Transcriptional Regulation in Cell Differentiation and Development

Front cover: circadian accumulation of mRNAs from putative transcriptional activation (DBP) target genes. Taken from the paper by Wuarin et al.

ISBN: 0 948601 37 X

Transcriptional Regulation in Cell Differentiation and Development

Proceedings of the joint British Society for Cell Biology –
Company of Biologists Symposium
University of Sussex, April 1992

Organized and Edited by

P. W. J. Rigby, R. Krumlauf and F. Grosveld
MRC National Institute for Medical Research,
London

SUPPLEMENT 16 1992
JOURNAL OF CELL SCIENCE
Published by THE COMPANY OF BIOLOGISTS LIMITED, Cambridge

The Company of Biologists Limited is a non-profit-making organization whose directors are active professional biologists. The Company, which was founded in 1925, is the owner and publisher of this and *The Journal of Experimental Biology* and *Development* (formerly *Journal of Embryology and Experimental Morphology*).

Journal of Cell Science is devoted to the study of cell organization. Papers will be published dealing with the structure and function of plant and animal cells and their extracellular products, and with such topics as cell growth and division, cell movements and interactions, and cell genetics. Accounts of advances in the relevant techniques will also be published. Contributions concerned with morphogenesis at the cellular and sub-cellular level will be acceptable, as will studies of micro-organisms and viruses, in so far as they are relevant to an understanding of cell organization. Theoretical articles and occasional review articles will be published.

Subscriptions

Journal of Cell Science will be published monthly in 1993 in the form of 3 volumes, each of 4 parts, and 1 Supplement. The subscription price of volumes 104, 105, 106 plus Supplement 17 is £565 (USA and Canada, US $988; Japan, £621) post free. Supplements may be purchased individually – prices on application to The Company of Biologists Ltd. Orders for 1993 may be sent to any bookseller or subscription agent, or to The Company of Biologists Ltd, Bidder Building, 140 Cowley Road, Cambridge CB4 4DL, UK, Fax (44) 223 423353. Copies of the journal for subscribers in the USA and Canada are sent by air to New Jersey for delivery with the minimum delay.

Back numbers of the *Journal of Cell Science* may be ordered through The Company of Biologists Ltd. This journal is the successor to the *Quarterly Journal of Microscopical Science*, back numbers of which are obtainable from Messers William Dawson & Sons, Cannon House, Park Farm Road, Folkestone, Kent CT19 5EE, UK.

Typeset, Printed and Published by
THE COMPANY OF BIOLOGISTS LIMITED
Department of Zoology, University of Cambridge, Downing Street,
Cambridge CB2 3EJ

ISBN: 0 948601 37 X

Preface

The last five years have seen enormous advances in our knowledge of both the mechanics of eukaryotic transcription and of the mechanisms by which it is regulated. We now know the names of many, perhaps most, of the general transcription factors that act, together with the polymerases, to assemble initiation complexes, and a considerable number of these are available as pure proteins. Indeed, factors that five years ago were nothing more than relatively poorly defined activities present in crude column fractions are now crystalline proteins of known three-dimensional structure. While much remains to be learned about how transcription is regulated during development and in response to extracellular signals, many general principles are now clear and in favourable cases proper regulation can be reconstituted in vitro using pure proteins.

The Spring Meeting of the British Society for Cell Biology, held in Brighton in April 1992, focused on the regulation of transcription during cell differentiation and development. The research presented at that meeting, which is summarised in the papers in this volume, covered such topics as tissue-specific expression, the regulation of transcription during vertebrate and invertebrate development, and how transcription is integrated with the cell cycle. One of the most striking, and pleasing, features of the meeting was that it made abundantly clear how rapidly studies of transcription are being integrated into the broad framework of biology. Those who would call themselves biochemists, geneticists, developmental biologists and cell biologists are all contributing their particular expertise to integrated approaches to the roles of transcriptional regulation in many of the most important areas of biology. As an example of how rapid progress has been, we doubt that many students of the subject would have predicted, had they been questioned at the beginning of the decade, that by 1992 we would be able to listen to a clear molecular exposition of how the transcriptional machinery responds to a biological phenomenon as complex as a circadian rhythm.

We hope that this volume will convey some of the excitement that was so clearly felt by those who attended that meeting and that it will serve to give readers interested in many aspects of biology a feeling for our present knowledge of this focal aspect of the subject.

Peter Rigby
Robb Krumlauf
Frank Grosveld
MRC National Institute for Medical Research,
London

November 1992

CONTENTS

Journal of Cell Science, Supplement 16, 1-7 (1992)
Printed in Great Britain © The Company of Biologists Limited 1992

The TATA-binding protein is a general transcription factor for RNA polymerase III

ROBERT J. WHITE[1], PETER W. J. RIGBY[2] and STEPHEN P. JACKSON[1]

[1]*Wellcome/CRC Institute, Tennis Court Road, Cambridge CB2 1QR, UK*
[2]*Laboratory of Eukaryotic Molecular Genetics, Medical Research Council, National Institute for Medical Research, The Ridgeway, Mill Hill, London NW7 1AA, UK*

Summary

The TATA-binding protein (TBP) is a principal component of the general factor TFIID and is required for specific transcription by RNA polymerase II. We have shown that TBP is also a general factor for RNA polymerase III.

Key words: TATA-binding protein, TBP, TFIID, transcription, RNA polymerase III.

Introduction

RNA polymerase III is responsible for the synthesis of a variety of small RNA molecules, including tRNA and 5 S rRNA. This transcription is coordinately regulated in response to a number of environmental stimuli, such as cell growth (Mauck and Green, 1974), differentiation (White et al., 1989) and transformation (White et al., 1990). In each of these cases, coordinate regulation is achieved by modulation of the activity of one or more of the general class III transcription factors (Carey and Singh, 1988; Hoeffler et al., 1988; Tower and Sollner-Webb, 1988; White et al., 1989; White et al., 1990). (Class III genes are those transcribed by RNA polymerase III, class II and I genes are those transcribed by RNA polymerases II and I, respectively). In order to achieve a clear understanding of such regulation, it is important to obtain a detailed knowledge of how many factors are involved in the basal transcription machinery.

The first estimate of the complexity of the class III transcription apparatus was gained as a result of fractionating extracts of human KB cells (Segall et al., 1980). The initial stage in this fractionation scheme involved chromatography on phosphocellulose and is schematised in Fig. 1. Two step fractions, PC-B and PC-C, were found to be necessary and sufficient to reconstitute transcription of a tRNA gene and an adenovirus VA gene in the presence of purified RNA polymerase III. Synthesis of 5 S rRNA also required these two fractions, as well as an additional fraction, PC-A. On this basis it was concluded that the PC-A fraction contains a gene-specific factor, which was named TFIIIA, whereas the PC-B and PC-C fractions each contain a general factor, which were named TFIIIB and TFIIIC, respectively (Segall et al., 1980). Since these experiments employed crude step fractions, the possibility of additional class III factors within the various fractions had not been excluded. Indeed two further factors have since been identified in silkworms (Ottonello et al., 1987; Young et al., 1991). However, in no other system were any additional general class III initiation factors identified for over a decade, despite many fractionation studies.

TBP is required for TATA-dependent transcription of U6 snRNA genes by RNA polymerase III

Extensive purification of a yeast TFIIIB preparation resolved two components, neither of which was sufficient on its own to support transcription of a U6 snRNA template by RNA polymerase III (Margottin et al., 1991). Whereas one of these components possessed TFIIIB activity, the other resembled the TATA-binding protein (TBP) in terms of its size, TATA-binding properties and ability to direct transcription of the adenovirus major late gene by RNA polymerase II in a TBP-dependent system. Furthermore, cloned TBP could substitute for this component in reconstituting RNA polymerase III transcription of the U6 snRNA gene. The involvement of TBP in U6 gene transcription is consistent with the presence of a TATA box ~30 bp upstream of the transcription start site. In the case of human and *Xenopus* U6 genes, this sequence has been shown to be important for polymerase III transcription (Carbon et al., 1987; Mattaj et al., 1988; Lobo and Hernandez, 1989; Simmen and Mattaj, 1990; Simmen et al., 1991). Indeed, the TATA box elements of several class II genes can functionally substitute for the U6 TATA sequence in directing transcription by RNA polymerase III (Lobo et al., 1991). As was the case for the yeast gene, addition of purified or cloned TBP was found to activate polymerase III transcription of the human U6 gene in a vari-

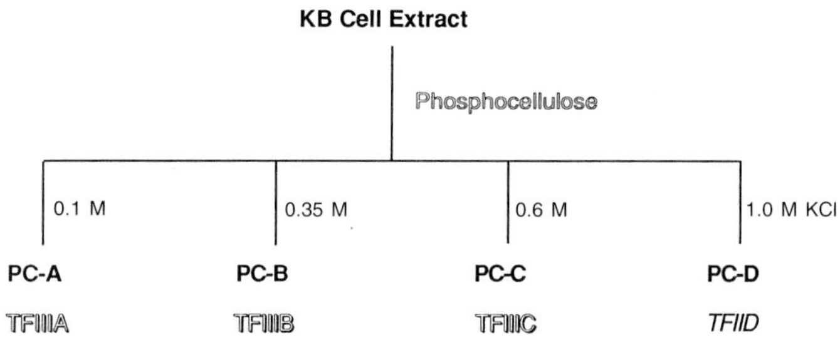

Fig. 1. Simplified version of fractionation scheme used by Segall et al. (1980) to separate class III transcription factors from an S100 extract of cultured human KB cells. The KCl concentrations at which the fractions eluted are indicated.

ety of reconstituted systems (Lobo et al., 1991; Simmen et al., 1991; Waldschmidt et al., 1991).

TBP: an unlikely candidate for a general class III factor

The presence of an essential TATA box in the promoters of U6 genes had suggested the possibility of a role for TBP, although a distinct TATA-binding factor for RNA polymerase III had, for a time, been considered likely (Carbon et al., 1987; Mattaj et al., 1988; Murphy et al., 1989; Simmen and Mattaj, 1990). Functional TATA boxes are also found in several other polymerase III promoters, such as those of the human 7SK gene (Murphy et al., 1987) and the Epstein-Barr virus EBER2 gene (Howe and Shu, 1989). In contrast, most class III genes have no TATA homology and generally have much less reliance on specific sequences upstream of the transcription start site (reviewed by Murphy et al., 1989 and Gabrielsen and Sentenac, 1991). The absence of a TATA box in most class III genes made TBP seem an unlikely candidate for a general class III transcription factor. However, transcription of a TATA-less template by RNA polymerase II was shown, nevertheless, to require the presence of TBP (Pugh and Tjian, 1991). Since this was the case for RNA polymerase II, it could clearly also be so for RNA polymerase III.

A second argument against TBP as a general class III factor was the fact that TFIID, of which TBP is a principal component, elutes from phosphocellulose in the PC-D fraction, which is not required for polymerase III transcription (Fig. 1; Matsui et al., 1980). However, western blotting with affinity-purified, anti-TBP antibodies has shown that TBP is, in fact, also present in significant amounts in both PC-B and PC-C, the fractions which are generally required by class III genes (R. J. White and S. P. Jackson, unpublished results). Therefore, a subpopulation of TBP molecules cofractionates with the established polymerase III general factors. So TBP could no longer be excluded, a priori, as a general class III factor.

A TATA-binding factor is required for transcription of all class III genes tested

We tested for an involvement of TBP in general polymerase III transcription using TATA-containing oligonucleotides. Preincubation of a HeLa nuclear extract with an oligonucleotide containing the sequence of the TATA box region of the adenovirus major late (ML) promoter inhibits transcription of a subsequently added human U6 snRNA gene; this effect is TATA-specific, since point-mutating the TATA box whilst leaving the remainder of the oligonucleotide unchanged abolishes the inhibition (Fig. 2, lanes 9 and 10). The sensitivity of U6 transcription to competing TATA sequences reflects its known requirement for TBP (Lobo et al., 1991; Margottin et al., 1991; Simmen et al.,

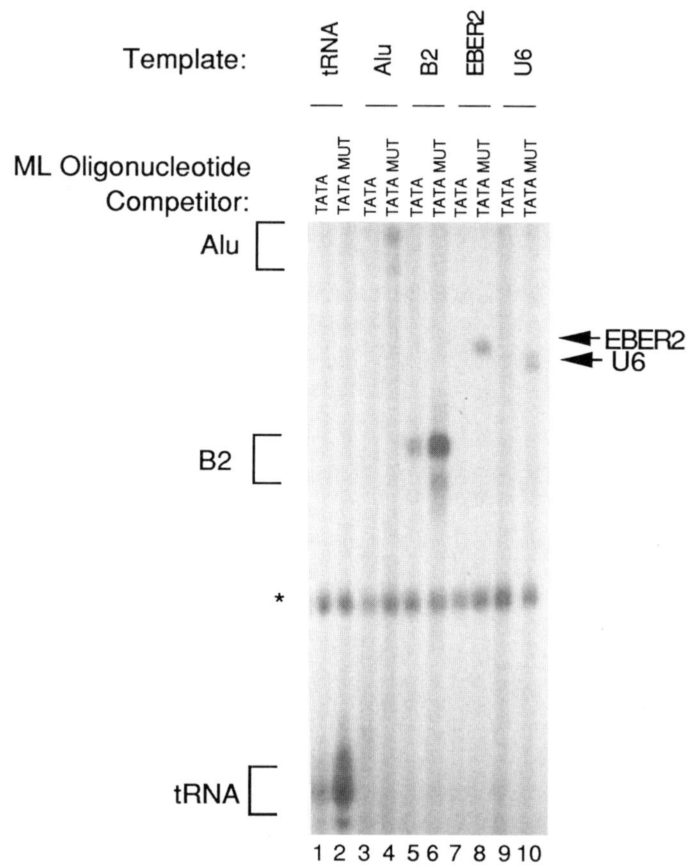

Fig. 2. Effect of oligonucleotide competitors upon RNA polymerase III transcription. tRNA^{Pro} (lanes 1-2), Alu (lanes 3-4), B2 (lanes 5-6), EBER2 (lanes 7-8) and U6 (lanes 9-10) genes were transcribed using HeLa nuclear extract that had been preincubated for 15 min at 30°C with 400 ng of ML TATA (lanes 1, 3, 5, 7 and 9) or ML TATA mutant (lanes 2, 4, 6, 8 and 10) oligonucleotides. The additional band in all lanes, that is marked by an asterisk and does not respond to competition, is due to end-labelling of endogenous small RNAs (G. Kunkel, personal communication). The ML TATA and ML TATA mutant oligonucleotides are as previously (White et al., 1992). The tRNA^{Pro} template is plasmid Mcet1 (Ciliberto et al., 1982). The Alu template is pRH5.7, a 5.7 kb fragment of human genomic DNA from the apoA1-CIII locus inserted into the *Sma*I site of pAT153. The B2 template is pTB14 (White et al., 1989). The EBER2 template is E2,-160 (Howe and Shu, 1989). The U6 template is pU6/Hae/RA.2 (Lobo et al., 1991). Extracts were prepared as previously (White et al., 1992). Transcription reactions were performed and processed as previously (White et al., 1992).

1991; Waldschmidt et al., 1991). Transcription of the Epstein-Barr virus EBER2 gene shows the same response (Fig. 2, lanes 7 and 8). A TATA box positioned 28 bp upstream of the start site of this gene has been shown to make a major contribution to EBER2 expression in vivo (Howe and Shu, 1989). Transcription of a member of the B2 middle-repetitive gene family is also specifically inhibited by the TATA sequence (Fig. 2, lanes 5 and 6). This gene has an internal promoter, but also has a TATA homology 15 bp upstream of the transcription start site (White, 1990). However, transcription of a human Alu gene and a *Caenorhabditis elegans* $tRNA^{Pro}$ gene, neither of which have TATA boxes (Ciliberto et al., 1982; C. Sharpe, personal communication), is also specifically inhibited by the ML TATA sequence (Fig. 2, lanes 1-4).

The fact that a TATA sequence specifically competes for transcription of these genes indicates that a TATA-binding factor is required for their expression. This is clearly a general effect, since it is shown by five different class III genes, with a range of promoter structures (Fig. 2). Strikingly, two of these genes have no TATA boxes and yet still require a TATA-binding factor for transcription. We have obtained the same response with six other TATA-less polymerase III templates, including the well-characterised adenovirus VA_I and *Xenopus borealis* somatic 5 S rRNA genes (White et al., 1992). In fact, transcription of all of the class III genes that we have tested is specifically inhibited by competition with TATA sequences. Therefore, there is a general polymerase III transcription factor that recognizes TATA boxes.

A general class III transcription factor has DNA-binding specificity and antigenic epitopes in common with TBP

It was important to exclude the possibility that the ML TATA sequence represents a cryptic binding site for one of the established class III general factors, TFIIIB and TFIIIC. TFIIIC has well-characterised DNA-binding properties, whereas TFIIIB by itself shows no sequence-specific interaction with DNA (reviewed by Gabrielsen and Sentenac, 1991). We therefore tested whether TFIIIC could recognize TATA boxes in a gel retardation assay and found that the ML TATA oligonucleotide was unable to compete for TFIIIC binding (R. J. White and S. P. Jackson, unpublished results). Therefore, the factor that recognizes TATA boxes and is required for transcription of all polymerase III templates is unlikely to be one of the established class III general factors.

Cloned human TBP binds to the ML TATA box and a point-mutation that abolishes the inhibitory effect of this sequence on polymerase III transcription (TATAAAA→TAGAGAA) also abolishes TBP binding (Hoffmann et al., 1990). This correlation was extended further by the observation that functional TATA boxes from two other well-characterised polymerase II promoters also inhibit transcription of all polymerase III templates tested (White et al., 1992). Thus, polymerase III transcription is inhibited by the CATATAA sequence that serves as a TATA box in the mouse β-globin promoter and by the CATAAAA sequence that serves as a TATA box in the human β-globin promoter. Point-mutation of either of these sequences (CATATAA→CATAGCA and CATAAAA→CATCGCC) again abolishes the ability to inhibit transcription of class III genes (White et al., 1992). Therefore, the TATA-binding general factor required for transcription by RNA polymerase III specifically recognizes three different TATA boxes that serve as functional targets for TBP.

We also tested the effect on polymerase III transcription of physically removing TBP by immunodepletion. An extract that had been immunodepleted using affinity-purified antisera against cloned TBP was found to be severely impaired relative to a mock-immunodepleted control extract, in its ability to transcribe all polymerase III templates tested (R. J. White, B. F. Pugh and S. P. Jackson, unpublished results). These templates included the *Xenopus borealis* somatic 5 S rRNA gene, the adenovirus VA_I gene and a human $tRNA^{Leu}$ gene, all of which lack TATA sequences.

Cloned TBP restores RNA polymerase III transcription to extracts that have been preincubated with TATA competitors

The general class III factor that recognizes TATA boxes has similar or identical DNA-binding specificity to TBP and is recognized by affinity-purified antibodies against TBP. These observations strongly implicate TBP as this general factor. However, the alternative of a distinct class III TATA-binding factor with epitopes and DNA-binding specificity in common with TBP is not excluded by this data. It was therefore necessary to test whether TBP itself could function as a general factor for RNA polymerase III. TBP had already been shown to activate TATA-dependent transcription of the U6 snRNA gene (Lobo et al., 1991; Margottin et al., 1991; Simmen et al., 1991; Waldschmidt et al., 1991). However, none of the classical class III genes, without TATA boxes, such as VA, tRNA and 5 S rRNA, had been shown to respond to TBP.

Cloned human TBP (Peterson et al., 1990), expressed in bacteria and purified to >95% homogeneity, restores transcription of TATA-less polymerase III templates in extracts preincubated with TATA-containing oligonucleotides (White et al., 1992). This is shown in Fig. 3 for the VA_I gene and a B2 gene. The inclusion of 2 μg/ml α-amanitin in this experiment excluded the possibility that TBP-induced transcription of these class III genes was by RNA polymerase II, since this concentration of α-amanitin selectively inhibits the polymerase II enzyme but not RNA polymerase III. Whereas cloned TBP can activate polymerase III transcription following preincubation of extracts with ML or β-globin TATA sequences, extracts preincubated with a TFIIIC binding site do not respond (White et al., 1992). This demonstrates the specificity of the effect and also confirms that TFIIIC is not the target for TATA box competitors.

The ability of cloned TBP to activate transcription of TATA-less polymerase III templates in extracts that have been depleted of endogenous TATA-binding factor(s) is strong evidence in support of the hypothesis that TBP is a general class III transcription factor. However, if an alter-

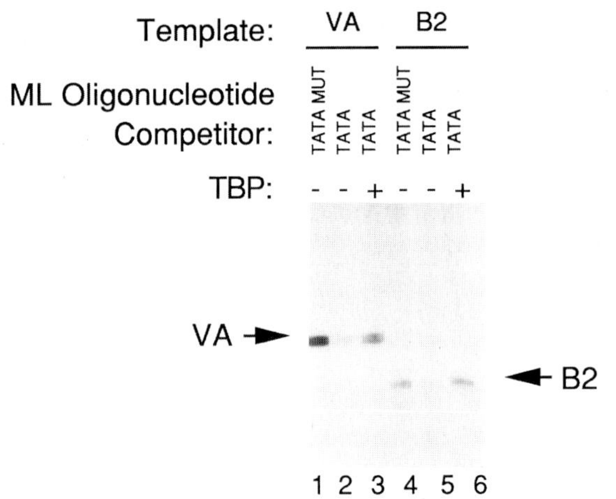

Fig. 3. Cloned TBP restores RNA polymerase III transcription to extracts depleted of TATA-binding proteins by preincubation with TATA oligonucleotides. VA_I (lanes 1-3) and B2 (lanes 4-6) genes were transcribed in the presence of 2 μg/ml α-amanitin using HeLa extract preincubated for 15 min at 30°C with 200 ng of the ML TATA mutant (lanes 1 and 4) or the ML TATA (lanes 2, 3, 5 and 6) oligonucleotides. In lanes 3 and 6, template was preincubated separately with 8 ng of >95% pure cloned human TBP for 15 min at 30°C before being added to the depleted extract. The VA_I template was pBRVA_I (White et al., 1989). Purified bacterially expressed cloned human TBP was a generous gift from B. F. Pugh and R. Tjian (Peterson et al., 1990).

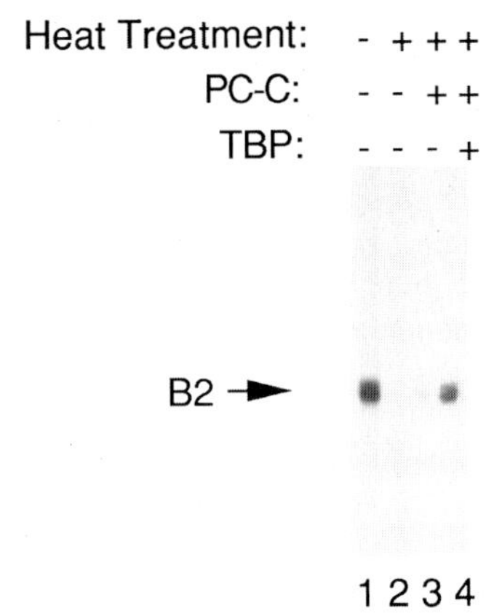

Fig. 4. Cloned TBP restores RNA polymerase III transcription to heat-inactivated extracts. A B2 gene was transcribed in the presence of 2 μg/ml α-amanitin using either untreated (lane 1) or heat treated (lanes 2-4) HeLa extract in the presence (lanes 3-4) or absence (lanes 1-2) of 4 μl of a PC-C fraction. Lane 4 contained 6.7 ng of cloned TBP. PC-C phosphocellulose column fraction was prepared according to the method of Segall et al. (1980). Heat treatment was for 15 min at 47°C according to Nakajima et al. (1988).

native TATA-binding factor were employed by RNA polymerase III, then addition of exogenous cloned TBP might simply serve to displace this factor from the TATA oligonucleotides, thereby allowing it to mediate transcription of class III genes. To exclude this possibility, we tested the effect of cloned TBP on polymerase III transcription in extracts in which endogenous TBP had been inactivated by heat treatment.

Cloned TBP restores RNA polymerase III transcription to heat-treated extracts

Mild heat treatment (47°C for 15 minutes) has been shown to inactivate TBP whilst leaving the other components of the class II general transcription apparatus unaffected (Nakajima et al., 1988). Such treatment abolishes transcription of all class III templates tested, consistent with a general requirement for TBP in polymerase III transcription (Fig. 4; Simmen et al., 1991; White et al., 1992; R. J. White and S. P. Jackson, unpublished observations). As well as TBP, a component of the PC-C fraction, presumably TFIIIC, is also inactivated in this way (Simmen et al., 1991). Addition of PC-C to the heat-treated extract elicits only a very slight restoration of transcription, but if cloned TBP is also included, polymerase III transcription is dramatically stimulated. This is shown in Fig. 4 for a B2 gene. It is also the case for U6 (Simmen et al., 1991), VA_I and tRNA genes (White et al., 1992). Therefore, TBP can directly activate transcription of both TATA-containing and TATA-less templates. Indeed, under appropriate conditions, the PC-C fraction and cloned TBP are sufficient to restore fully levels of transcription in a heat-treated extract to the same level as those occurring in an unheated extract (Simmen et al., 1991; White et al., 1992). Heat treating the cloned TBP abolishes its ability to activate transcription of class III templates (White et al., 1992). The transcription reconstituted in a heat-inactivated extract using PC-C and TBP is abolished by 100 μg/ml but not by 2 μg/ml α-amanitin, which is the α-amanitin sensitivity diagnostic of mammalian RNA polymerase III (White et al., 1992). Therefore, TBP can direct polymerase III transcription of a variety of class III genes.

TBP functions as a general class III transcription factor in vivo

Transcription of a broad range of class III genes therefore requires a factor with the same TATA-binding specificity as TBP, the same heat-sensitivity as TBP, with antigenic epitopes in common with TBP, and which can be functionally substituted for by cloned TBP. This strongly suggests that TBP is a general factor for RNA polymerase III. However, the above experiments were all performed in vitro. It therefore remained to confirm that TBP plays the same role in vivo. This has been achieved using yeast genetics.

Expression of U6 snRNA, $tRNA^{Trp}$ and $tRNA^{Ile}$ is rapidly inhibited following transfer of strains bearing temperature-sensitive mutations in TBP to the non-permissive temperature (Cormack and Struhl, 1992). A mutation in the gene encoding the largest subunit of RNA polymerase II does not produce a similar effect, suggesting that the response of U6 and tRNA transcription to TBP mutations is not an indirect consequence of a defect in polymerase II transcription. The rapid response of U6 and tRNA synthesis to inactivation of TBP is also consistent with a direct involvement of TBP in the expression of these genes (Cormack and Struhl, 1992). Furthermore, cell-free extracts made from yeast strains bearing mutated TBP are unable to transcribe 5 S rRNA or $tRNA^{Leu3}$ genes (Schultz et al., 1992). Addition of cloned TBP restores 5 S rRNA and tRNA transcription by these extracts, thereby proving that the TBP mutations are not acting indirectly (Schultz et al., 1992). These experiments conclusively demonstrate that TBP is a general transcription factor for RNA polymerase III in vivo.

Discussion

TBP is a general transcription factor for RNA polymerase III. We have tested eleven different class III genes, including representatives of all the types of promoter organisation known to occur in polymerase III templates (reviewed by Gabrielsen and Sentenac, 1991), and in all cases we have observed a requirement for TBP. Furthermore, each of the yeast genes tested has also shown the same TBP-dependence (Cormack and Struhl, 1992; Schultz et al., 1992). Therefore, the TBP requirement for RNA polymerase III transcription is a general one.

Two general class III initiation factors, TFIIIB and TFIIIC, were defined in 1980 (Segall et al., 1980). Since that time, two new factors, TFIIID and TFIIIR, have been discovered in silkworms (Ottonello et al., 1987; Young et al., 1991). However, it remains to be determined whether these factors are ubiquitous or silkworm-specific. Furthermore, since TFIIID and TFIIIR were only tested on two templates (5 S rRNA and $tRNA^{Ala}$ genes), it is not yet clear that these are truly general factors. Indeed, TFIIIC can no longer be rigorously included in the general category, since it is not required for transcription of U6 genes (Moenne et al., 1990; Lobo et al., 1991; Margottin et al., 1991; Waldschmidt et al., 1991). At present, TFIIIB and TBP are the only class III initiation factors that can be considered to be truly general.

It has been reported that little or no free TBP exists in HeLa cell extracts (Timmers and Sharp, 1991). Instead, human TBP is found in large, multi-component complexes such as TFIID, which contain additional polypeptides that are tightly bound to TBP, and are consequently referred to as TBP-associated factors or TAFs (Pugh and Tjian, 1991; Tanese et al., 1991). It therefore seems likely that the TBP population that is active in RNA polymerase III transcription also exists in association with other proteins. The fact that a population of TBP molecules copurifies with TFIIIB activity in yeast (Margottin et al., 1991) suggests that TBP and TFIIIB may interact in solution, which further implies that they may interact in a functional transcription complex. It also seems likely that there are class III-specific TAFs associated with the TBP subpopulation that mediates RNA polymerase III transcription. If so, then these represent additional components of the class III transcription machinery that are awaiting discovery.

Since many class III promoters do not have TATA boxes, it is not clear how they recruit TBP. TBP is capable of specifically recognizing sequences that are quite different from a consensus TATA element (Singer et al., 1990). It is therefore possible that TBP binds directly to at least some TATA-less class III genes. However, since the only sequences that are obviously conserved between tRNA, VA and B2 genes are the A- and B-block internal promoter elements that are bound by TFIIIC, TBP would presumably have to recognise different sequences in different genes, were direct binding to be the rule. It seems far more probable that TBP is recruited to such genes by protein-protein interactions. Template-commitment experiments support this conclusion (R. J. White and S. P. Jackson, unpublished observations). This is in keeping with results for a TATA-less RNA polymerase II template, where TBP recruitment is achieved by interaction with a protein component of the TFIID complex, referred to as a 'tethering' factor (Pugh and Tjian, 1991).

TBP is also required for specific transcription by RNA polymerase I (Comai et al., 1992; Cormack and Struhl, 1992; Schultz et al., 1992). In this case the essential class I factor SL1 is composed of TBP and three TAFs (Comai et al., 1992). Polymerase I templates do not have TATA homologies, but murine SL1 can bind directly to the mouse rRNA gene promoter (Bell et al., 1990). Which component of SL1 is responsible for this DNA binding remains to be determined. In contrast, human SL1 cannot bind independently to the human rRNA gene promoter, but needs to be tethered via protein-protein interactions with an upstream-binding factor termed UBF (Bell et al., 1988; Bell et al., 1990). Therefore, the human class I system provides a good example of TBP recruitment to a TATA-less promoter by protein-protein interactions.

It will be interesting to determine whether the different classes of transcription machinery utilise the same or different regions of the TBP polypeptide. The C-terminal 180 amino acids of TBP are at least 80% identical in all organisms from which it has been cloned; these include human, yeast (*Saccharomyces cerevisiae* and *Schizosaccharomyces pombe*), the fruit fly *Drosophila melanogaster* and the plant *Arabidopsis thaliani* (see Greenblatt, 1991, for a review). This conserved region has been shown to be sufficient to allow basal transcription in vitro from a TATA-containing promoter by RNA polymerase II (Hoey et al., 1990; Horikoshi et al., 1990; Peterson et al., 1990). We have found that it is also sufficient for RNA polymerase III transcription of a TATA-less tRNA gene (R. J. White, B. F. Pugh and S. P. Jackson, unpublished observations). Three different point-mutations within this region abolish transcription by each of the nuclear RNA polymerases in vivo, suggesting that much of the protein is important for all classes of transcription (Cormack and Struhl, 1992; Schultz et al., 1992). However, two other point-mutations produce class-specific effects: one abolishes transcription by RNA polymerases II and III but allows transcription by RNA polymerase I; another, which inactivates TATA binding, still functions with TATA-less class I and III genes but not with a TATA-containing class II promoter (Schultz et al., 1992). Therefore, components of the different classes of transcription machinery may contact overlapping but distinct regions of TBP.

For over a decade the TATA-binding protein was considered to be specifically required for TATA-dependent transcription by RNA polymerase II. Recently, its known sphere of influence has rapidly expanded to include TATA-dependent transcription by RNA polymerase III (Margottin et al., 1991), TATA-independent transcription by RNA polymerase II (Pugh and Tjian, 1991), TATA-independent transcription by RNA polymerase III (White et al., 1992) and, finally, transcription by RNA polymerase I (Comai et al., 1992). These dramatic discoveries (reviewed by White and Jackson, 1992) have established TBP as the central component of nuclear transcription. It is unique in being a general factor required for transcription by RNA polymerases I, II and III. The fact that it has maintained this position from yeast to man suggests that it will prove to be the case in all eukaryotes. If so, then transcription of all nuclear genes is likely to require this small, remarkable polypeptide.

We thank Frank Pugh and Robert Tjian for purified TBP, Colin Sharpe for pRH5.7, Nouria Hernandez for pU6/Hae/RA.2 and Gregory Howe for E2,-160. This work was funded by the Cancer Research Campaign and the Medical Research Council of Great Britain. R. J. W. and S. P. J. are members of the Zoology Department of Cambridge University. S. P. J. is also supported by a Leukaemia Society of America Senior Fellowship.

References

Bell, S. P., Jantzen, H.-M. and Tjian, R. (1990). Assembly of alternative multiprotein complexes directs rRNA promoter selectivity. *Genes Dev.* **4**, 943-954.

Bell, S. P., Learned, R. M., Jantzen, H.-M. and Tjian, R. (1988). Functional cooperativity between transcription factors UBF1 and SL1 mediates human ribosomal RNA synthesis. *Science* **241**, 1192-1197.

Carbon, P., Murgo, S., Ebel, J.-P., Krol, A., Tebb, G. and Mattaj, I. W. (1987). A common octamer motif binding protein is involved in the transcription of U6 snRNA by RNA polymerase III and U2 snRNA by RNA polymerase II. *Cell* **51**, 71-79.

Carey, M. F. and Singh, K. (1988). Enhanced B2 transcription in Simian virus 40-transformed cells is mediated through the formation of RNA polymerase III transcription complexes on previously inactive genes. *Proc. natl. Acad. Sci. U.S.A.* **85**, 7059-7063.

Ciliberto, G., Castagnoli, L., Melton, D. A. and Cortese, R. (1982). Promoter of a eukaryotic tRNAPro gene is composed of three noncontiguous regions. *Proc. natl. Acad. Sci. U.S.A.* **79**, 1195-1199.

Comai, L., Tanese, N. and Tjian, R. (1992). The TATA-binding protein and associated factors are integral components of the RNA polymerase I transcription factor, SL1. *Cell* **68**, 965-976.

Cormack, B. P. and Struhl, K. (1992). The TATA-binding protein is required for transcription by all three nuclear RNA polymerases in yeast cells. *Cell* **69**, 685-696.

Gabrielsen, O. S. and Sentenac, A. (1991). RNA polymerase III (C) and its transcription factors. *Trends Biochem. Sci.* **16**, 412-416.

Greenblatt, J. (1991). Roles of TFIID in transcriptional initiation by RNA polymerase II. *Cell* **66**, 1067-1070.

Hoeffler, W. K., Kovelman, R. and Roeder, R. G. (1988). Activation of transcription factor IIIC by the adenovirus E1A protein. *Cell* **53**, 907-920.

Hoey, T., Dynlacht, B. D., Peterson, M. G., Pugh, B. F. and Tjian, R. (1990). Isolation and characterization of the *Drosophila* gene encoding the TATA box binding protein, TFIID. *Cell* **61**, 1179-1186.

Hoffmann, A., Sinn, E., Yamamoto, T., Wang, J., Roy, A., Horikoshi, M. and Roeder, R. G. (1990). Highly conserved core domain and unique N terminus with presumptive regulatory motifs in a human TATA factor (TFIID). *Nature, Lond.* **346**, 387-390.

Horikoshi, M., Yamamoto, T., Ohkuma, Y., Weil, P. A. and Roeder, R. G. (1990). Analysis of structure-function relationships of yeast TATA box binding factor TFIID. *Cell* **61**, 1171-1178.

Howe, J. G. and Shu, M.-D. (1989). Epstein-Barr virus small RNA (EBER) genes: unique transcription units that combine RNA polymerase II and III promoter elements. *Cell* **57**, 825-834.

Lobo, S. M. and Hernandez, N. (1989). A 7bp mutation converts a human RNA polymerase II snRNA promoter into an RNA polymerase III promoter. *Cell* **58**, 55-67.

Lobo, S. M., Lister, J., Sullivan, M. L. and Hernandez, N. (1991). The cloned RNA polymerase II transcription factor IID selects RNA polymerase III to transcribe the human U6 gene *in vitro*. *Genes Dev.* **5**, 1477-1489.

Margottin, F., Dujardin, G., Gerard, M., Egly, J.-M., Huet, J. and Sentenac, A. (1991). Participation of the TATA factor in transcription of the yeast U6 gene by RNA polymerase C. *Science* **251**, 424-426.

Matsui, T., Segall, J., Weil, P. A. and Roeder, R. G. (1980). Multiple factors required for accurate initiation of transcription by purified RNA polymerase II. *J. Biol. Chem.* **255**, 11992-11996.

Mattaj, I. W., Dathan, N. A., Parry, H. D., Carbon, P. and Krol, A. (1988). Changing the RNA polymerase specificity of U snRNA gene promoters. *Cell* **55**, 435-442.

Mauck, J. C. and Green, H. (1974). Regulation of pre-transfer RNA synthesis during transition from resting to growing state. *Cell* **3**, 171-177.

Moenne, A., Camier, S., Anderson, G., Margottin, F., Beggs, J. and Sentenac, A. (1990). The U6 gene of *Saccharomyces cerevisiae* is transcribed by RNA polymerase C (III) *in vivo* and *in vitro*. *EMBO J.* **9**, 271-277.

Murphy, S., Di Liegro, C. and Melli, M. (1987). The *in vitro* transcription of the 7SK RNA gene by RNA polymerase III is dependent only on the presence of an upstream promoter. *Cell* **51**, 81-87.

Murphy, S., Moorefield, B. and Pieler, T. (1989). Common mechanisms of promoter recognition by RNA polymerases II and III. *Trends Genet.* **5**, 122-126.

Nakajima, N., Horikoshi, M. and Roeder, R. G. (1988). Factors involved in specific transcription by mammalian RNA polymerase II: purification, genetic specificity, and TATA box-promoter interactions of TFIID. *Molec. cell. Biol.* **8**, 4028-4040.

Ottonello, S., Rivier, D. H., Doolittle, G. M., Young, L. S. and Sprague, K. U. (1987). The properties of a new polymerase III transcription factor reveal that transcription complexes can assemble by more than one pathway. *EMBO J.* **6**, 1921-1927.

Peterson, M. G., Tanese, N., Pugh, B. F. and Tjian, R. (1990). Functional domains and upstream activation properties of cloned human TATA binding protein. *Science* **248**, 1625-1630.

Pugh, B. F. and Tjian, R. (1991). Transcription from a TATA-less promoter requires a multisubunit TFIID complex. *Genes Dev.* **5**, 1935-1945.

Schultz, M. C., Reeder, R. H. and Hahn, S. (1992). Variants of the TATA-binding protein can distinguish subsets of RNA polymerase I, II, and III promoters. *Cell* **69**, 697-702.

Segall, J., Matsui, T. and Roeder, R. G. (1980). Multiple factors are required for the accurate transcription of purified genes by RNA polymerase III. *J. Biol. Chem.* **255**, 11986-11991.

Simmen, K. A., Bernues, J., Parry, H. D., Stunnenberg, H. G., Berkenstam, A., Cavallini, B., Egly, J.-M. and Mattaj, I. W. (1991). TFIID is required for *in vitro* transcription of the human U6 gene by RNA polymerase III. *EMBO J.* **10**, 1853-1862.

Simmen, K. A. and Mattaj, I. W. (1990). Complex requirements for RNA polymerase III transcription of the *Xenopus* U6 promoter. *Nucl. acids Res.* **18**, 5649-5657.

Singer, V. L., Wobbe, C. R. and Struhl, K. (1990). A wide variety of DNA sequences can functionally replace a yeast TATA element for transcriptional activation. *Genes Dev.* **4**, 636-645.

Tanese, N., Pugh, B. F. and Tjian, R. (1991). Coactivators for a proline-rich activator purified from the multisubunit human TFIID complex. *Genes Dev.* **5**, 2212-2224.

Tower, J. and Sollner-Webb, B. (1988). Polymerase III transcription factor B activity is reduced in extracts of growth-restricted cells. *Molec. cell. Biol.* **8**, 1001-1005.

Timmers, H. Th. M. and Sharp, P. A. (1991). The mammalian TFIID protein is present in two functionally distinct complexes. *Genes Dev.* **5**, 1946-1956.

Waldschmidt, R., Wanandi, I. and Seifart, K. H. (1991). Identification of transcription factors required for the expression of mammalian U6 genes *in vitro*. *EMBO J.* **10**, 2595-2603.

White, R. J. (1990). Regulation of transcription of the B2 gene family by RNA polymerase III. *PhD Thesis. University of London.*

White, R. J. and Jackson, S. P. (1992). The TATA-binding protein plays a central role in transcription by RNA polymerases I, II and III. *Trends Genet.* **8**, 284-288.

White, R. J., Jackson, S. P. and Rigby, P. W. J. (1992). A role for the TATA-box-binding protein component of the transcription factor IID complex as a general RNA polymerase III transcription factor. *Proc. natl. Acad. Sci. U.S.A.* **89**, 1949-1953.

White, R. J., Stott, D. and Rigby, P. W. J. (1989). Regulation of RNA polymerase III transcription in response to F9 embryonal carcinoma stem cell differentiation. *Cell* **59**, 1081-1092.

White, R. J., Stott, D. and Rigby, P. W. J. (1990). Regulation of RNA polymerase III transcription in response to Simian virus 40 transformation. *EMBO J.* **9**, 3713-3721.

Young, L. S., Dunstan, H. M., Witte, P. R., Smith, T. P., Ottonello, S. and Sprague, K. U. (1991). A class III transcription factor composed of RNA. *Science* **252**, 542-546.

Journal of Cell Science, Supplement 16, 9-14 (1992)
Printed in Great Britain

Transcriptional repression by methylation of CpG

RICHARD MEEHAN[1], JOE LEWIS[1], SALLY CROSS[1], XINSHENG NAN[1], PETER JEPPESEN[2] and ADRIAN BIRD[1]

[1]*Institute of Cell and Molecular Biology, University of Edinburgh, Kings Buildings, Edinburgh EH9 3JR, UK*
[2]*MRC, Human Genetics Unit, Western General Hospital, Crewe Road, Edinburgh EH4 2XU, UK*

Summary

Methylated DNA in mammals is associated with transcriptional repression and nuclease resistant chromatin. In this review we discuss how these effects may be mediated by proteins that bind to methylated DNA.

Key words: DNA methylation, transcriptional repression, methyl-CpG binding proteins.

Methylation of DNA is essential in mammals

In mammals approximately 70% of all CpGs are methylated at the 5 position of cytosine. It is not clear what the functional role of this modification is, but it has been recently shown by gene targeting experiments that methylation of DNA is necessary for mouse development (Li et al., 1992). Mutant mouse embryos that have reduced levels (1/3 of wild type) of genomic m^5C were generated by introducing a mutation into the DNA methyltransferase (MTase) gene. Homozygous mutant embryos were stunted and died at midgestation. A similar reduction of m^5C in the DNA of embryonic stem (ES) cells had no effect on their ability to survive in tissue culture. This suggests that reduced levels of MTase cause abnormal development.

How might decreased levels of DNA methylation cause embryonic lethality? There is much evidence that methylation near promoters leads to stable inactivation of the associated gene, and in one case (X-inactivation in placental mammals) we know that the stability of repression in the organism is due in part to methylation (see Riggs and Pfeifer, 1992). For example the CpG island-containing genes on the X chromosome become methylated following X chromosome inactivation in eutherian mammals (Grant and Chapman, 1988). The equivalent CpG island genes remain non-methylated on the active X chromosome. A similar process of methylation associated inactivation occurs to retroviral proviruses following infection of early mouse embryos (Jähner et al., 1982). In both of these cases there is some evidence that the genes become inactivated prior to or at the onset of methylation (Gautsch and Wilson, 1983; Lock et al., 1987; Singer-Sam et al., 1990). Although DNA methylation may not be directly responsible for the initial inactivation event, it is seen as integral to the maintenance of gene repression (Pfeifer et al., 1990b). There may well be other cases where the repressive effects of methylation have been harnessed for the benefit of controlling gene expression during development. If we assume that methylation is functioning as a transcriptional repressor in the developing embryo then one possibility is that high levels of MTase are required to repress genes stably whose inappropiate expression could cause embryonic lethality. These 'genes' may correspond to normally inactivated retroviruses (Jähner et al., 1982), retroposons and genes that are usually tissue-specific in their expression.

At this point it is also worth considering that DNA methylation has been proposed to have numerous other roles in the regulation of chromatin activity with respect to DNA repair, recombination and replication. Perturbation of these processes could also have lethal consequences for the developing embryo. For example, the mouse major satellite sequence is undermethylated in the embryonal carcinoma-derived F9 cells and is shifted in its replication timing from late to early S-phase in comparison with its methylated counterpart (Selig et al., 1988). This may implicate methylation in the maintainance of satellite DNA sequences in a late replicating mode in differentiated cells. It has also recently been shown that CpG methylation inhibits recombination between V(D)J substrates maintained on minichromosomes (Hsieh and Lieber, 1992). Likewise transcriptionally active DNA promoter regions are preferentially repaired after exposure to DNA damaging agents in comparison with inactive DNA regions (Mellon et al., 1986). Thus a more liberal interpretation of the function of DNA methylation would be as a chromatin repressor rather than solely as a transcriptional repressor.

The observation that ES cells can survive in tissue culture with reduced amounts of m^5C in their DNA implies that embryonic cells are less dependent on DNA methylation than somatic cells. It would of be interest to know whether somatically derived cell lines carrying the same MTase mutation could also maintain viability in culture. In fact many established cell lines have high levels of de novo methylation at the promoter regions of genes that are probably not required by the cell for growth on a plastic dish (Jones et al., 1990; Antequera et al., 1990). A somatically derived cell line may find it selectively more advantageous to utilize the inhibitory effects of DNA methylation on genes that are dispensable in culture.

How does methylation inhibit transcription?

There is a variety of evidence in the literature which suggests that the presence of DNA methylation at gene promoters can inhibit transcription in two possible ways (Watt and Molloy, 1988; Boyes and Bird, 1991). One model proposes that CpG methylation can interfere with transcription directly by modifying the binding site (through methylation) of transcription factors so that they can no longer bind their cognate sequences. Alternatively there are factors in the nucleus which specifically bind methylated DNA and thereby deny transcription factors access to gene promotors.

In the first case we know of some transcription factors that are sensitive to methylation in this way (Kovesdi et al., 1987; Watt and Molloy,1988; Iguchi-Ariga and Schaffner, 1989; Shen and Whitlock, 1989; Comb and Goodman, 1990). Two factors detected in HeLa cells are unable to bind to sites containing methyl-CpG and are thus unable to stimulate transcription from the adenovirus late and E2 promoters (Kovesdi et al., 1987; Watt and Molloy, 1988). On the other hand the general transcription factor Sp1 binds equally well to methylated and non-methylated sites and stimulates transcription from both kinds of template (Harrington et al., 1988; Höeller et al., 1988). It should also be recognised that many transcription factors do not contain the dinucleotide CpG in their recognition sequence, so it is difficult to envisage how methylation could directly inhibit the binding of such factors.

Current evidence strongly favours the second idea that repression is mediated by proteins that bind to DNA containing methyl-CpG (Boyes and Bird, 1991; Levine et al., 1991). Early evidence supporting such a mechanism came from microinjection studies with the methylated *Herpes simplex* virus thymidine kinase (tk) gene (Buschausen et al., 1987). In this case the methylated tk gene was transcribed normally for about 8 hours until repressed. The timing of inhibition coincided with assembly of the methylated tk gene into chromatin. This suggested that the inhibition was indirect and may involve proteins that bind to methylated DNA. Additional evidence for the existence of such factors in mammalian nuclei was that m^{5}C (but not thymidine) in chromatin is refractory to digestion by micrococcal nuclease (Solage and Cedar, 1978) and to nucleases that can cleave at CpG (Hansen et al., 1988; Antequera et al., 1989). Sequences artificially methylated in vitro are transcriptionally repressed when transfected into cells and also adopt a nuclease-insensitive chromatin structure (Keshet et al., 1986). A refinement of the indirect model which can accommodate both gene repression and altered chromatin would be that methyl-CpG binding proteins, or MeCPs, bind to a methylated gene leading to an altered chromatin structure which would in turn deny access to the transcription machinery (Lewis and Bird, 1991). Binding of MeCPs in this model could also account for the other chromatin suppressor effects associated with DNA methylation such as late replication and inhibition of recombination.

Methyl-CpG binding proteins (MeCPs)

We have detected two such proteins (MeCPs 1 and 2) by conventional biochemical techniques in mammalian and avian nuclei. We have purified, cloned and sequenced one of these (MeCP2). MeCP1 binds in vitro to DNA containing at least 12 symmetrically methylated CpGs (Meehan et al., 1989), while MeCP2 can bind to a single methylated CpG pair (Lewis et al., 1992). Neither protein will bind significantly to DNA that contains m^{5}C in a sequence other than CpG (Meehan et al., 1989, 1992). Furthermore, neither protein shows an affinity for TpG, which also carries a methyl group at the 5 position of the pyrimidine ring preceding a G (but is not self complementary). The relaxed sequence specificity and widespread tissue distribution of these proteins makes them likely candidates for mediators of the effects of methylation. MeCP1 is of relatively low abundance (about 5000 molecules per nucleus), and is loosely bound, whereas MeCP2 is comparatively abundant (about 200,000 molecules per nucleus) and is only released from chromatin by high salt concentrations. Both proteins are deficient in embryonal carcinoma (EC) and stem (ES) cells.

A third methylated DNA binding protein has been detected in human placenta but this protein is highly sequence-specific (Huang et al., 1984; Wang et al., 1986) and may actually correspond to a methylation-insensitive transcription factor. This protein is therefore unlikely to be involved in general methylation-mediated inhibition. It has also been reported that histone H1 binding to methylated DNA in vitro can inhibit the normally methylation-insensitive restriction enzyme *Msp*I (Higurashi and Cole, 1991).

Biological significance of MeCPs

Studies of MeCP1 have implicated it in methylation-associated gene inactivation. The expression of the X-linked mouse PGK1 gene promoter can be inhibited by methylation in transcription extracts and in cell lines which contain MeCP1 (Boyes and Bird, 1991). Expression can be restored by competition with methylated DNA substrates which are specific for MeCP1. In addition, methylated genes are not efficiently repressed in ES and EC cell lines or extracts which lack MeCPs (Boyes and Bird, 1991; Levine et al., 1991). Histone H1 is present in ES and EC cell lines and so can be ruled out as a methylation-specific transcriptional repressor.

Methylation can also inhibit transcription directly by interference of site-specific methylation within a recognition site for a transcription factor with the binding of that factor. Although we know of some transcription factors that are blocked in this way (see earlier), it is striking that out of seven fully methylated promoters studied so far, none are inhibited strongly under conditions where MeCP1 is absent (Boyes and Bird, 1991, 1992; Levine et al., 1991).

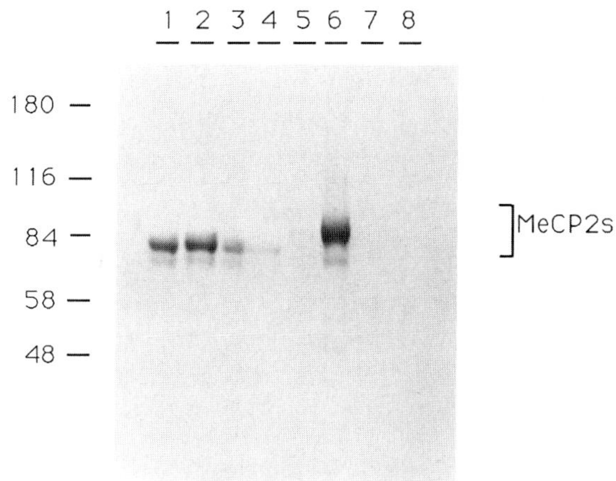

Fig. 1. Detection of MeCP2 in rodent nuclear extracts by western blotting. Nuclear proteins equivalent to 20 µg of DNA were transferred to nitrocellulose membranes after separation by SDS/PAGE and incubated with antibody (Ab76) raised against purified rat MeCP2 (Lewis et al., 1992). The nuclei were from the following sources: lane1, mouse kidney; lane 2, mouse brain; lane 3, mouse spleen; lane 4, mouse liver; lane 5, rat testes; lane 6, rat brain; lane 7, PC13 a mouse embryonal carcinoma cell line; lane 8, D2 an SV40 transformed mouse embryo stem cell line. The numbers on the left correspond to the molecular mass (in kDa) of protein size markers. Pre-immune serum failed to give any signal under these conditions (data not shown).

Thus the predominant repression mechanism appears to work via MeCP1. In keeping with this, several studies have shown that transcription is sensitive to any methylation near the promoter, not just to specific methyl groups at key sites (Murray and Grosveld, 1987).

Although purified MeCP2 has a preference for DNA symmetrically methylated at the dinucleotide CpG, it cannot selectively inhibit transcription from CpG-rich methylated DNA templates in vitro (Meehan et al., 1992). Thus the biological significance of MeCP2 is presently uncertain. In addition to its high specificity for methyl-CpG pairs, the protein contains domains that can interact with low affinity with non-methylated DNA sequences. The latter are readily competed out in the presence of excess non-methylated DNA, implying that they do not overlap the protein domain responsible for methyl-CpG binding (Meehan et al., 1992). It would be premature to conclude, however, that MeCP2 is not involved in transcriptional repression. Nuclease treatment of chromatin indicates that MeCP2 is bound to chromatin, but the transcription experiments were carried out in histone-free extracts in which chromatin cannot form (Meehan et al., 1992). If the natural ligand for MeCP2 is chromatin, as is the case for histone H1 (Wolffe, 1990), this could explain our failure to observe specific effects.

Evidence that MeCP2 is associated with methyl-CpGs in vivo relies upon in situ localisation of the protein using antibodies. A polyclonal antibody raised against purified, denatured rat MeCP2 cross reacts with the homologous protein from mouse (Fig. 1). The mobility of the reacting protein is identical to that of the mouse MeCP2-like activity (80 kD) detected by southwestern assay (Meehan et al., 1992). Antibodies against MeCP2 localise preferentially to centromeric heterochromatin regions, although euchromatic chromosome arms are also stained (Fig. 2; see also Lewis et al., 1992). Pre-immune serum did not stain the chromosomes significantly (Fig. 2). More than half of the methyl-CpGs in the mouse genome are located in the major satellite DNA which is concentrated in centromeric heterochromatin (Manuelides, 1981; Horz and Altenberger, 1981), as calculated from the known distribution of CpGs in satellite and bulk DNA. The asymmetrical distribution of methyl-CpG can also be visualised directly using antibodies against 5-methylcytosine, which preferentially stain regions of pericentromeric heterochromatin (Miller et al., 1974). Thus MeCP2 colocalizes with chromosomal regions that are known to be rich in methyl-CpG. We could also demonstrate that the methylated form of the 234-bp satellite monomer DNA was a good substrate for MeCP2 by southwestern analysis (Lewis et al., 1992) but did not bind MeCP1 in the bandshift assay (S. Cross, unpublished data).

One speculation for MeCP2 function may be in the genome-wide protection of methyl-CpGs against nucleases, as it is much more abundant and more tightly bound in the nucleus than MeCP1. Brain nuclei, which have the highest levels of MeCP2, show particularly striking protection of methyl-CpGs against nucleases (Antequera et al., 1989). Conversely PC13 cells, which have very reduced levels of MeCP1 and MeCP2, show markedly reduced levels of protection (Antequera et al., 1989; Levine et al., 1991). As stated earlier, it has been reported that histone H1 binding to methylated DNA in vitro can inhibit the normally methylation-insensitive restriction enzyme *Msp*I, implying that H1 could be responsible for nuclease protection in nuclei (Higurashi and Cole, 1991). However, this observation does not explain the different levels of protection seen between mouse brain and PC13 nuclei, which appear to have comparable levels of histone H1, unless histone H1 is very different between PC13 cells and brain. As yet, we have been unable to detect a preference for binding to methylated DNA by histone H1 in either the bandshift or southwestern assays.

Genomic sequencing studies on the human PGK1 promoter have not provided any evidence for binding by MeCPs in vivo (Pfeifer et al., 1990a; Pfeifer and Riggs, 1991). However differences were observed between the PGK1 promoters on the active X chromosome (Xa) and the inactive X chromosome (Xi). The unmethylated Xa promoter was free of nucleosomes but did show several footprints indicative of transcription factors (including Sp1), whereas the methylated DNA (60 out of a possible 61 CpG sites) of the Xi promoter was wrapped around positioned particles, which are presumed to be nucleosomes. Interestingly the phased nucleosomes on the Xi promoter covered most of the *Msp*I sites tested for nuclease sensitivity.

The lack of footprints attributable to MeCPs over the methylated PGK1 promoter does not prove their absence. If MeCPs bind randomly and weakly with methyl-CpGs then it may be difficult to detect their interaction with methylated DNA by present methods (Pfeifer and Riggs, 1991). We know that MeCP2 binding to chromatin is very sensitive to nuclease treatment (Meehan et al., 1992). This may account for the failure to detect MeCPs interacting with methyl-CpGs of PGK1 on the Xi using DNase I. There

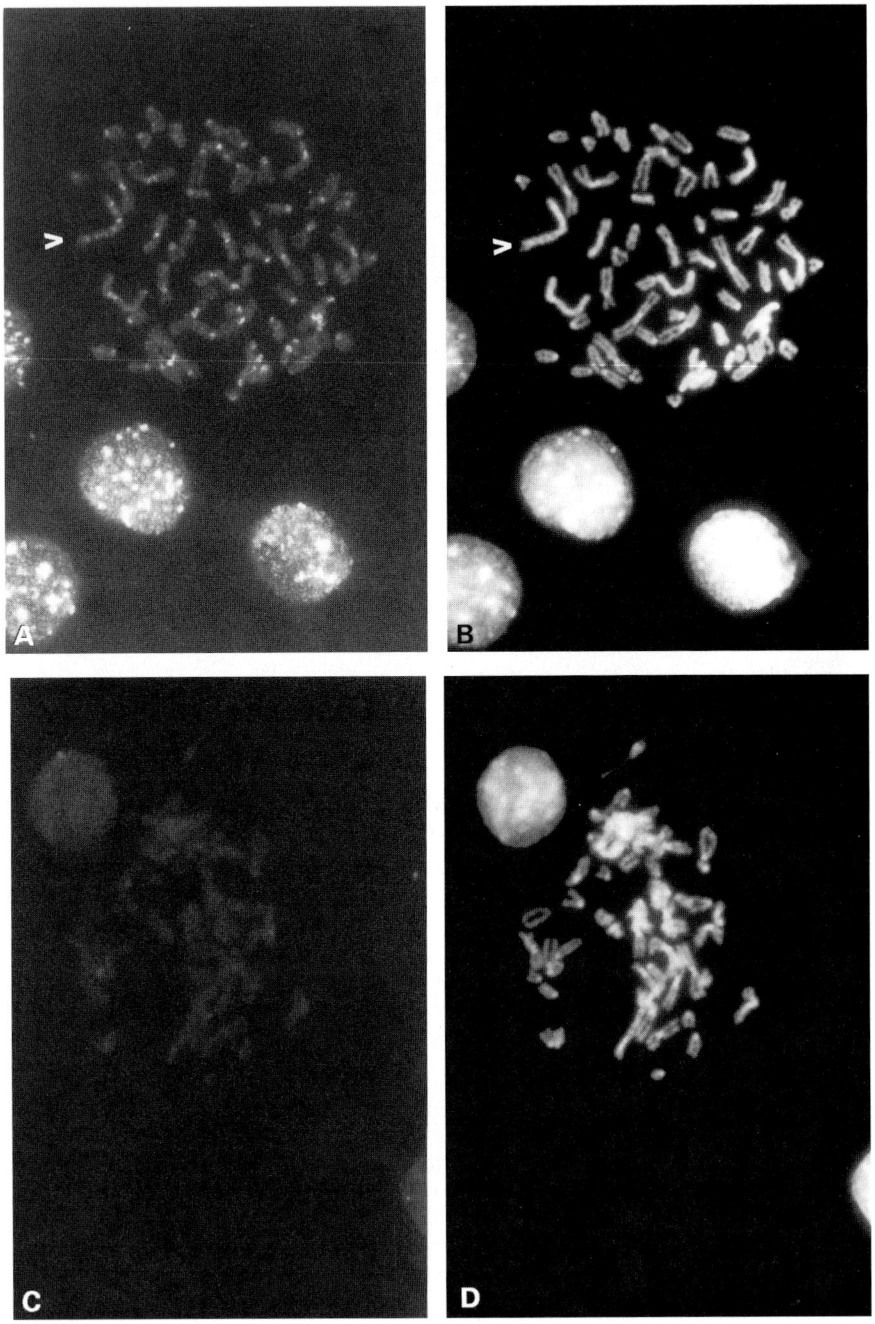

Fig. 2. Distribution of MeCP2 in mouse nuclei and metaphase chromosomes. MeCP2 was detected by indirect immunofluorescence of nuclei and metaphase chromosomes from mouse fibroblasts (cell line L929) using Ab76. Antibody labelling was carried out as described by Jeppesen et al. (1992) and modified by Lewis et al. (1992). Anti-MeCP2 immunofluorescence (A) is confined to nuclei and metaphase chromosomes, where it is particularly enriched in the heterochromatic domains but there is also significant staining on the euchromatic arms. (B) The same field is observed using the DNA fluorochrome Hoechst 33258 and shows the brightly fluorescent heterochromatin in both nuclei and chromosomes. The fluorescent heterochromatin corresponds with the concentration of anti-MeCP2 immunofluorescence. This is especially striking in the case of a marker multicentric chromosome (arrowhead) where each block of residual pericentromeric heterochromatin is immunofluorescently labelled. (D,E) A similar preparation of mouse L929 cells was labelled under identical conditions to those described for A and B using pre-immune rabbit serum instead of AB76. Only background non-specific FITC fluorescence is evident (C). The same field observed by Hoechst 33258 fluorescence is shown in D.

is no evidence that methylation of DNA favours nucleosome formation or determines positioning (Felsenfeld et al., 1982; Drew and McCall, 1987). Without MeCPs it is difficult to account for the reduced *Msp*I resistance in MeCP deficient cells (Antequera et al., 1989; Levine et al., 1991). This problem may be overcome if MeCPs guide nucleosome formation so that methyl-CpGs are rendered nuclease resistant (see Riggs and Pfeifer, 1992, and below).

Mechanisms of methylation-mediated repression

The similarities and differences between MeCP1 and MeCP2 suggest a working model for their roles in methylation-mediated repression. MeCP1 is of relatively low abundance (about 5000 molecules per nucleus), and is loosely bound, whereas MeCP2 is comparatively abundant (about 200,000 molecules per nucleus) and is only released from chromatin by high salt concentrations. MeCP1 may thus compete with transcription factors in the nucleoplasm for binding to methylated DNA, the outcome depending on the density of methylation and the affinity of the factors (Boyes and Bird, 1992; Bird, 1992). We hypothesise that binding to MeCP1 then guides the DNA into a heterochromatic structure involving stable association with MeCP2. How this might happen is unknown, but it probably occurs at DNA replication, since the resistance of methylated DNA to nucleases is known to increase dramatically at this time. Hsieh and Lieber (1992) showed that high-density CpG methylation prevents the V(D)J joining reaction in lymphocytes, but only after DNA replication has taken place. Resistance to *Msp*I was much higher following replication, consistent with the idea that methylation guides the replicating DNA into a 'heterochromatic' structure.

Although direct interference of methyl-CpG with transcription factor binding may not be the primary mechanism of methylation-mediated repression, it may contribute to the MeCP-mediated repression described above. The three parameters which determine the effects of DNA methylation on gene expression are: (1) location of the methylated sites relative to the promoter; (2) density of methylated sites; (3) promoter strength (Boyes and Bird, 1992; Bird, 1992). Direct blockage of a factor which contributes to promoter strength may weaken certain promoters, and may therefore bias the competition between MeCP1 binding and transcription factor binding in favour of MeCP1. In this hypothetical case, direct and indirect inhibition mechanisms, which were thought to be mutually exclusive alternatives, would work together to bring about repression.

Conclusions

MeCPs have been detected in organisms which maintain a large fraction of their genomes in a methylated state, including mammals, birds and plants (Meehan et al., 1992; Zhang et al., 1989). Thus the interaction of MeCPs with methylated DNA may be a common strategy in regulating chromatin activity in these types of organisms. This idea is reinforced by the fact that we have not detected MeCPs in organisms with genomes which have either very reduced methylated DNA fractions or no detectable methylated DNA (eg. *Drosophila*) (Meehan et al., 1992). Our ability to isolate and clone these proteins should allow us to reconstruct in vitro the interaction of chromatin and MeCPs, and investigate how this could influence chromatin structure and activity. Of prime importance will be the generation of mutations in the MeCP loci of mice and the comparison of the phenotype of such mutants with the MTase mutation (Li et al., 1992). We might expect that loss of MeCPs and loss of MTase should give similar phenotypes.

We thank Jillian Charlton for expert technical assistance, Peri Tate and Paco Antequera for critical reading of the manuscript through many drafts and Christine Struthers for help with the manuscript. This work was supported by the Imperial Cancer Research Fund and the Wellcome trust. R. M. and J. L. are members of the ICRF epigenetics laboratory; S. C. and X. N. are supported by the Wellcome trust.

References

Antequera, F., MacLeod, D. and Bird, A. P. (1989). Specific protection of methylated CpGs in mammalian nuclei. *Cell* **58**, 509-517.

Antequera, F., Boyes, J. and Bird, A. (1990). High levels of *de novo* methylation and altered chromatin structure at CpG islands in cell lines. *Cell* **62**, 503-514.

Bird, A. P. (1992). The essentials of DNA methylation. *Cell* **70**, 5-8.

Boyes, J. and Bird, A. (1991). DNA methylation inhibits transcription indirectly via a methyl-CpG binding protein. *Cell* **64**, 1123-1134.

Boyes, J. and Bird, A. (1992). Repression of genes by DNA methylation depends on CpG density and promoter strength: evidence for involvement of a methyl-CpG binding protein. *EMBO J.* **11**, 327-333.

Buschhausen, G., Wittig, B., Graessman, M. and Graessman, H. (1987). Chromatin structure is required to block transcription of the methylated herpes simplex virus thymidine kinase gene. *Proc. Nat. Acad. Sci. USA* **84**, 1177-1181.

Comb, M. and Goodman, H. M. (1990). CpG methylation inhibits proenkephalin gene expression and binding of the transcription factor AP-2. *Nucl. Acids Res.* **18**, 3975-3982.

Drew, M. R. and McCall, M. J. (1987). Structural analysis of a reconstituted DNA containing three histone octamers and histone H5. *J. Mol. Biol.* **197**, 485-511.

Felsenfeld, G., Nickol, J., Behe, M., McGhee, J. D. and Jackson, D. (1982). Methylation and chromatin structure. *Cold Spring Harbor Symp. Quant. Biol.* **47**, 577-584.

Gautsch, J. W. and Wilson, M. C. (1983). Delayed *de novo* methylation in tetracarcinoma cells suggests additional tissue-specific mechanisms for controlling gene expression. *Nature* **301**, 32-37.

Grant, S. G. and Chapman, V. M. (1988). Mechanisms of X-chromosome regulation. *Annu. Rev. Genet.* **22**, 199-233.

Hansen, R. S., Ellis, N. A. and Gartler, S. M. (1988). Demethylation of specific sites in the 5′ region of the inactive X-linked human phosphoglycerate kinase gene correlates with appearance of nuclease sensitivity and gene expression. *Mol. Cell Biol.* **8**, 4692-4699.

Harrington, M. A., Jones, P. A., Masayoshi, I. and Karin, M. (1988). Cytosine methylation does not effect binding of transcription factor Sp1. *Proc. Nat. Acad. Sci. USA* **85**, 2066-2070.

Higurashi, M. and Cole, R. D. (1991). The combination of DNA methylation and H1 histone binding inhibits the action of a restriction nuclease on plasmid DNA. *J. Biol. Chem.* **266**, 8619-8625.

Höeller, M., Westin, G., Jiricny, J. and Schaffner, W. (1988). Sp1 transcription factor binds DNA and activates transcription even when the binding site is CpG methylated. *Genes Dev.* **2**, 1127-1135.

Horz, W. and Altenberger, W. (1981). Nucleotide sequence of mouse satellite DNA. *Nucl. Acids Res.* **9**, 683-676.

Hsieh, C-L. and Lieber, M. R. (1992). CpG methylated minichromosomes become inaccessible for V(D)J recombination after undergoing replication. *EMBO J.* **11**, 315-325.

Huang, L-H., Wang, R., Gama-Sosa, M. A., Shenoy, S. and Ehrlich, M. (1984). A protein from human placental nuclei binds preferentially to 5-methylcytosine rich DNA. *Nature* **308**, 293-295.

Iguchi-Ariga, S. M. M. and Schaffner, W. (1989). CpG methylation of the cAMP-responsive enhancer/promoter sequence TGACGTCA abolishes specific factor binding as well as transcriptional activation. *Genes Dev.* **3**, 612-619.

Jähner, D., Stihlmann, H., Stewart, L. L., Harkens, K., Löhlen, J., Simon, I. and Jaenisch, R. (1982). *De novo* methylation and expression of retroviral genomes during mouse embryogenesis. *Nature* **298**, 632-628.

Jeppesen, P. N., Mitchell, A., Turner, B. and Perry, P. (1991). Antibodies to defined histone epitopes reveal variations in chromatin conformation and underacetylation of centric heterochromatin in human metaphase chromosomes. *Chromosoma* **101**, 322-332.

Jones, P., Wolkowicz, M., Rideout, W., Gonzales, F., Marziasz, C., Coetzee, G. and Tapscott, S. (1990). *De novo* methylation of the MyoD1 CpG island during the establishment of immortal cell lines. *Proc. Nat. Acad. Sci. USA* **87**, 6117-6121.

Keshet, I., Lieman-Hurwitz, J. and Cedar, H. (1986). DNA methylation affects the formation of active chromatin. *Cell* **44**, 535-543.

Kovesdi, I., Reichel, R. and Nevins, J. R. (1987). Role of adenovirus E2 promoter binding factor in E1A-mediated coordinate gene control. *Proc. Nat. Acad. Sci. USA* **84**, 2180-2184.

Levine, A., Cantoni, G-L. and Razin, A. (1991). Inhibition of promoter activity by methylation: possible involvement of protein mediators. *Proc. Nat. Acad. Sci. USA* **88**, 6515-6518.

Lewis, J. D. and Bird, A. P. (1991). DNA methylation and chromatin structure. *FEBS Lett.* **285**, 155-159.

Lewis, J. D., Meehan, R. R., Henzel, W. J., Maurer-Fogy, I., Jeppesen, P., Klein, F. and Bird, A. (1992). Purification, sequence and cellular localization of a novel chromosomal protein that binds to methylated DNA. *Cell* **69**, 905-914.

Li, E., Bestor, T. H. and Jaenisch, R. (1992). Targeted mutation of the

DNA methyltransferase gene results in embryonic lethality. *Cell* **61**, 915-926.

Lock, L. F., Takagi, N. and Martin, G. R. (1987). Methylation of the HPRT gene on the inactive X chromosome occurs after chromosome inactivation. *Cell* **48**, 39-46.

Manuelides, L. (1981). Consensus sequence of mouse satellite DNA indicates it is derived from tandem 116 base pair repeats. *FEBS Lett.* **129**, 25-28.

Meehan, R. R., Lewis, J. D., McKay, S., Kleiner, E. L. and Bird, A. P. (1989). Identification of a mammalian protein that binds specifically to DNA containing methylated CpGs. *Cell* **58**, 499-507.

Meehan, R. R., Lewis, J. D. and Bird, A. P. (1992). Characterization of MeCP2, a vertebrate DNA binding protein with affinity for methylated DNA. *Nucl. Acids Res.* **20**, 5085-5092.

Mellon, I., Bohr, V. A., Smith, L. A. and Hanawalt, P. C. (1986). Preferential DNA repair of an active gene in human cells. *Proc. Nat. Acad. Sci. USA* **83**, 8878-8882.

Miller, L. L., Schnoedl, W., Allen, J. and Erlanger, B. F. (1974). 5-methylcytosine localized in mammalian constitutive heterochromatin. *Nature* **251**, 636-637.

Murray, E. J. and Grosveld, F. (1987). Site specific demethylation in the promoter of human gamma globin gene does not alleviate methylation mediated repression. *EMBO J.* **6**, 2329-2335.

Pfeifer, G. P., Tanguay, P. L. Steigerwald, S. D. and Riggs, A. D. (1990a). In vivo footprint and methylation analysis by PCR-aided genomic sequencing: comparison of active and inactive X-chromosomal DNA at the CpG island and promoter of PGK-1. *Genes Dev* **4**, 1277-1287.

Pfeifer, G. P., Steigerwald, B. S. D., Hansen, R. S., Gartler, S. M. and Riggs, A. D. (1990b). Polymerase chain reaction aided genomic sequencing of an X-chromosome linked CpG island: methylation patterns suggest clonal inheritance, CpG site autonomy and an explanation of activity state stability. *Proc. Nat. Acad. Sci. USA* **87**, 8252-8256.

Pfeifer, G. P. and Riggs, A. D. (1991). Chromatin differences between active and inactive X chromosomes revealed by genomic footprinting of permeabilized cells using DNaseI and ligation mediated PCR. *Genes Dev* **5**, 1102-1113.

Riggs, A. D. and Pfeifer, G. P. (1992). X-chromosome inactivation and cell memory. *Trends Genet.* **8**, 169-174.

Selig, S., Ariel, M., Goitein, R., Marcus, M. and Cedar, H. (1988). Regulation of mouse satellite DNA replication time. *EMBO J.* **7**, 419-426.

Shen, E. S. and Whitlock, J. P. Jr. (1989). The potential role of DNA methylation in the response to 2,3,7,8-Tetrachlorodibenzo-*p*-dioxin. *J. Biol. Chem.* **264**, 17754-17758.

Singer-Sam, J., Grant, M., LeBon, J. M., Okuyama, K., Chapman, V., Monk, M. and Riggs, A. D. (1990). Use of *Hpa*II-polymerase chain reaction assay to study DNA methylation in the Pgk-1 CpG island of mouse embryos at the time of X-chromosome inactivation. *Mol. Cell Biol.* **10,** 4987-4989.

Solage, H. and Cedar, H. (1978). Organization of 5-methylcytosine in chromosomal DNA. *Biochemistry* **17**, 2934-2938.

Wang, R., Zhang, X. Y., Khan, R., Zhou, Y., Huang, L. H. and Ehrlich, M. (1986). Methylated DNA-binding protein from human placenta recognizes specific methylated sites on several prokaryotic DNAs. *Nucl. Acids Res.* **14**, 9843-9859.

Watt, F. and Molley, P. L. (1988). Cytosine methylation prevents binding to DNA of a HeLa cell transcription factor required for optimal expression of the adenovirus major late promoter. *Genes Dev.* **2**, 1136-1143.

Wolffe, A. P. (1990). New approaches to chromatin function. *New Biologist* **2**, 211-218.

Zhang, D., Ehrlich, K. C., Supakar, P. C. and Ehrlich, M. (1989). A plant DNA-binding protein that recognizes 5-methylcytosine residues. *Mol. Cell. Biol.* **9**, 1351-1356.

Journal of Cell Science, Supplement 16, 15-20 (1992)
Printed in Great Britain © The Company of Biologists Limited 1992

Developmental regulation of globin gene expression

MARK MINIE[1], DAVID CLARK[1], CECELIA TRAINOR[1], TODD EVANS[1,*], MARC REITMAN[1], ROBERT HANNON[2], HANNAH GOULD[2] and GARY FELSENFELD[1]

[1]*Laboratory of Molecular Biology, NIDDK, National Institutes of Health, Bethesda, MD 20892, USA* and [2]*Biomolecular Sciences Division, King's College, London WC2B 5RL, England*

*Present address: Department of Biological Sciences, University of Pittsburgh, Pittsburgh, PA 15260, USA

Summary

We have used the globin family of genes in chicken to study developmental regulation of gene expression, both at the level of individual interaction of *trans*-acting factors with local promoters and enhancers, and at the level of chromatin structure. Regulation of all members of the α- and β-globin clusters is affected by the erythroid regulatory factor GATA-1. Separate mechanisms exist for regulation of individual members of the family. As an example, we describe the control mechanisms that play a role in the expression of the ρ-globin gene, which is expressed only in primitive lineage erythroid cells. In addressing the involvement of chromatin structure in gene activation, we have examined the role of locus control elements, and also considered the way in which RNA polymerase molecules might accommodate to the presence of nucleosomes on transcribed genes.

Key words: erythroid development, GATA-1, chromatin, transcription.

Introduction

The globin gene family has served as a model for the study of developmentally regulated gene expression (Evans et al., 1990), allowing investigators to address the questions of how the entire family is activated in an erythroid-specific manner, and how individual family members are turned on and off at successive stages of erythroid development. We have focussed our attention on erythroid development in the chicken, in part because embryonic erythroid development in chickens is well described, and because cells at each developmental stage can be obtained in abundance. Furthermore, since avian erythrocytes retain their nuclei, many regulatory factors can be isolated readily from cells found in the adult circulation.

Two distinct erythroid lineages are produced in the developing chick embryo (Dieterlen-Lievre, 1988; Nikinmaa, 1990). The primitive lineage predominates in the circulation until about day 5 after fertilization; at that point the definitive lineage appears. Primitive lineage cells express the embryonic β-globin genes ρ and ε, and the α-globin genes, α^{π}, α^{A} and α^{D}. Definitive cells on the other hand express β^{H} and β^{A}, as well as α^{A} and α^{D} (Fig. 1). The steps leading to the activation of these genes require the participation of typical *trans* acting factors that bind to promoters and enhancers, as well as the modification of chromatin structure over the α and β domains. Understanding of the globin regulatory pathways thus requires an analysis of both local and long-range interactions.

Our earliest studies of globin gene expression were concentrated on the β^{A}-globin gene, and were greatly aided by the development of a method for transfection of plasmids into primary erythroid cells (obtained from the embryonic circulation) with little or no damage to the cells (Hesse et al., 1986; Lieber et al., 1987). The method permits testing of the activity of globin promoters and enhancers in cells of both the primitive and definitive lineages. This led to the identification of a strong, erythroid-specific enhancer element just 3′ of the β^{A}-globin gene (Hesse et al., 1986; Choi and Engel, 1986), which preferentially activates that gene in definitive lineage cells. As was subsequently shown (Nickol and Felsenfeld, 1988; Choi and Engel, 1988), the enhancer also functions in primitive lineage cells to activate the embryonic ε gene, located downstream of the enhancer.

The general erythroid factor GATA-1

DNase I footprinting in vitro and site mutagenesis allowed us to identify essential elements of the enhancer (Evans et al., 1988; Reitman and Felsenfeld, 1988). Of particular interest was a pair of DNA sequence motifs of the form (A/T)GATA(A/G), which we had already identified in the promoter regions of the ρ and α^{D} genes (Kemper et al., 1987). Related binding sites for this factor (now called GATA-1, originally named Eryf1 in chicken, and NF-E1 or GF-1 in human and mouse) were identified in the promot-

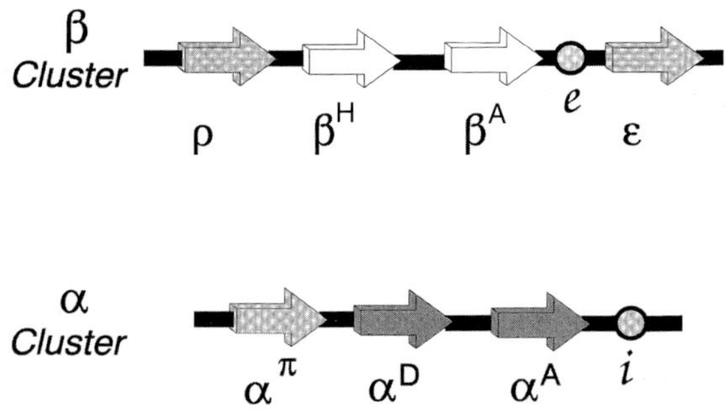

Fig. 1. Organization of the α-globin and β-globin clusters in chicken. The letters *e* and *i* mark the β- and α-cluster enhancers respectively.

ers and enhancers of all the globin genes (Evans et al., 1988). The cDNA from mouse (Tsai et al., 1989) and both the cDNA and gene from chicken, cGATA-1 (Evans and Felsenfeld, 1989; Hannon et al., 1991), have been cloned. The human GATA-1 cDNA and gene (Trainor et al., 1990; Zon et al., 1990; Zon and Orkin, 1992) have also been cloned. cGATA-1 is a 304 amino acid protein with two metal ion 'finger' motifs, which binds typically as a monomer to an asymmetric site (Evans and Felsenfeld, 1989). *Trans*-activation studies in chick embryo fibroblasts (Evans and Felsenfeld, 1991) show that some promoters carrying a single copy of a GATA-1 binding site are strongly stimulated by expression of GATA-1. When plasmids carrying such a GATA-1 expression vector are co-transfected with an appropriate CAT reporter plasmid, a 360-fold stimulation of CAT activity is observed relative to antisense controls (Fig. 2). Remarkably, the presence of additional GATA-1 sites in the promoter does not increase this response in fibroblasts. A quite different effect is seen when the same group of reporters is introduced into primary embryonic erythroid cells: the reporter carrying multiple GATA-1 sites is much more active (Evans and Felsenfeld, 1991). This result has led us to suggest that cGATA-1 is interacting either with an activator present in fibroblasts, or an erythroid cell-specific repressor.

It seems clear that cGATA-1 must play an important role in the activation of the erythroid program. We do not know how cGATA-1 itself is first activated in the early stages of development. Examination of the cGATA-1 promoter (Hannon et al., 1991; Fig. 3) reveals the presence of a cluster of three strong GATA-1-binding sites upstream of the site of transcription initiation. *Trans*-activation studies in primary chick embryo fibroblasts confirm that these sites contribute to the expression of GATA-1. Although this autoregulatory function may well serve to maintain expression of the gene, it clearly cannot initiate it. The first steps in activation of GATA-1 expression may involve either the binding to these sites of other members of the GATA family, or interaction of other factors with putative sites that we have identified in or near the GATA-1 promoter. Notable among these is a *c-myb* site and a site for an unidentified member of the steroid hormone receptor superfamily in the first intron (Fig. 3). We are presently investigating the possible role of these sites in control of GATA-1 expression in early erythroid progenitors.

Control of individual globin genes

GATA-1 clearly can serve as part of a general erythroid 'switch' mechanism, but how can it contribute to lineage-specific expression of individual globin family members? We choose as an example the ρ-globin gene, a member of the β-globin family that is expressed only in cells of the primitive lineage. When plasmid constructions containing the ρ-globin promoter are transiently transfected into primary erythroid cells, the same lineage specificity is observed (Fig. 4). The construction ρCAT is active in primitive lineage cells, but not in the definitive lineage. As shown in Fig. 4, constructions carrying the β^A promoter (β^ACAT**E**) or the α^A promoter (α^ACAT**I**) also are expressed in a way consistent with their behavior in vivo. These results indicate that the information for lineage-specific ('stage'-specific) control of ρ-globin gene expression is contained in the 456 bp-promoter segment of ρCAT. To identify the controlling elements, we have made a series of deletion and site mutations in the ρCAT promoter (Minie et al., 1992). The effect of the site mutations is summarized in Fig. 5. Two elements critical for expression from the ρ promoter in primitive cells are a GATA-1 motif at −221 bp from the cap site, and an Sp1-binding site 50 bp 5′ of the transcriptional start site.

Since GATA-1 and Sp1 are present in both primitive and definitive lineages, it is perhaps difficult to see how they

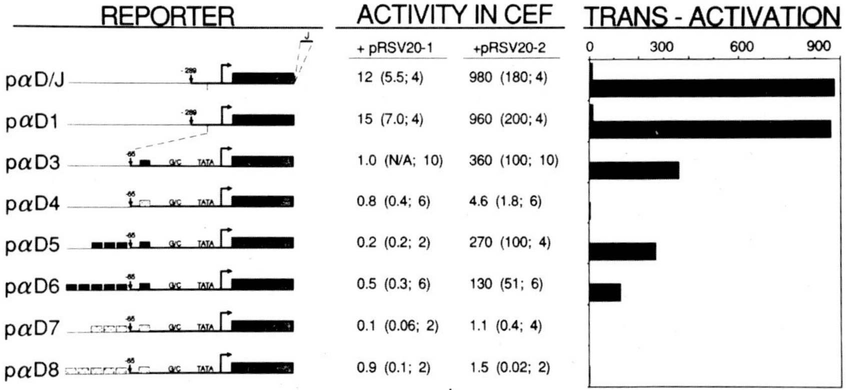

Fig. 2. Effect of expressing cGATA-1 in primary chicken embryo fibroblasts on transcription from various promoters containing GATA-1-binding sites. GATA-1 expression vectors were transiently co-transfected with CAT reporter plasmids carrying various promoters. Results were normalized for transfection efficiency by co-transfection of a β-galactosidase expression vector. The solid black boxes represent GATA-1-binding sites. (From Evans and Felsenfeld, 1991).

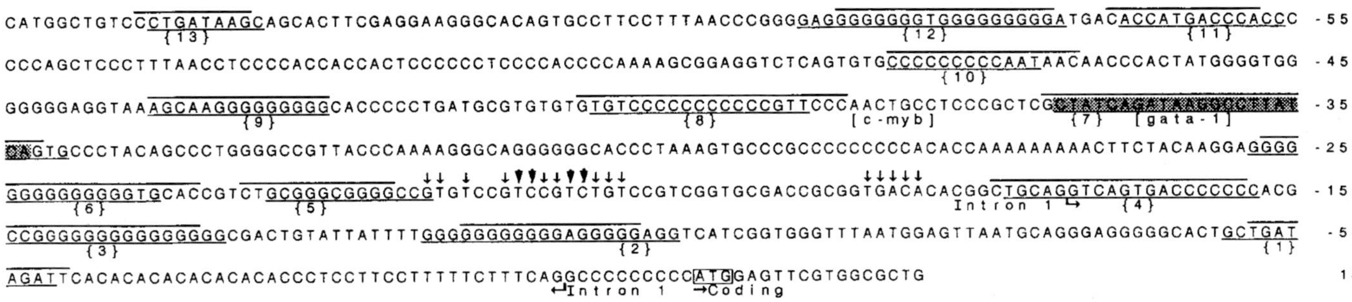

Fig. 3. Protein-binding sites on the promoter of the cGATA-1 gene. Lines represent sequences protected in DNase I footprinting experiments with extracts from erythroid cells. The steroid hormone receptor superfamily binding site is the palindromic sequence GGTCA.. at the 5′ end of the first intron. (From Hannon et al., 1991).

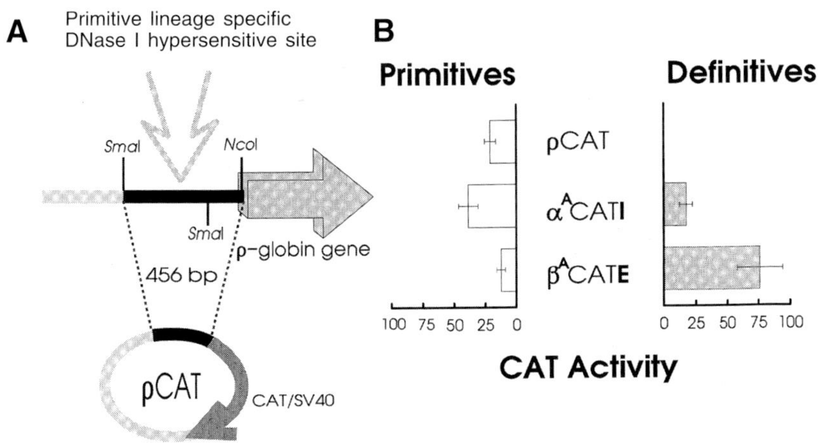

Fig. 4. (A) Plasmid construction used in transient expression studies of the ρ-globin promoter. (B) Lineage specific expression of *cat* under the control of the ρ-, α^A- and β^A-globin promoters. (From Minie et al., 1992).

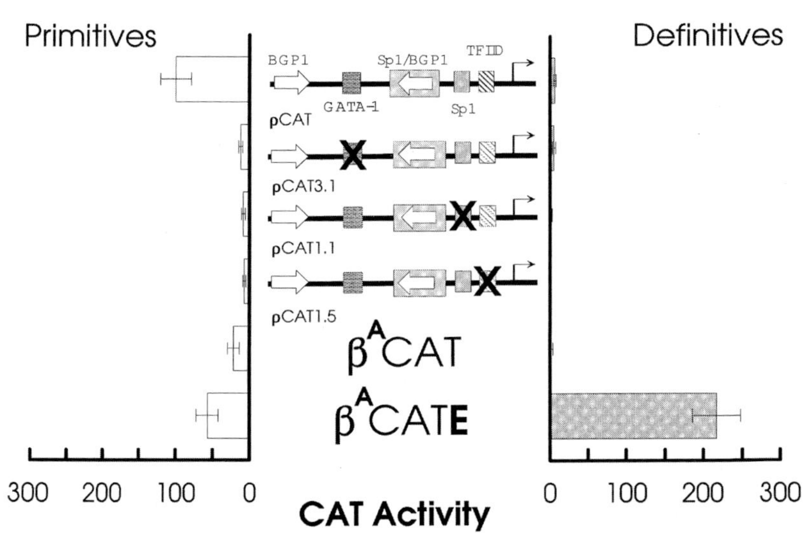

Fig. 5. Effect of mutation in GATA-1-, Sp1- and TFIID-binding sites of the ρ-globin promoter on expressed CAT activity. (Minie et al., 1992).

could be responsible for lineage-specific expression. However, measurement of the nuclear concentrations of both factors in primitive and definitive lineages reveals that there are major differences between the two lineages. As shown in Fig. 6, the concentrations of GATA-1 and Sp1 (as measured by their ability to bind to their specific DNA sites in a gel shift assay) decrease by almost an order of magnitude as the primitive cells are replaced by the definitive lineage. We have also measured GATA-1 mRNA levels: these decrease by 3- to 4-fold.

Since we know the approximate nuclear concentrations of the factors, and their specific binding constants, we can estimate the effect of the decrease in nuclear abundance of factors on the occupancy of sites on the ρ promoter within the nucleus. We calculate that the observed concentration decrease might result in a reduction in the number of doubly

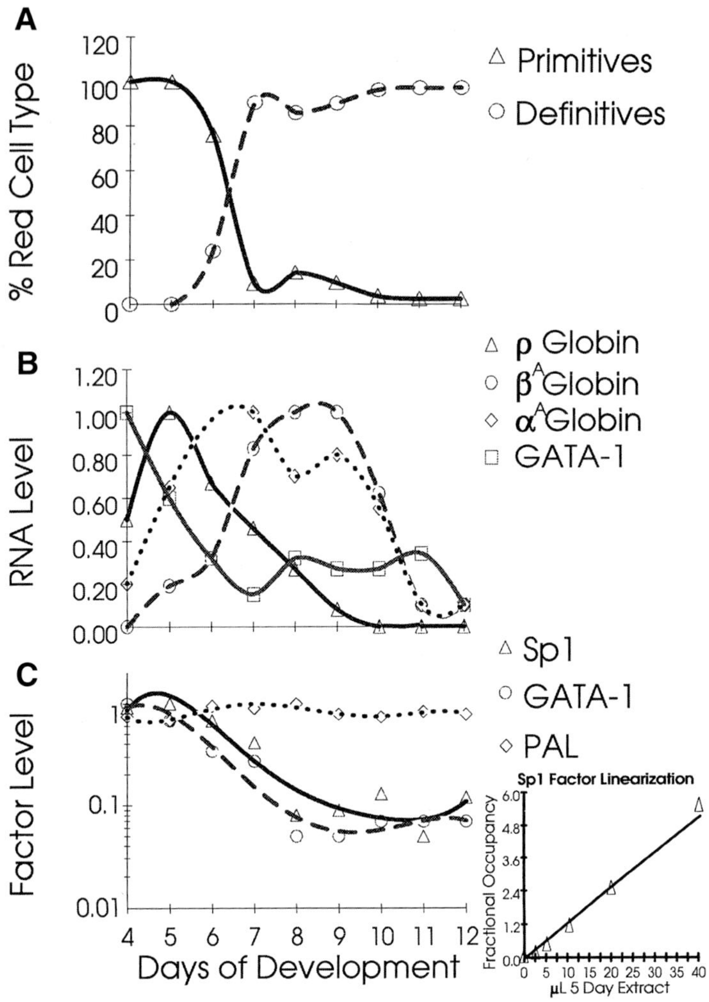

Fig. 6. (A) Abundance of cell lineages in the embryonic circulation during chicken development. (B) Levels of globin and GATA-1 mRNAs during development. (C) Developmental variation in the abundance of Sp1, GATA-1 and the NF-1 family member PAL, measured by gel shift assay. The inset shows a calibration curve for the assay. (From Minie et al., 1992).

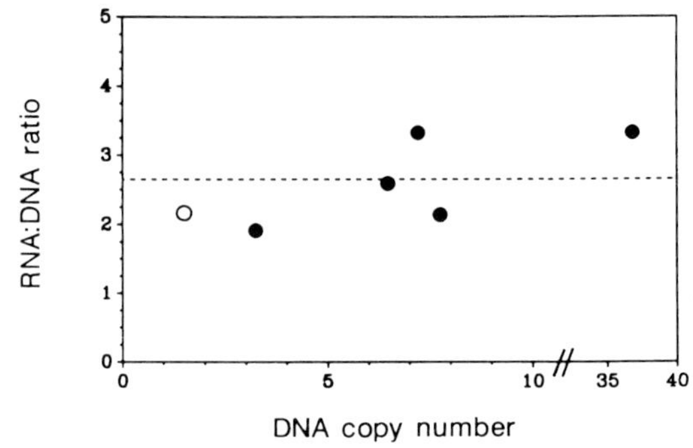

Fig. 7. Levels of chicken β^A-globin mRNA in mice made transgenic with a DNA segment containing the β^A-globin promoter, gene and 3′ β/ε enhancer. The RNA:DNA ratio is the amount of transgene RNA, per transgene copy, expressed as a percentage of the mouse β^{maj} RNA, per gene copy. (From Reitman et al., 1990).

occupied promoter sites by almost two orders of magnitude (Minie et al., 1992). The binding affinities of Sp1 and GATA-1 sites in the control elements of individual globin genes vary considerably. We have suggested that this combination of varying affinity with varying concentration is important in differential regulation of individual globin family members. Although we cannot exclude the possibility of contributions to lineage-specific regulation by other factors and elements not included in the ρCAT construction, sufficient information for such regulation appears to be contained within the proximal promoter region of the ρ-globin gene. Thus, regulation of ρ-globin expression can be controlled by factors that are not confined to a particular lineage. A similar concentration-dependent mechanism (involving additional factors) appears to account for lineage-specific expression of the α^{π}-globin gene (J. Knezetic and G. Felsenfeld, unpublished). On the other hand, control of the β^A-globin gene appears to depend upon the presence of a lineage-specific factor, NF-E4 (Gallarda et al., 1989).

Role of chromatin structure

Distant regulatory elements also exert control over globin gene expression. The locus control region (LCR) of the human β-globin domain is 10 to 20 kb away from the globin gene cluster (Grosveld et al., 1987; reviewed by Evans et al., 1990). Our search for equivalent elements in the chicken β locus has led us to a further examination of the properties of the β/ε enhancer described above. We find that transgenic mice carrying the enhancer, coupled to the β^A-globin gene and its promoter, express the chicken gene in a copy-number-dependent fashion, i.e. the amount of chicken β-globin mRNA made is directly proportional to the number of integrated copies of the gene (Fig. 7; Reitman et al., 1990). This result shows that LCR elements need not be located at the ends of a putative loop domain: the β/ε enhancer is situated in the middle of the β cluster.

Locus control regions are distinguishable from other kinds of regulatory elements that may affect chromatin structure. The above results suggest that LCRs are not likely to mark domain boundaries. Although matrix attachment regions (MARs) may occur nearby, there is no evidence that LCR activity depends on the presence of an MAR. The LCR is also different in properties from DNA sequences, such as the *scs* elements in *Drosophila* (Kellum and Schedl, 1991, 1992) and the A elements in the chicken lysozyme locus (Stief et al., 1989), that appear to isolate genes from the effects of *cis*-acting sequences or chromatin structures which lie outside the region they flank. To a first approximation, when a pair of *scs* sequences surrounds a gene and its local regulatory elements, the gene is expressed at levels governed only by those elements, regardless of where it is inserted in the genome. The *scs* functions passively, perhaps by forming closed loop domains that sterically restrict access from the outside. Locus control regions appear to be more closely related to conventional enhancers, though many enhancers do not function as LCRs, and there is some evidence that LCR elements may not always be enhancers.

It is possible that LCRs serve to establish locally 'open' chromatin domains at the promoters of their target genes (see Felsenfeld, 1992). In any case, the LCR is probably an 'active' control region that exerts its effect by perturbing chromatin structure and stabilizing interactions of *trans*-acting factors with sites on local promoters and enhancers.

Transcription on chromatin templates

It is well established that (with the exception of promoters and enhancers) transcriptionally active eukaryotic genes to a large extent retain their bound histones (Ericsson et al., 1990; Postnikov et al., 1991; Tazi and Bird, 1990; Felsenfeld, 1992). Except at very high rates of transcription, nucleosome cores and even some higher order chromatin structure are likely to be present (Kimura et al., 1983; Caplan et al., 1987; Fisher and Felsenfeld, 1986). How does RNA polymerase transcribe through such structures? We have investigated the possibility that histone octamer displacement accompanies a transient wave of positive supercoiling generated by the advance of the polymerase. The generation of positive supercoils ahead of the polymerase, and negative supercoils behind, has been observed in vivo and in vitro (Liu and Wang, 1987; Tsao et al., 1989; Brill and Sternglanz, 1988). Since formation of nucleosome cores requires the introduction of negative supercoils, it might be expected that positive supercoiling would disfavor nucleosome core formation. We found, however, that in reconstitution experiments in vitro, histone octamers form normal nucleosome structures on positively supercoiled DNA, as judged by their circular dichroic properties, and the chemical reactivity and cross-linking behavior of the histones (Clark and Felsenfeld, 1991). Thus the presence of positive supercoiling does not automatically result in the displacement of a histone octamer. On the other hand, if positive and negative supercoiled plasmids are present together in solution, added histone octamers, under reversible binding conditions, will always prefer the negatively supercoiled DNA. This has led us to propose a model (Fig. 8) in which histone octamers are transferred, one at a time, from a position directly in front of the advancing polymerase to a position directly behind. We have not yet been able to test this hypothesis, but we have undertaken experiments to determine what happens to individual histone octamers during transcription. Our results (Clark and Felsenfeld, 1992) show that nucleosome cores are displaced from their original binding site to other sites on the same plasmid. Although the latter experiments were carried out with a prokaryotic RNA polymerase (SP6), it seems likely that transcription by eukaryotic polymerases will take advantage of the same mechanism, which appears to make use of an intrinsic property of the histones.

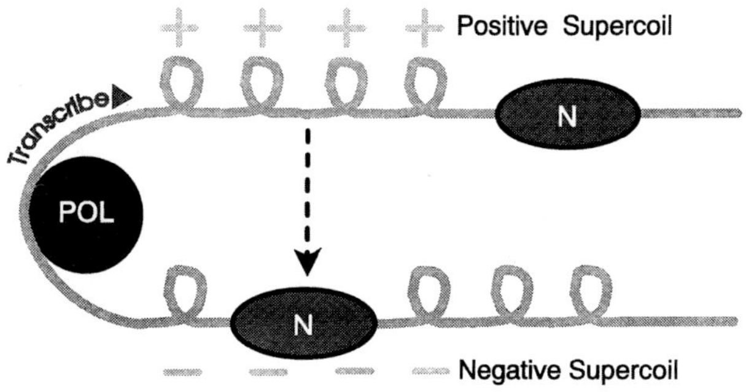

Fig. 8. A hypothetical model for the passage of RNA polymerase through DNA covered with histone octamers, in which one octamer at a time is transferred from the path of the advancing polymerase, driven in part by local supercoiling differences. (From Clark and Felsenfeld, 1991).

Conclusion

The globin gene system provides a model for the study of developmental regulation of gene expression. Within this system it is possible to explore mechanisms of regulation at every level of organization, from the relatively simple interactions of *trans*-acting factors with their sites in local promoters and enhancers, to the still poorly understood effects of distant elements that appear to mediate chromatin structure. The role of chromatin structure is now being addressed in many laboratories. We do not yet know how promoters and enhancers are cleared of histone octamers when genes are activated; activation is likely to involve different mechanisms for different kinds of genes (Felsenfeld, 1992). We also have only a vague idea of how chromatin structure of transcriptionally active domains differs from that of compact chromatin, or how the structure is generated. The globin system of genes can be used to address all of these questions.

References

Brill S. J. and Sternglanz, R. (1988). Transcription-dependent DNA supercoiling in yeast DNA topoisomerase mutants. *Cell* **29**, 403-11.

Caplan, A., Kimura, T., Gould, H. and Allan, J. (1987). Perturbation of chromatin structure in the region of the adult beta-globin gene in chicken erythrocyte chromatin. *J. Mol. Biol.* **193**, 57-70.

Choi, O.-R., and Engel, J. D. (1986). A 3′ enhancer is required for temporal and tissue-specific transcriptional activation of the chicken adult beta-globin gene. *Nature* **323**, 731-734.

Choi O.-R. and Engel, J. D. (1988). Developmental regulation of beta-globin gene switching. *Cell* **55**, 17-26.

Clark, D. J. and Felsenfeld, G. (1991). Formation of nucleosomes on positively supercoiled DNA. *EMBO J.* **10**, 387-395.

Clark, D. and Felsenfeld, G. (1992). A nucleosome core is transferred out of the path of a transcribing polymerase. *Cell* **71**, 11-22.

Dieterlen-Lievre, F. (1988). Birds. In *Vertebrate Blood Cells*, (ed. A. F. Rowley and N. A. Ratcliffe) pp. 257-336. Cambridge University Press.

Ericsson C., Grossbach, U., Bjorkroth, B. and Daneholt, B. (1990). Presence of histone H1 on an active Balbiani ring gene. *Cell* **60**, 73-83.

Evans, T., Reitman, M. and Felsenfeld, G. (1988). An erythrocyte-specific DNA-binding factor recognizes a regulatory sequence common to all chicken globin genes. *Proc. Nat. Acad. Sci. USA* **85**, 5976-5980.

Evans, T. and Felsenfeld, G. (1989). The erythroid-specific transcription factor Eryf1: a new finger protein. *Cell* **58**, 877-885.

Evans, T., Felsenfeld, G. and Reitman, M. (1990). Control of globin gene transcription. *Annu. Rev. Cell Biol.* **6**, 95-124.

Evans, T. and Felsenfeld, G. (1991). *trans*-Activation of a globin promoter in nonerythroid cells. *Mol. Cell. Biol.* **11**, 843-853.

Felsenfeld, G. (1992). Chromatin as an essential part of the transcriptional mechanism. *Nature* **355**, 219-224.

Fisher, E. A. and Felsenfeld, G. (1986). A comparison of the folding of globin and ovalbumin gene-containing chromatin isolated from chicken oviduct and erythrocytes. *Biochemistry* **25**, 8010-8016.

Gallarda, J. L., Foley, K. P., Zhuoying, Y. and Engel, J. D. (1989). The β-globin stage selector element factor is erythroid-specific promoter/enhancer binding protein NF-E4. *Genes Dev.* **3**, 1845-1859.

Grosveld, F., van Assendelft, G. B., Greaves, D. R. and Kollias, G. (1987). Position-independent, high-level expression of the human beta-globin gene in transgenic mice. *Cell* **51**, 975-985.

Hannon, R., Evans, T., Felsenfeld, G. and Gould, H. (1991). Structure and promoter activity of the gene for the erythroid transcription factor GATA-1. *Proc. Nat. Acad. Sci. USA* **88**, 3004-3008.

Hesse, J. E., Nickol, J. M., Lieber, M. R. and Felsenfeld, G. (1986). Regulated gene expression in transfected primary chicken erythrocytes. *Proc. Nat. Acad. Sci. USA* **83**, 4312-4316.

Kellum, R. and Schedl, P. (1991). A position-effect assay for boundaries of higher order chromosomal domains. *Cell* **64**, 941-50.

Kellum, R. and Schedl, P. (1992). A group of scs elements function as domain boundaries in an enhancer-blocking assay. *Mol. Cell. Biol.* **12**, 2424-2431.

Kemper, B., Jackson, P. D. and Felsenfeld, G. (1987). Protein binding sites within the 5′ DNase I hypersensitive region of the chicken α^D-globin gene. *Mol. Cell. Biol.* **7**, 2059-2069.

Kimura, T., Mills, F. C., Allan, J. and Gould, H. (1983). Selective unfolding of erythroid chromatin in the region of the active beta-globin gene. *Nature* **306**, 709-712.

Lieber, M. R., Hesse, J. E., Nickol, J. M and Felsenfeld, G. (1987). The mechanism of osmotic transfection of avian embryonic erythrocytes: analysis of a system for studying developmental gene expression. *J. Cell Biol.* **105**, 1055-1065.

Liu, L. F. and Wang, J. C. (1987). Supercoiling of the DNA template during transcription. *Proc. Nat. Acad. Sci. USA* **84**, 7024-7027.

Minie, M., Kimura, T. and Felsenfeld, G. (1992). The developmental switch in embryonic ρ-globin expression is correlated with erythroid lineage-specific differences in transcription factor levels. *Development* **115**, 1149-1164.

Nickol, J. M. and Felsenfeld, G. (1988). Bidirectional control of the chicken beta- and epsilon-globin genes by a shared enhancer. *Proc. Nat. Acad. Sci. USA* **85**, 2548-2552.

Nikinmaa, M. (1990). *Vertebrate Red Blood Cells: Adaptations of Function to Respiratory Requirements.* New York: Springer-Verlag.

Postnikov Y. V., Shick, V. V., Belyavsky, A. V., Khrapko, K. R., Brodolin, K. L., Nikolskaya, T. A. and Mirzabekov, A. D. (1991). Distribution of high mobility group proteins 1/2, E and 14/17 and linker histones H1 and H5 on transcribed and non-transcribed regions of chicken erythrocyte chromatin. *Nucleic Acids Res.* **25**, 717-25.

Reitman, M. and Felsenfeld, G. (1988). Mutational analysis of the chicken beta-globin enhancer reveals two positive-acting domains. *Proc. Nat. Acad. Sci. USA* **85**, 6267-6271.

Reitman, M., Lee, E., Westphal, H. and Felsenfeld, G. (1990). Site-independent expression of the chicken β^A-globin gene in transgenic mice. *Nature* **348**, 749-752.

Stief, A., Winter, D. M., Stratling, W. H. and Sippel, A. E. (1989). A nuclear DNA attachment element mediates elevated and position-independent gene activity. *Nature* **341**, 343-345.

Tazi, J. and Bird, A. (1990). Alternative chromatin structure at CpG islands. *Cell* **60**, 909-920.

Trainor, C. D., Evans, T., Felsenfeld, G. and Boguski, M. (1990). The sequence, modular structure and evolution of a human erythroid transcription factor. *Nature* **343**, 92-96.

Tsai, S.-F., Martin, D. I. K., Zon, L. I., D'Andrea, A. D., Wong, G. G. and Orkin, S. H. (1989). Cloning of cDNA for the major DNA-binding protein of the erythroid lineage through expression in mammalian cells. *Nature* **339**, 446-451.

Tsao, Y. P., Wu, H. Y., and Liu, L. F. (1989). Transcription-driven supercoiling of DNA: direct biochemical evidence from in vitro studies. *Cell* **56**, 111-118.

Zon, L. I., Tsai, S.-F., Burgess, S., Matsudaira, P., Bruns, G. A. P. and Orkin, S. H. (1990). The major human erythroid DNA-binding protein (GF-1): primary sequence and localization of the gene to the X chromosome. *Proc. Nat. Acad. Sci. USA* **87**, 668-672.

Zon, L. I. and Orkin, S. H. (1992). Sequence of the human GATA-1 promoter. *Nucleic Acids Res.* **20**, 1812.

Journal of Cell Science, Supplement 16, 21-31 (1992)
Printed in Great Britain © The Company of Biologists Limited 1992

cis and *trans* regulation of tissue-specific transcription

J. D. ENGEL[1,*], H. BEUG[2], J. H. LaVAIL[3], M. W. ZENKE[2], K. MAYO[1], M. W. LEONARD[1], K. P. FOLEY[1], Z. YANG[1], J. M. KORNHAUSER[1], L. J. KO[1], K.-C. LIM[1], K. M. GEORGE[1] and K. BREIGEL[2]

[1]*Department of Biochemistry, Molecular Biology and Cell Biology, Northwestern University, Evanston, IL 60208-3500, USA*
[2]*Institute of Molecular Pathology, A1030 Vienna, Austria*
[3]*Department of Anatomy, UCSF, San Francisco, California, USA*

*Author for correspondence

Summary

Analysis of both the *cis*-regulatory sequences which control globin gene switching as well as the *trans*-acting factors which bind to these sequences to elicit a differential, developmentally regulated response has lent insight into the general mechanisms responsible for tissue-specific gene regulation. We show here that the chicken adult β-globin gene promoter sequences are intimately involved in competitive interaction with the β/ε-globin enhancer to regulate differentially ε- versus β-globin gene transcription. Secondly, we show that the family of GATA transcription factors directs gene regulation in a variety of discrete cell types, and describe potential cellular target genes for each member of the GATA factor family, as well as potential mechanisms whereby multiple GATA factors expressed in a single cell might be used to elicit differential transcriptional activities.

Key words: globin gene switching, GATA, transcription factor, T lymphocytes, HIV.

Promoter competition regulates avian β/ε-globin gene switching

During vertebrate ontogeny, erythroid cells undergo a series of morphological and biosynthetic modifications in parallel with changes in lineage and sites of erythropoiesis (Bruns and Ingram, 1973). Transcriptional regulation of this process, particularly with respect to the α- and β-like globin genes, has been extensively studied as a paradigm for cellular commitment, differentiation and embryonic development. In chickens, the *cis*-linked, β-like globin genes are arranged in the order 5′-ρ-β^H-β-ε-3′ (Dolan et al., 1981) and their expression is restricted to either the primitive or definitive erythroid compartments. Primitive cells form by 36 hours of development and express embryonic ρ- and ε-globin (Bruns and Ingram, 1973). Beginning at day 5, erythroid cells of the definitive lineage rapidly replace these primitive cells and express adult β^H- and β-globin (Bruns and Ingram, 1973). This changing pattern of globin isotypes during development is referred to as hemoglobin switching.

On the basis of transient transfection experiments, we proposed that the ε- to β-globin switch is mediated at the transcriptional level by mutually exclusive competition between the promoters of each gene for interaction with a single enhancer (Choi and Engel, 1988). Central to this model was the finding that correct developmental regulation of the ε-globin gene is observed only when it is linked in *cis* to both the shared β/ε enhancer and the adult gene; either duplication of the enhancer or deletion of the adult gene resulted in ε-globin expression in both primitive and definitive erythroid cells (Choi and Engel, 1988). Further analysis demonstrated that definitive suppression of ε-globin transcription requires a stage selector element (SSE) located between −112 and −12 bp of the adult promoter (Choi and Engel, 1988). This is in contrast to β-globin gene transcription, which is regulated autonomously, and is not dependent upon the linked embryonic gene for definitive-stage specificity (Choi and Engel, 1988). Biochemical analysis of proteins interacting with the enhancer and SSE strongly suggested that NF-E4 and βCTF (factors that bind the SSE only in definitive cells; Fig. 1B) are the principle determinants controlling the two alternate promoter/enhancer interactions at different developmental stages (Engel et al., 1989; Gallarda et al., 1989a,b).

As a further test of this model, a cotransfection RNA/PCR assay was used to analyze the effects of individual SSE mutations on transcription of linked β- and ε-globin genes (Foley and Engel, 1992). Specific SSE mutations (see below) were introduced into a test construct containing the genomic β/ε-globin locus which was modified by insertion of short oligonucleotides into the third exon of each gene [pβ′ε′(+).βP(−112)] (Fig. 1A); these inserted sequences serve to differentiate transcripts derived from the transfected and endogenous globin genes. The assay also employs (as a control for transfection and RNA/PCR efficiency) a second construct with genes marked by slightly larger inserts [pβ″ε″(++)] (Fig. 1A).

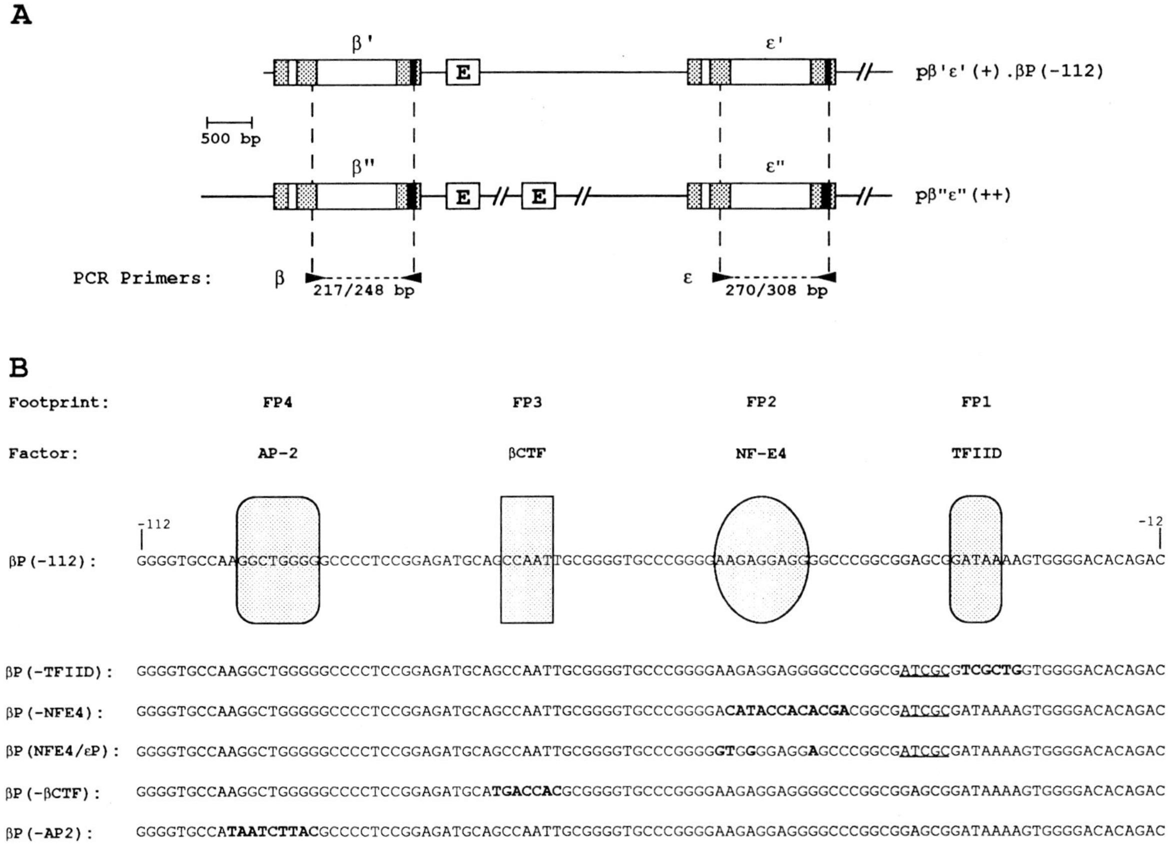

Fig. 1. RNA/PCR assay and β-globin SSE mutations. (A) Structures of the test [pβ′ε′(+).βP(−112)] and control [pβ″ε″(++)] constructs. Shaded and black regions represent exons and third exon inserts, respectively. The positions of enhancer elements are depicted by open boxes labeled E. Arrowheads separated by dashed lines represent β- and ε-globin specific PCR primers and the expected sizes of amplification products derived from cDNA templates are indicated. (B) Sequences of mutated SSEs used to replace the wild-type sequence [βP(−112)] of the test construct. Mutations in individual transcription factor-binding sites are indicated in bold lettering. The identities of the transcription factors that are presumed to bind to these sequences are represented above the wild-type SSE (and are used to identify each mutation).

This control construct contains a duplication of the β/ε enhancer to allow expression of both genes in definitive cells (Choi and Engel, 1988). Two PCR primer pairs were designed that specifically amplify *both* test and control transcripts from either the marked β- or ε-globin genes (Fig. 1A). RNA/PCR (incorporating α[^{32}P]dCTP) was then performed on total RNA isolated from definitive erythroid cells transiently transfected with the test and control constructs. Subsequent analysis by denaturing polyacrylamide gel electrophoresis demonstrated that the ratios between β- and ε-globin test and control signals were constant within the exponential phase of amplification and corresponded to the template ratios present in the starting RNA samples (data not shown; Foley and Engel, 1992).

Since the SSE is required for adult β-globin gene expression as well as for suppression of ε-globin transcription in definitive cells (Choi and Engel, 1988), we individually mutated all four characterized protein-binding sites in this region (Fig. 1B). These modified SSEs were introduced into the test construct and adult β-globin expression was analyzed in definitive cells as described above. Mutation of the TATA box was found to decrease β-globin transcription by 10-fold, while mutation of the distal AP-2 site reduced expression by 2-fold (Fig. 2). Disruption of either the NF-E4 or βCTF-binding sites significantly decreased transcription to within approximately 2-fold of the level observed after deletion of the entire SSE (Fig. 2). Given that NF-E4 and βCTF are definitive stage-specific factors (Gallarda et al., 1989a), these results are consistent with their being principally responsible for β-globin gene activation in the definitive lineage. While βCTF is expressed in definitive cells both before and after overt β-globin transcription, NF-E4 is restricted to the mature definitive stage (Gallarda et al., 1989a). It therefore seems likely that NF-E4 may serve as a key control point regulating β-globin induction during erythroid cell maturation.

The SSE mutations were also analyzed for their effects on expression of the linked ε-globin gene in definitive cells. Mutations in the TATA, NF-E4 and βCTF sites, as well as deletion of the entire SSE, resulted in a 10- to 20-fold activation of ε-globin expression, while the AP-2 mutation resulted in a 4-fold increase (Fig. 2). Comparison of the changes in β- versus ε-globin transcription for each individual mutated site clearly indicates that in definitive cells the expression of each gene is reciprocally related to the other and dependent upon the level of SSE activity (Fig. 2).

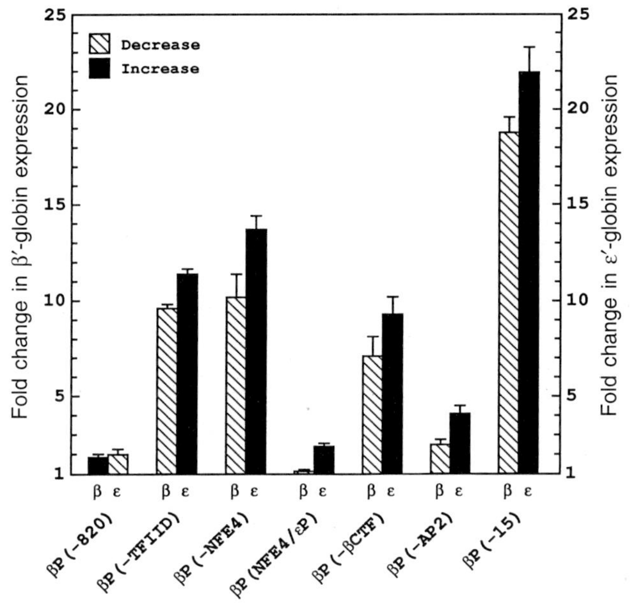

Fig. 2. SSE mutations result in reciprocal changes in β- versus ε-globin expression. The test construct and its derivatives were cotransfected with the control construct into HD3 cells (Beug et al., 1982) and analyzed by RNA/PCR as described in the text. SSE mutations are identified as in Fig. 1, while βP(−820) and βP(−15) refer to deletions of the β-globin promoter to −820 and −15 bps, respectively. The βP(NF-E4/εP) construct is not discussed in the text. The results presented are the fold changes in β′/β″ (β) and ε′/ε″ (ε) ratios relative to that of the wild-type SSE and represent the means (±1 s.d.) from three independent experiments. Increased and decreased expression are indicated by solid and hatched bars, respectively.

These experiments demonstrate that the key predictions of the promoter competition model are correct, and we can draw several conclusions from these results: firstly, that definitive stage-specific transcription of the β-globin gene is dependent on expression of the SSE-binding factor NF-E4; secondly, that β-globin expression is absent in primitive cells due to the lack of NF-E4 and βCTF; and thirdly, that the SSE element is required for suppression of ε-globin transcription in definitive cells. In the first two instances, NF-E4 and βCTF appear to act as transcriptional activators directly on the β-globin promoter; in the third instance, however, they likely function indirectly by mediating preferential association between the enhancer and the adult promoter, thereby preventing ε-globin gene transcription. These results strongly support the existence of a competitive promoter-enhancer equilibrium.

Analysis of the GATA genes and transcription factors

We originally cloned (from a chicken cDNA library) a family of transcription factors capable of binding to a DNA motif with the core sequence GATA and showed that each member of this family is a potent transcriptional activator in vivo (Yamamoto et al., 1990). The three family members (cGATA-1, cGATA-2 and cGATA-3) exhibit greater than 90% amino acid homology to one another within the DNA-binding domain, but are less highly conserved outside of this region. Indeed, the duplicated zinc finger motif that forms the DNA-binding domain is highly conserved among all members of the GATA family from all species examined so far (Zon et al., 1991).

The cGATA factors exhibit distinct tissue and temporal patterns of expression (Yamamoto et al., 1990; M. W. Leonard, K.-C. Lim and J. D. Engel, unpublished observations). The various GATA family members have overlapping but distinct patterns of expression; GATA-1 expression is restricted to erythroid cells, mast cells and megakaryocytes (Tsai et al., 1989; Martin et al., 1990; Romeo et al., 1990). GATA-2 is expressed in a wide variety of cell types and GATA-3 is abundantly expressed in T-lymphocytes, mature erythrocytes and the developing brain (Yamamoto et al., 1990). In some cases, multiple GATA factors are coexpressed and may potentially bind the same regulatory sites of downstream target genes. For example, GATA-1, -2 and -3 are all expressed in definitive chicken erythroid cells (Yamamoto et al., 1990); similarly, GATA-2 and GATA-3 are expressed in identical neurons within the chicken central nervous system (J. M. Kornhauser, M. W. Leonard, M. Yamamoto, J. H. LaVail, K. E. Mayo and J. D. Engel, unpublished observations).

(A) Binding site specificity of the cGATA factor family

The GATA family of transcription factors are all related by their very similar zinc finger DNA-binding domains. The consensus recognition sequence for these factors, WGATAR (Evans et al., 1988; Wall et al., 1988; W = A/T, R = A/G), contains inherent ambiguity, and although each family member, cGATA-1, -2 and -3, has been shown to bind to the same GATA site found in the mouse α-globin promoter (TGATAA), we proposed that each factor might have a slightly different, distinguishable binding specificity still encompassed by the GATA core consensus. The determination of the precise recognition sequences for each factor may be critical to understanding the role these factors play, individually as well as relative to each other, in the tissues in which they are coexpressed. Defining which factor is bound to any given promoter or enhancer region will be necessary to discern which of the GATA factors has a functional role at that given site. In addition to overlapping expression patterns, the cell types with highest expression levels differ for the various GATA factors. It is anticipated that the target genes for GATA-3 in T-cells would differ from those of GATA-1 in red blood cells. Because the WGATAR-binding consensus for the GATA factors was compiled on the basis of erythroid-specific genes, the recognition element of the GATA factors mediating transcriptional regulation in non-erythroid cells could well be different, and must, therefore, be directly assessed.

In order to investigate the binding specificities of the GATA factors, an oligonucleotide was synthesized with a central (GAT) core sequence bordered by randomized nucleotides: 3 nucleotides 5′ (positions −1, −2 and −3) and 4 nucleotides 3′ (positions 1, 2, 3 and 4) from the fixed GAT sequence. The double-stranded oligonucleotide was radiolabeled and then used as a probe in electrophoretic gel mobility shift assays (EGMSA) with purified, bacterially

canonical:			W	**GATA**	R		
cGATA-1:	N	A/C$_{.81}$	A$_{.63}$	**GAT**A$_{1.0}$	A$_{.86}$	C$_{.59}$	A$_{.59}$
cGATA-2:	A$_{.52}$	G/C$_{.69}$	A/T$_{.86}$	**GAT**A$_{.96}$	A/T$_{1.0}$	G/C$_{.71}$	A$_{.47}$
cGATA-3:	A$_{.54}$	G$_{.62}$	A/T$_{.75}$	**GAT**A$_{.83}$	A/T$_{.97}$	T$_{.43}$	A/C$_{.73}$

Fig. 3. Optimum binding sites determined for the cGATA-1, -2 and -3 transcription factors. The consensus preferred binding site for each of the cloned chicken GATA factors is shown. Sites were selected in three rounds on the basis of binding by bacterially expressed, purified glutathione S-transferase fusion proteins of cDNA clones encoding cGATA-1, -2 and -3 (Yamamoto et al., 1990). Selected sites were cloned and sequenced, and then analyzed for the frequency of encountering each nucleotide at each randomized position. Nucleotides shown are those most frequently recovered at each position when sequencing the following number of independently recovered binding sites: cGATA-1, 22 clones; cGATA-2, 52 clones, cGATA-3, 79 clones.

expressed glutathione-S-transferase fusion proteins (Smith and Johnson, 1988) of each cGATA-1, -2 and -3. Selected sites with a high affinity for each GATA protein were then recovered by gel isolating the reduced mobility band and elution from the gel. The selected oligonucleotides were then amplified by PCR. This process was repeated to complete four rounds of selection, and the pool of oligonucleotides was cloned and individual sites were sequenced (Blackwell and Weintraub, 1990).

The frequency of encountering either G, A, T or C was determined for each randomized nucleotide position. Fig. 3 compares, for the three chicken GATA proteins, the most highly favored nucleotide at each position, and the frequencies with which it is recovered in a number of sequenced, selected sites. All the GATA proteins strongly selected for GAT<u>AA</u>, with position 1 being more highly selected than position 2. This result was quite unanticipated, since an A or G (at position +2) is part of the traditional WGATA<u>R</u> consensus (Evans et al., 1988; Wall et al., 1988).

Beyond the 5-nucleotide similarity, at the positions further from the core, the individual factors begin to show different optimal sites. At position −1, GATA-2 and GATA-3 selected a T as frequently as an A, consistent with the GATA consensus; however, GATA-1 strongly preferred an A at that position. Not surprisingly, the further away from the central GAT core, the weaker the selection for a particular nucleotide at that position. Each of the factors seems to have unique preferences at positions −2 and +3; these positions might then be those anticipated to be the most important in discriminating amongst the three proteins. The furthest positions analyzed, −3 and +4, showed relatively weak selection, but, once again, there are subtle, consistent differences between the three proteins.

Surprisingly, the canonical GATA site, WGATAR, was not found to be the most favored site for any of the GATA factors. At the +2 (R) position, not only is G not selected, it seems to be selected <u>against</u>; none of the sequenced binding sites (of >150) obtained with any of the three factors had a G at this position, whereas T and C were found, albeit very infrequently. The WGATAR consensus was determined to include A/G at the +2 position because there are GATAG sites found in the chicken β-globin enhancer (Evans et al., 1988), the chicken α-globin enhancer (Knezetic and Felsenfeld, 1989), the chicken β^{H} promoter (Perkins et al., 1989) and the human $^{A}\gamma$ globin promoter (Martin et al., 1989); however, the site in the chicken β-globin enhancer is the only one of these which has been shown to be functionally significant by mutational analysis (Reitman and Felsenfeld, 1988).

(B) Cloning and characterization of the mGATA-3 gene

Functionally important consensus GATA-binding sites have been identified in transcriptional regulatory regions of a number of murine and human T-cell specific genes (Ko et al., 1991) and so we sought to determine whether T-lymphocytes from species other than chickens similarly express high levels of a GATA-binding factor. We isolated murine and human homologues of cGATA-3 cDNA (mGATA-3 and hGATA-3 respectively; Ko et al., 1991), and demonstrated that all three factors are highly conserved in their amino acid sequences (greater than 90% overall identity between hGATA-3 and cGATA-3 compared with approximately 40% identity between hGATA-1 and cGATA-1; Trainor et al., 1990), have similar tissue distributions (being abundantly expressed in T-lymphocytes and brain) and are capable of activating transcription in vivo (Ko et al., 1991).

As an initial step towards understanding how this complex pattern of GATA-3 expression is achieved, we have isolated and characterized the mGATA-3 locus. Four overlapping clones isolated from a genomic DNA library encompass the entire coding region of mGATA-3 with an additional 18 kb of 5′ and 15 kb of 3′ flanking sequences. The gross organization of the mGATA-3 gene (Fig. 4) is similar to that of mGATA-1 (Tsai et al., 1991) and cGATA-1 (Hannon et al., 1991). Thus, the gene comprises six exons, the first being entirely untranslated and the second containing the initiation methionine codon. Each of the duplicated $C\text{-}X_2\text{-}C\text{-}X_{17}\text{-}C\text{-}X_2\text{-}C$ zinc fingers that define the GATA family is located in a separate exon (exons 4 and 5) and the carboxy terminus of the protein, 3′ untranslated region and polyadenylation site are located in exon 6.

Primer extension and RNase protection assays were performed to determine the transcription initiation site (data not shown). Transcription of the mGATA-3 gene initiates at a unique site 188 nucleotides upstream of the first intron, in contrast to the situation for mGATA-1 and cGATA-1, where extensive mRNA 5′-end heterogeneity was found with multiple transcription initiation sites mapping over approximately 75 nucleotides. As is the case for mGATA-1, the mGATA-3 promoter lacks an identifiable TATA box (although motifs with the sequence TTAA and TAGAA are located at −27 and −37, respectively) and no consensus for the transcription initiator element (Smale and Baltimore, 1989) is present. A number of consensus binding sites for a variety of transcription factors (particularly SP1 and AP2) are present within the proximal 310 bp 5′ to the cap site, but the functional importance of individual elements for mGATA-3 promoter activity has yet to be ascertained. No consensus GATA factor-binding sites exist within this region.

Quantitative RT/PCR was used to assess the level of

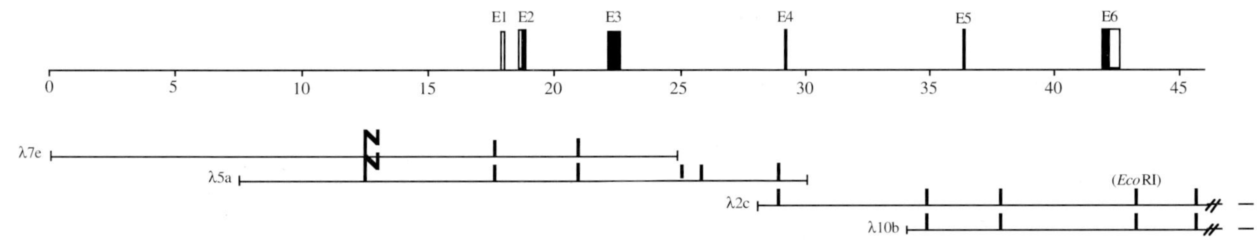

Exon 1 (noncoding 5' end, 188 nt): TCAG/gttggt..Intron 1 (419 bp)...ttgcag/GTGA

Exon 2 (ATG, 630 nt)...CACG/gtgagt...Intron 2 (2.5 kbp)...tcacag/GGAGC...

Exon 3 (536 nt)...ACAG/gtgagt...Intron 3 (7.0 kbp)...ttccag/AAGG...

Exon 4 (amino finger; 145 nt)...GCTG/gtaagt... Intron 4 (6.5 kbp)...atnccag/TCGG...

Exon 5 (carboxy finger; 125 nt)...TAAT/gtaagc...Intron 5 (4.5 kbp)...tttag/ATT...

Exon 6 (3' end; 828 nt).

Chromosome localization: centromere-proximal end of chromosome 2 (10p13); N. Jenkins, personal comunication)

Fig. 4. Organization of the mGATA-3 gene. Recombinant λ bacteriophage DNAs describing the mGATA-3 locus (top line, markers every 5 kbp) are shown. The vertical bars within the clones describe the position of restriction sites for *Eco*RI or the single *Not*I site (N). The sequence organization of the gene, the splice boundaries and the chromosomal location of mGATA-3, are shown in abbreviated form below the genomic clones.

GATA-3 expression in transformed and primary T-lymphocytes from a number of sources. cGATA-3 is expressed in v-*rel* transformed pre-B/T cells, is approximately 5-fold less abundant in RP9 mature B-cells and greater than 10-fold more abundant in MSB mature T-cells (M. W. Leonard, K.-C. Lim and J. D. Engel, unpublished observations). hGATA-3 and mGATA-3 are similarly expressed at some 50- to 100-fold greater abundance in normal mature T-lymphocytes than in B-cells. Furthermore, FACS sorted $CD4^-CD8^-$, $CD4^+CD8^+$, $CD4^+$ and $CD8^+$ murine thymocyte fractions all express high levels of mGATA-3 (the lowest expression being in the $CD4^+CD8^+$ cells and the highest (8-fold higher) in the $CD4^+$ single positive population; M. W. Leonard, D. Kouissis, F. Grosveld and J. D. Engel, unpublished observations). Taken together these data strongly suggest that the GATA-3 transcription factor is likely to play an important role in the regulation of T-cell specific transcription from the earliest stages of T-lymphocyte development.

(C) GATA factors that conditionally regulate transcription

As mentioned above, one enigmatic aspect of GATA factor expression is that multiple members of the family, which ostensibly share very similar DNA-binding site specificity, are expressed in the same cells during erythroid and neuronal development (Yamamoto et al., 1990). The observed changing ratios of cGATA-1, -2 and -3 factors with respect to one another during erythroid differentiation (Yamamoto et al., 1990; M. W. Leonard, K.-C. Lim and J. D. Engel, unpublished observations) suggested that the ratio of these factors might play a role in determining the decision pathway in erythroid progenitor cells: i.e., that expression of one GATA factor at greater or lesser abundance might lead to an altered propensity for regeneration or differentiation in the erythroid developmental program. In order to address this question empirically, conditional mutants of the cGATA factors were prepared. By fusing the hormone-binding domain of the human estrogen receptor (hER; Kumar et al., 1986) to the carboxy termini of the chicken GATA factors (Yamamoto et al., 1990; Fig. 5A), we hoped to generate cGATA regulatory proteins whose activities could be conditionally regulated by the addition or removal of the ligand β-estradiol (E_2).

As anticipated, the presence or absence of estradiol did not influence the ability of the three wild-type cGATA factors to *trans*-activate the C3βGH reporter plasmid (Fig. 5B; Yamamoto et al., 1990). The cGATA/ER proteins, on the other hand, *trans*-activate the reporter plasmid only in the presence of E_2 (Fig. 5B). cGATA-1/ER was the best transcriptional activator in this assay and, perhaps surprisingly, activates the reporter plasmid to a greater extent than the parent factor, cGATA-1. The level of *trans*-activation by cGATA-2/ER is approximately half that of cGATA-2. cGATA-3/ER appears to be the poorest activator of the chimeric proteins, although its ability to stimulate transcription is quite comparable to that of the parental wild-type cGATA-3 factor. In the presence of the estrogen antagonist tamoxifen, the degree of *trans*-activation of cGATA-1/ER is reduced to a level comparable to that of native cGATA-1, whereas treatment of cells with antagonist ICI results in rather poor *trans*-activation by cGATA-1/ER. cGATA-2/ER and cGATA-3/ER are both completely inactive in the presence of either antagonist.

While the *trans*-activating activities of cGATA-2/ER and cGATA-3/ER do not differ markedly from their wild-type counterparts, cGATA-1/ER *trans*-activates a reporter construct significantly better than the native cGATA-1 factor itself. As there is a hormone-responsive *trans*-activation domain within the hormone-binding domain of the estrogen receptor (Webster et al., 1988; Lees et al., 1989), it is possible that this hER *trans*-activation domain contributes to the observed increase in *trans*-activation potential of this chimera.

By fusing individual cGATA transcription factor cDNAs to a segment of cDNA encoding the hormone-binding domain of the human estrogen receptor, we generated

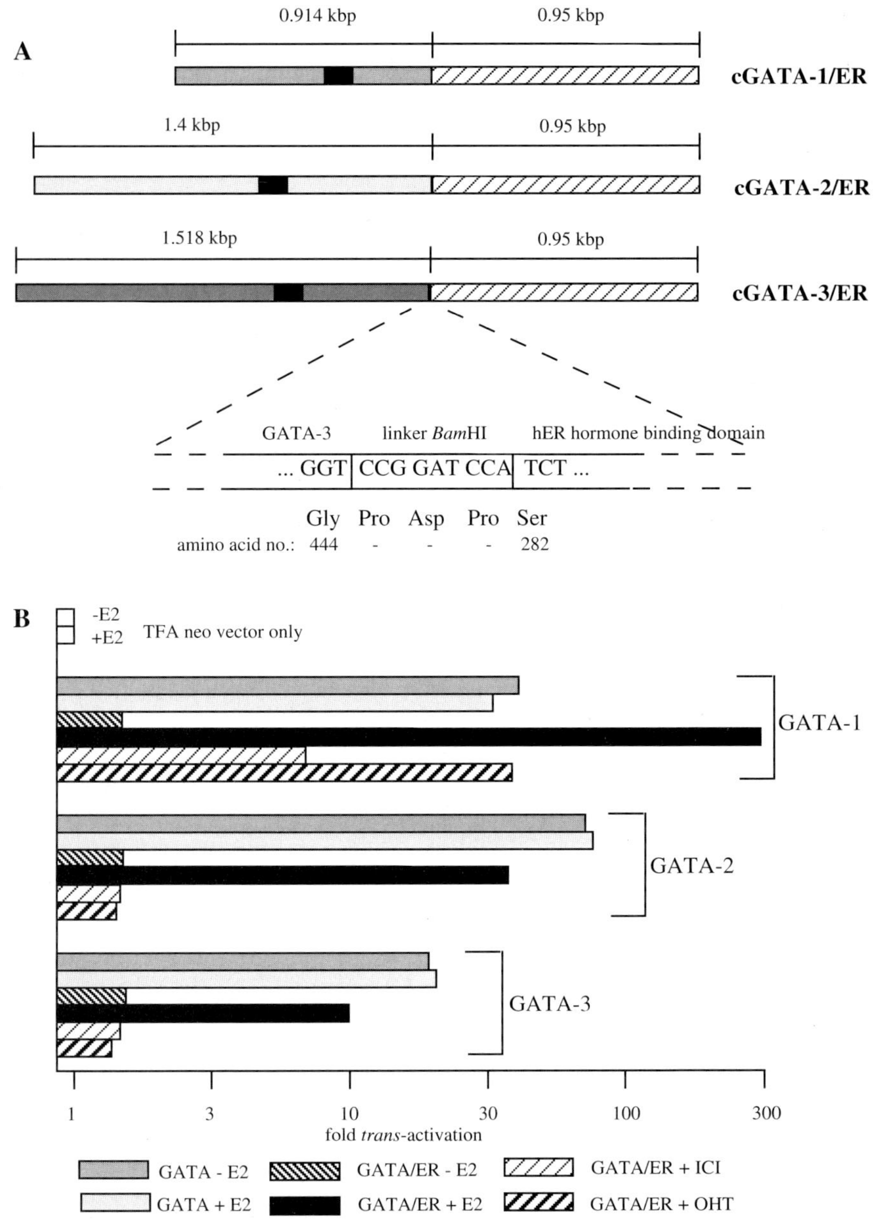

Fig. 5. Conditional GATA factors. (A) Constructions. Full-length cGATA factor cDNAs, lacking termination codons, were generated using PCR. These were then joined in frame to a 0.95 kbp *Bam*HI/*Eco*RI cDNA fragment encoding the hormone-binding domain (amino acids 282 to 595) of the human estrogen receptor (hER) using a PCR-introduced *Bam*HI restriction site at the 3′ end of each of the cGATA cDNA clones. The introduction of the *Bam*HI linker resulted in the insertion of 3 amino acids (Pro-Asp-Pro) in the joining region of the chimeric proteins between the last encoded amino acid of the cGATA factors and Ser_{282} of hER. The resulting chimeric cDNAs (1.9 kb cGATA-1/ER, 2.35 kb cGATA-2/ER and 2.45 kb cGATA-3/ER) were then subcloned into the eukaryotic expression vector TFAneo, in which synthesis of the hybrid cGATA/ER mRNAs is directed by the RSV LTR. (B) cGATA/ER chimeric proteins *trans*-activate GATA-directed reporter plasmids in hormone dependent manner. The optimal ratio of activator (wild-type cGATA or chimeric cGATA/ER) and reporter (C3βGH) plasmids was cotransfected into QT6 quail fibroblast cells, as described (Yamamoto et al., 1990). Sixteen hours post-transfection, β-estradiol (10^{-6} M) or an estrogen antagonist (ICI 164,384 at 10^{-6} M or tamoxifen at 10^{-7} M) in ethanol was added to the cells. Thirty-six hours later, the media was assayed for secreted human growth hormone using the Allegro hGH kit (Nichols Institute Diagnostics). *Trans*-activation activity was calculated from at least two independent experiments as described (Yamamoto et al., 1990).

cGATA/ER chimeric proteins. Using an artificial promoter containing GATA factor-binding sites, we showed that cGATA/ER chimeric proteins are indeed hormone responsive in cotransfection, *trans*-activation assays. In the absence of E_2, the chimeric GATA/ER proteins fail to activate transcription; only the ligand-bound species are able to modulate gene expression.

(D) hGATA-3 regulates HIV-1 transcription

Human immunodeficiency virus type 1 (HIV-1) is the etiologic agent of acquired immune deficiency syndrome (AIDS), and infects human $CD4^+$ T cells and myeloid cells of the monocytic lineage using the CD4 receptor for viral entry (see Cullen and Green, 1989; McCune, 1991; Vaishnav and Wong-Staal, 1991). The retroviral long terminal repeat (LTR) of HIV-1 has been suggested to play a number of roles in the regulation of the life cycle of the virus after it infects cells (Rosen et al., 1985). The viral LTR U3 region contains a variety of *cis*-acting regulatory sequences responsible for modulating viral gene expression. Little is known, at present, regarding the identities of factors that may govern the cell type-restricted expression pattern of HIV-1.

Since the expression profile of hGATA-3 is consistent with the documented tropism of HIV-1 replication, one might anticipate that hGATA-3 could play a role in the T cell-specific regulation of HIV-1 transcription. Indeed, initial inspection of the sequences in the viral LTR indicated the presence of four potential GATA factor-binding sites (Fig. 6A). Both strands of HIV-1 U3 region DNA were examined for their ability to bind to hGATA-3 by DNase I footprinting analysis (data not shown). The footprinting results actually revealed six specific hGATA-3-binding sites, all localized within the 5′ domain of the LTR. The four consensus GATA sites were protected by hGATA-3 factor, as well as two additional non-canonical

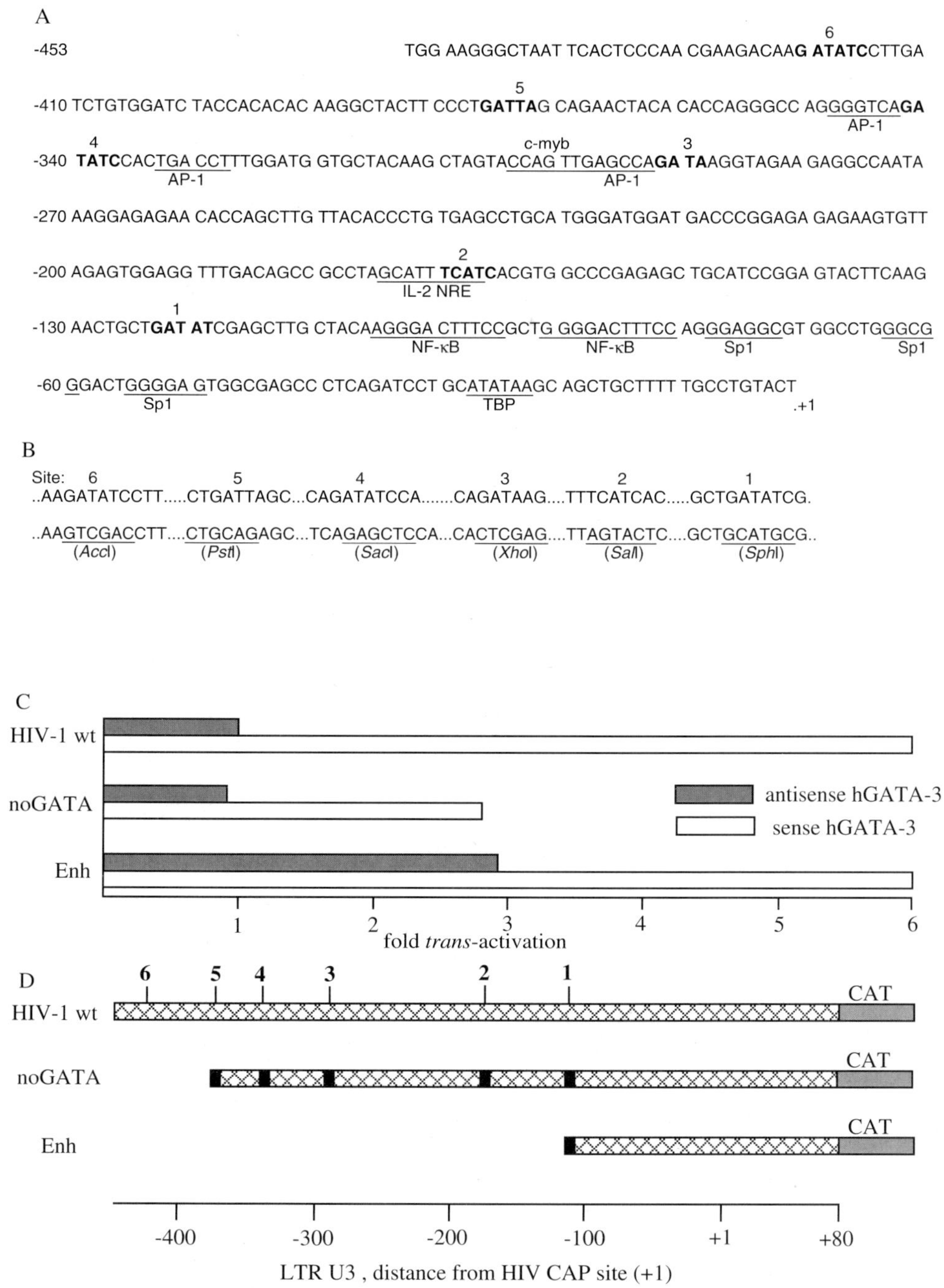

Fig. 6. (A) Nucleotide sequence of, and transcription factor-binding sites in, the HIV-1 LTR. Nucleotide sequence of the HIV-1 LTR from the CAP site (nucleotide +1) extending 5′ to the boundary of the U3 region is shown. The demonstrated and/or putative (consensus sequence) transcription factor-binding sites are underlined and the identity of the corresponding proteins are shown either below or above the sequence (Jones et al., 1986; Nabel and Baltimore, 1987; Hauber and Cullen, 1988; Jakobovits et al., 1988; Shaw et al., 1988; Harrich et al., 1989; Smith and Greene, 1989; Dasgupta et al., 1990; Kato et al., 1991; Berkhout and Jeang, 1992). The binding sites for hGATA-3 identified by in vitro footprinting are shown in outlined letters and are numbered 1 to 6 consecutively, beginning nearest to the HIV-1 CAP site. The numbers at left indicate the nucleotide number relative to the transcriptional start site. (B) GATA-binding sites and mutations. All six hGATA-3-binding sites and surrounding sequences are listed on the top line with the GATA sequence shown in bold. Each GATA site was mutated to a specific restriction enzyme site by PCR mutagenesis (underlined) as shown on the bottom line. (C) hGATA-3 *trans*-activates HIV-1 LTR-driven transcription in HeLa cells. HeLa cells were transfected using the calcium phosphate procedure (Graham and van der Eb, 1973) with 10 μg of the wild-type or mutated HIV-1 LTR constructs plus 5 μg of RSV/hGATA-3 (Ko et al., 1991). 40% of the cells from a confluent plate were used for each transfection. Cell lysates were then prepared by repeated freeze/thaw. Protein concentrations were determined and CAT assays were performed using aliquots of extract containing the equal quantities of recovered protein (Gorman et al., 1982). The result was quantified by determining the amount of ^{14}C-chloramphenicol produced in the enzymatic assay and the conversion was quantitated on a Molecular Dynamics PhosphorImager. The results shown are calculated relative to the conversion by cotransfection with antisense hGATA-3, and are the average of four independent experiments. The transfections were done with the HIV/CAT constructs indicated at the left and with sense strand hGATA-3 (open bar) or antisense strand hGATA-3 (shaded bar). (D) HIV-1 LTR wild type and mutations. Line 1: the wild-type sequence of the HIV-1 LTR −453 to +80. The ‘noGATA’ construct (line 2) represents a plasmid in which each of the GATA-binding sites was mutated by PCR into unique restriction enzyme sites. All of the mutated LTRs were then subcloned 5′ to a promoterless chloramphenicol acetyltransferase (CAT) gene (pCATbasic; Promega) and were confirmed by DNA sequencing.

GATA-binding sites: GATTA (site 5) and GATGA (site 2; Fig. 6A).

We next investigated the possibility that hGATA-3 might *trans*-activate HIV-1 LTR-mediated transcription by transfection into HeLa cells, where only low levels of hGATA-2, but no hGATA-3, are expressed. The HIV-1 LTR

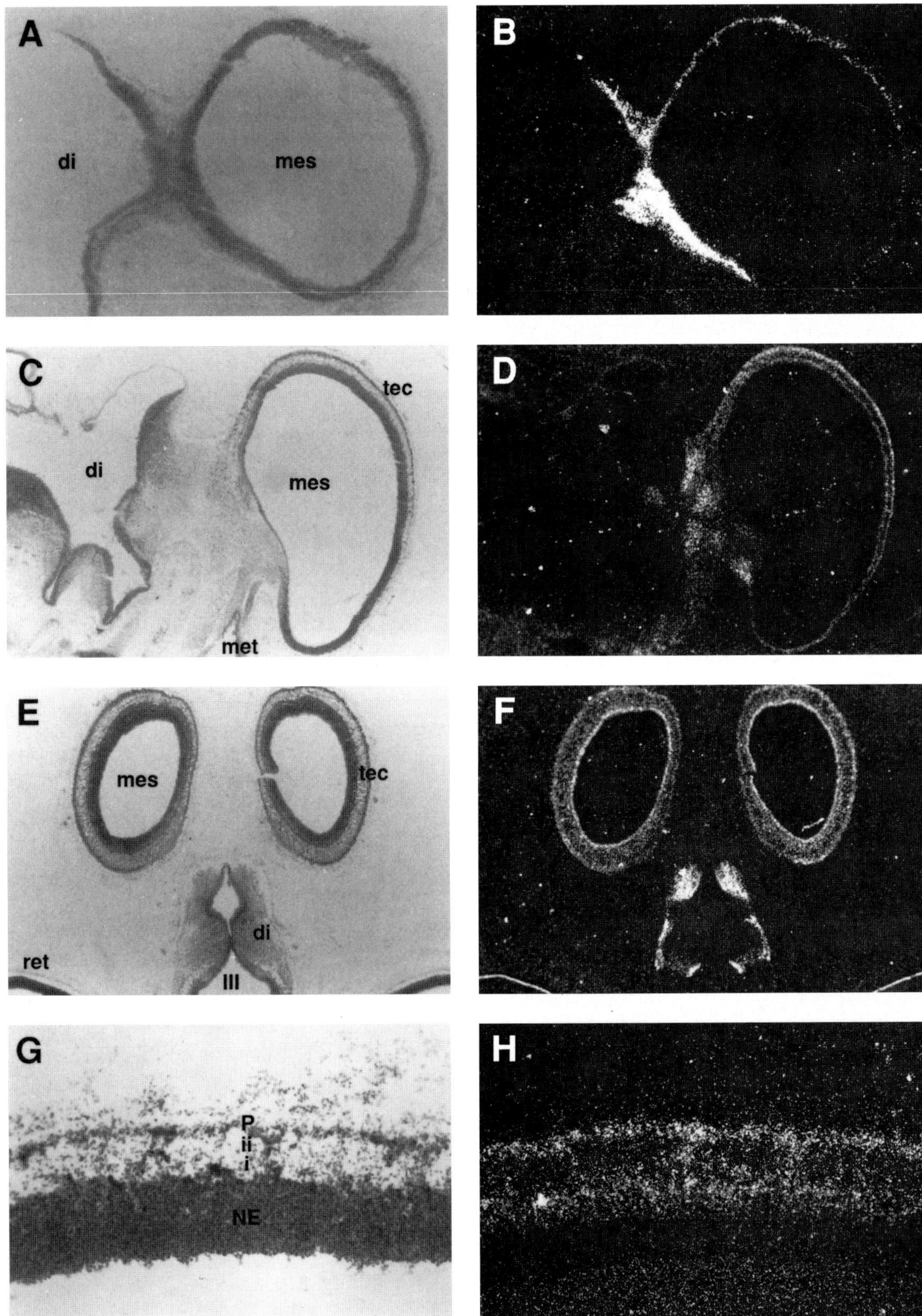

(nucleotides –453 to +80, Fig. 6A) was subcloned into pCAT basic (Promega) to generate HIVwt/CAT; this reporter construct was cotransfected into HeLa cells with an activator plasmid constitutively expressing hGATA-3 protein (RSV/hGATA-3; Ko et al., 1991). As shown in Fig. 6C, hGATA-3 indeed stimulates HIVwt/CAT expression approximately six-fold. These data show that hGATA-3 activates HIV-1 transcription in vivo.

To examine whether the increase in expression was a direct effect of hGATA-3 binding to the identified GATA

Fig. 7. cGATA-2 and cGATA-3 expression in the chicken optic tectum. (A,B) Sagittal section of E4 embryo hybridized to a cGATA-2 probe; (C,D) sagittal section of E6 embryo hybridized with cGATA-3 probe; (E,F) coronal section of E6 embryo hybridized with cGATA-2 probe; (G,H) higher magnification view of a coronal section of E6 brain hybridized with cGATA-3 probe. Bright-field photomicrographs in A, C, E and G show the morphology of the tissue; the same fields are visualized by dark-field microscopy in B, D, F and H to show the autoradiographic silver grains indicating hybridization. di, diencephalon; mes, mesencephalon; met, metencephalon; tec, optic tectum; III, third ventricle; ret, retina; NE, neural epithelium; P, pial layer. Chicken embryos were frozen and stored at −80°C until sectioning. 20 μm-thick coronal or sagittal sections were cut and mounted on gelatin/poly-L-lysine-coated slides. Plasmid cDNA clones encoding cGATA-2 or cGATA-3 (Yamamoto et al., 1990) were digested with restriction enzymes prior to in vitro transcription using either SP6 or T7 RNA polymerase, to generate mRNA-sense or -antisense ^{35}S-labeled RNA probes, respectively. These probes did not contain the sequences encoding the DNA-binding domain to ensure that the hybridization was factor-specific.

sites within the LTR of HIV-1, a series of mutations were generated by PCR in which the GATA sites (Fig. 6B) were individually changed into unique restriction enzyme recognition sequences and then ligated into a single HIV-1 LTR lacking all of the GATA-binding sites (noGATA; Fig. 6D). DNase I footprint analysis confirmed that the mutations indeed eliminated hGATA-3 binding to the GATA sites (data not shown). A 5′ deletion to −120 was also constructed in which only the enhancer region lying 3′ to all the GATA-binding sites remained (Enh; Fig. 6D); this construct still includes the NF-κB, Sp1, TF-IID and LBP sites. The individual and multiple GATA-binding site mutations were then functionally analyzed for their effect upon the ability of hGATA-3 to direct transcription from the HIV-1 LTR. Mutagenesis of the individual GATA-binding sites resulted in only a slight decrease in the transcriptional activation by hGATA-3 (data not shown). However, when all of the mutations were combined into a single plasmid (noGATA/CAT; Fig. 6C), its ability to be *trans*-activated by cotransfected hGATA-3 was significantly reduced. These data taken together demonstrate that hGATA-3 stimulates HIV-1 transcription by binding to the GATA sites within the HIV-1 LTR U3 region.

In summary, the binding of hGATA-3 is required for a 6-fold increase in HIV-1 LTR-mediated transcriptional activation in non-lymphoid cells, and mutations which abolish all of the hGATA-3-binding sites within the LTR result in a quantitatively similar decrease in HIV-1 expression upon cotransfection into non-lymphoid cells (Fig. 6). The reduction in activity seen upon transfection of a mutant HIV-1 LTR bearing mutations in all of the GATA-binding sites into Jurkat cells (data not shown) strongly supports the hypothesis that hGATA-3 may be one of the factors mediating basal expression of HIV 1, and is indeed required for optimal expression of the virus.

(E) Expression of GATA-2 and GATA-3 in the brain

To determine if the expression of the GATA-2 and GATA-3 transcription factor genes is restricted to a subset of cells within the chicken brain, we performed in situ hybridization studies using ^{35}S-labeled cGATA complementary RNA probes. At E3.5 and E4, the earliest embryonic stages examined, strong specific hybridization to GATA-2 and -3 mRNAs is detected at the rostro-ventral boundary of the mesencephalon in the region of the constriction between the mesencephalon (the developing optic lobe) and the diencephalon (Fig. 7). Specific expression of GATA-2 and -3 is also observed within the rostral optic tectum, and most of this hybridization is localized to the outermost portion of the neural epithelium. During development, formation of the tectal layers proceeds in a rostro-ventro-lateral to caudo-dorso-medial direction (LaVail and Cowan, 1971a), and so this rostral portion of the tectum which shows more intense labeling is more mature. GATA mRNAs are also detected along the ventral surface of the metencephalon (data not shown).

By day E6, GATA mRNA expression is more prominent within the optic tectum, although labeling is still visible in the developing diencephalon (Fig. 7). In the day E12 optic tectum, twelve distinct cell layers have been described (LaVail and Cowan, 1971a). GATA-2 and -3 mRNAs are found mainly in layers vi through ix, with the highest expression in layers vi and viii (data not shown). At this stage, the neuroepithelium (NE) has shrunk, and many neuronal progenitor cells have migrated peripherally. No GATA mRNA is detected in the NE at E12, suggesting that the cells expressing GATA-2 and -3 have indeed migrated out of this layer by this time. cGATA-2 and cGATA-3 are expressed in the same tectal cell layers. GATA-2 and -3 mRNAs continue to be expressed in the adult optic tectum. Throughout development the highest levels of cGATA mRNAs are in the mesencephalic region and the developing optic tectum. These data suggest that GATA mRNA expression is restricted to a limited number of cell lineages within the brain, specifically those of the mesencephalon, the mesencephalon-diencephalon junction region, and the ventral metencephalon.

These results suggest a role for cGATA-2 and -3 transcription factors in the regulated expression of specific genes in the developing chicken visual system. Qualitatively, we find that GATA mRNA expression is most prominent within, and is precisely localized to, discrete groups of cells in the developing brain which are physiologically associated with the visual system. These cells are generated during days 6 to 9 of embryogenesis (LaVail and Cowan, 1971b). Thus, the cGATA-2 and -3 genes become activated in neurons generated during this defined period of major neural and morphological organization, and continue to be expressed thereafter. The cells of the optic tectum expressing GATA mRNAs appear to be neuronal, although it is also possible that these factors may be expressed in non-neuronal cells. At all developmental stages examined, the spatial patterns of cGATA-2 and cGATA-3 expression in the brain appear to be identical.

The conservation between chicken and mouse GATA-3 amino acid sequence (Ko et al., 1991) prompted us to perform mRNA in situ hybridization to ask whether the analogous neural structure (the superior colliculus) in mice expresses GATA-2 and GATA-3. At day 14 of gestation, both mGATA-2 and -3 are expressed in the mesencephalic roof (this structure further differentiates to form the supe-

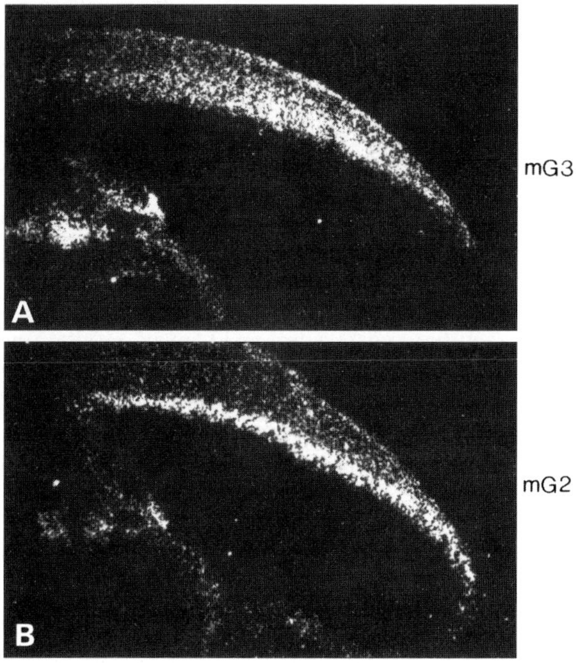

Fig. 8. mGATA-2 and mGATA-3 expression in the superior colliculus at E14.5. Adjacent sagittal sections of an E14.5 embryo were probed with riboprobes specific for each factor. (A) mGATA-3 is expressed in the outer cell layers of the mesecephalic roof. (B) mGATA-2 is expressed in the cell layer directly bordering the ventricular zone. Both factors also show hybridization to distinct cell layers in the tuberculum posterius.

rior colliculus, the primary receptive center for the optic tracts). Thus it appears that the factors may play an important role in the developing vision systems of both species. However, in the mouse, mGATA-2 and -3 are expressed in different cell layers of the developing superior colliculus. mGATA-2 is expressed in less mature cells directly bordering the ventricular zone, while mGATA-3 is expressed in the outer (more-mature) cell layers where pre-neurons are migrating (Fig. 8). This is quite clearly different from the localized expression pattern of these factors seen in the chicken optic tectum, and the significance of this disparity is not yet clear.

Other areas of the murine central nervous system were also analyzed for expression of the two factors. At day 11 of gestation, both mGATA-2 and -3 are expressed in the motor neuron pool and sympathetic ganglia. The pattern is similar at day 14, with the exception that mGATA-2 expression in the motor neuron pool is greatly diminished (data not shown). Taken together these results suggest specific and distinct functions for the two transcription factors during murine neural development.

This work was supported by research grants from the NIH (HL 24415, HL 45168 and GM 28896) and a postdoctoral fellowship from the Leukemia Society of America (M. W. L.).

References

Berkhout, B. and Jeang, K.-T. (1992). Functional roles for the TATA promoter and enhancers in basal and tat-induced expression of the human immunodeficiency virus type 1 long terminal repeat. *J. Virology* **66**, 139-149.

Beug, H., Palmieri, S., Freudenstein, C., Zentgraf, H. and Graf, T. (1982). Hormone-dependent terminal differentiation in vitro of chicken erythroleukemia cells transformed by ts mutants of avian erythroblastosis virus. *Cell* **28**, 907-919.

Blackwell, T. K. and Weintraub, H. (1990). Differences and similarities in DNA-binding preferences of myoD and E2a protein complexes revealed by binding site selection. *Science* **250**, 1104-1110.

Bruns, G. A. P. and Ingram, V. M. (1973). The erythroid cells and haemoglobins of the chick embryo. *Phil. Trans. R. Soc. London Ser. B.* **266**, 225-305.

Choi, O.-R. B. and Engel, J. D. (1988). Developmental regulation of β-globin gene switching. *Cell* **55**, 17-26.

Cullen, B. R. and Greene, W. C. (1989). Regulatory pathways governing HIV-1 replication. *Cell* **58**, 423-426.

Dasgupta, P., Saikumar, P., Reddy, C. D. and Reddy, E. P. (1990). Myb protein binds to human immunodeficiency virus 1 long terminal repeat sequences and trans-activates LTR-mediated transcription. *Proc. Nat. Acad. Sci. USA* **87**, 8090-8094.

Dolan, M., Sugarman, B. J., Dodgson, J. B. and Engel, J. D. (1981). Chromosomal arrangement of the chicken β-type globin genes. *Cell* **24**, 669-677.

Engel, J. D., Choi, O.-R., Endean, D. J., Foley, K. P., Gallarda, J. L. and Yang, Z. (1989). Genetics and biochemistry of the embryonic to adult switch in the chicken ε- and β-globin genes. In *Hemoglobin Switching part A: Transcriptional Regulation* (ed. G. Stamatoyannopoulos and A. Neinhuis), pp 89-103. New York: Alan R. Liss.

Evans, T., Reitman, M. and Felsenfeld, G. (1988). An erythrocyte-specific DNA-binding factor recognizes a regulatory sequence common to all chicken globin genes. *Proc. Nat. Acad. Sci. USA* **85**, 5976-5980.

Foley, K. P. and Engel, J. D. (1992). Individual stage selector element mutations lead to reciprocal changes in β- vs. ε-globin gene transcription: genetic confirmation of promoter competition during globin gene switching. *Genes Dev.* **6**, 730-744.

Gallarda, J. L., Foley, K. P., Yang, Z. and Engel, J. D. (1989a). The β-globin stage selector element factor is erythroid-specific promoter/enhancer binding protein NF-E4. *Genes Dev.* **3**, 1845-1859.

Gallarda, J. L., Yang, Z., Endean, D. J., Foley, K. P. and Engel, J. D. (1989b). *cis* and *trans* determinants of chicken β-globin transcription. In *Tissue Specific Gene Expression* (ed. R. Renkawitz), pp. 103-121. Weinheim: VCH Publishers.

Gorman, C. M., Moffat, L. F. and Howard, B. H. (1982). Recombinant genomes which express chloramphenicol acetyltransferase in mammalian cells. *Mol. Cell. Biol.* **2**, 1044-1051.

Graham, F. L. and van der Eb, A. J. (1973). A new technique for the assay of infectivity of human adenovirus 5 DNA. *Virol.* **52**, 456-467.

Hannon, R., Evans, T., Felsenfeld, G. and Gould, H. (1991). Structure and promoter activity of the gene for the erythroid transcription factor GATA-1. *Proc. Nat. Acad. Sci. USA* **88**, 3004-3008.

Harrich, D., Garcia, J., Wu, F., Mitsuyasu, R., Gonzalez, J. and Gaynor, R. (1989). Role of SP1-binding domains in in vivo transcriptional regulation of the human immunodeficiency virus type 1 long terminal repeat. *J. Virology* **63**, 2585-2591.

Hauber, J. and Cullen, B. R. (1988). Mutational analysis of the *trans*-activation-responsive region of the human immunodeficiency virus type 1 long terminal repeat. *J. Virology* **62**, 673-679.

Jakobovits, A., Smith, D. H., Jakobovits, E. B. and Capon, D. J. (1988). A discrete element 3′ of human immunodeficiency virus 1 (HIV-1) and HIV-2 mRNA initiation sites mediates transcriptional activation by an HIV *trans* activator. *Mol. Cell. Biol.* **8**, 2555-2561.

Jones, K. A., Kadonaga, J. T., Luciw, P. A. and Tjian, R. (1986). Activation of the AIDS retrovirus promoter by the cellular transcription factor, Sp1. *Science* **232**, 755-759.

Kato, H., Horikoshi, M. and Roeder, R. G. (1991). Repression of HIV-1 transcription by a cellular protein. *Science* **251**, 1476-1479.

Knezetic, J. A. and Felsenfeld, G. (1989). Identification and characterization of a chicken α-globin enhancer. *Mol. Cell. Biol.* **9**, 893-901.

Ko, L. J., Yamamoto, M., Leonard, M. W., George, K. M., Ting, P. and Engel, J. D. (1991). Murine and human T lymphocyte GATA-3 factors mediate transcription through a *cis*-regulatory element within the human T cell receptor δ gene enhancer. *Mol. Cell. Biol.* **11**, 2778-2784.

Kumar, V., Green, S., Staub, A. and Chambon, P. (1986). Localisation of the oestradiol-binding and putative DNA-binding domains of the human oestrogen receptor. *EMBO J.* **5**, 2231-2236.

LaVail, J. H. and Cowan, W. M. (1971a). The development of the chick optic tectum. I. Normal morphology and cytoarchitectonic development. *Brain Res.* **28**, 391-419.

LaVail, J. H. and Cowan, W. M. (1971b). The development of the chick optic tectum. II. Autoradiographic studies. *Brain Res.* **28**, 421-441.

Lees, J. A., Fawell, S. E. and Parker, M. G. (1989). Identification of two transactivation domains in the mouse oestrogen receptor. *Nucl. Acids Res.* **17**, 5477-5488.

Martin, D. I. K., Tsai, S.-F. and Orkin, S. H. (1989). Increased γ-globin expression in a nondeletion HPFH mediated by an erythroid-specific DNA-binding factor. *Nature* **338**, 435-438.

Martin, D. I. K., Zon, L. I., Mutter, G. and Orkin, S. H. (1990). Expression of an erythroid transcription factor in megakaryocytic and mast cell lineages. *Nature* **344**, 444-447.

McCune, J. M. (1991). HIV-1: The infective process in vivo. *Cell* **64**, 351-363.

Nabel, G. and Baltimore, D. (1987). An inducible transcription factor activates expression of human immunodeficiency virus in T cells. *Nature* **326**, 711-713.

Perkins, N. D., Nicolas, R. H., Plumb, M. A. and Goodwin, G. H. (1989). The purification of an erythroid protein which binds to enhancer and promoter elements of haemoglobin genes. *Nucl. Acids Res.* **17**, 1299-1314.

Reitman, M. and Felsenfeld, G. (1988). Mutational analysis of the chicken β-globin enhancer reveals two positive-acting domains. *Proc. Nat. Acad. Sci. USA* **85**, 6267-6271.

Romeo, P.-H., Prandini, M.-H., Joulin, V., Mignotte, V., Prenant, M., Vainchenker, W., Marguerie, G. and Uzan, G. (1990). Megakaryocytic and erythrocytic lineages share specific transcription factors. *Nature* **344**, 447-449.

Rosen, C. A., Sodroski, J. G. and Haseltine, W. A. (1985). The location of *cis*-acting regulatory sequences in the human T cell lymphotropic virus type III (HTLV-III/LAV) long terminal repeat. *Cell* **41**, 247-253.

Shaw, J.-P., Utz, P. J., Durand, D. B., Toole, J. J., Emmel, E. A. and Crabtree, G. R. (1988). Identification of a putative regulator of early T cell activation genes. *Science* **241**, 202-205.

Smale, S. T. and Baltimore, D. (1989). The 'initiator' as a transcriptional control element. *Cell* **57**, 1259-1273.

Smith, D. B. and Johnson, K. S. (1988). Single-step purification of polypeptides expressed in *Escherichia coli* as fusions with glutathionine S-transferase. *Gene* **67**, 31-40.

Smith, M. R. and Greene, W. C. (1989). The same 50-kDa cellular protein binds to the negative regulatory elements of the interleukin 2 receptor α-chain gene and the human immunodeficiency virus type 1 long terminal repeat. *Proc. Nat. Acad. Sci. USA* **86**, 8526-8530.

Trainor, C. D., Evans, T., Felsenfeld, G. and Boguski, M. S. (1990). Structure and evolution of a human erythroid transcription factor. *Nature* **343**, 92-96.

Tsai, S.-F., Martin, D. I. K., Zon, L. I., D'Andrea, A. D., Wong, G. G. and Orkin, S. H. (1989). Cloning of cDNA for the major DNA-binding protein of the erythroid lineage through expression in mammalian cells. *Nature* **339**, 446-451.

Tsai, S.-F., Strauss, E. and Orkin, S. H. (1991). Functional analysis and in vivo footprinting implicate the erythroid transcription factor GATA-1 as a positive regulator of its own promoter. *Genes Dev.* **5**, 919-931.

Vaishnav, Y. N. and Wong-Staal, F. (1991). The biochemistry of AIDS. *Ann. Rev. Biochem.* **60**, 577-630.

Wall, L., deBoer, E. and Grosveld, F. (1988). The human β-globin gene 3′ enhancer contains multiple binding sites for an erythroid-specific protein. *Genes Dev.* **2**, 1089-1100.

Webster, N. J. G., Green, S., Jin, J. R. and Chambon, P. (1988). The hormone-binding domains of the estrogen and glucocorticoid receptors contain an inducible transcription activation function. *Cell* **54**, 199-207.

Yamamoto, M., Ko, L. J., Leonard, M. W., Beug, H., Orkin, S. H. and Engel, J. D. (1990). Activity and tissue-specific expression of the transcription factor NF-E1 multigene family. *Genes Dev.* **4**, 1650-1662.

Zon, L. I., Mather, C., Burgess, S., Bolce, M. E., Harland, R. M. and Orkin, S. H. (1991). Expression of GATA-binding proteins during embryonic development in *Xenopus laevis*. *Proc. Nat. Acad. Sci. USA* **88**, 10642-10646.

Journal of Cell Science, Supplement 16, 33-38 (1992)
Printed in Great Britain © The Company of Biologists Limited 1992

The *bicoid* and *dorsal* morphogens use a similar strategy to make stripes in the *Drosophila* embryo

Y. TONY IP, MICHAEL LEVINE and STEPHEN J. SMALL

Biology Department, Center for Molecular Genetics, UCSD, 9500 Gilman Drive, La Jolla, CA 92093, USA

Summary

The anterior-posterior (A-P) and dorsal-ventral (D-V) axes of the early *Drosophila* embryo are established by two key maternal morphogens: *bicoid* (*bcd*) and *dorsal* (*dl*), respectively. The *bcd* protein is expressed in a broad concentration gradient along the A-P axis, with peak levels present at the anterior pole, while *dl* is expressed in a gradient along the D-V axis with peak levels along the ventral surface. The two morphogens are unrelated and their gradients are formed by distinct processes. Nonetheless, we have obtained evidence that they generate sharp on/off stripes of target gene expression through a similar mechanism. Both morphogens establish overlapping patterns of transcriptional activators and repressors in the early embryo. The activators and repressors bind to closely linked sites within short (300 to 500 bp) target promoter elements that have the properties of on/off switches. The activators act in concert with the morphogen to define a broad region where target genes can be initiated. Borders of target gene expression are established by the repressors, resulting in the formation of stripes.

Key words: *Drosophila*, embryogenesis, morphogen, stripes, repression.

Introduction

We are interested in understanding how morphogens control development. In particular, how do gradients of morphogens generate sharp patterns of gene expression in the early *Drosophila melanogaster* embryo? There are two key maternal morphogens in *Drosophila*, *bicoid* (*bcd*) and *dorsal* (*dl*) (for review, see St Johnston and Nusslein-Volhard, 1992). These are expressed during oogenesis and are present in the unfertilized egg. The *bcd* morphogen is important for the specification of cell identity along the anterior-posterior (A-P) axis. The protein is distributed in a broad concentration gradient in early embryos, with peak levels at the anterior pole and progressively lower levels in more posterior regions. In contrast, *dl* is uniformly expressed along the A-P axis, but is graded along the dorsal-ventral (D-V) axis, with peak levels along the ventral surface and progressively lower levels in more lateral and dorsal regions (Steward et al., 1988). This *dl* gradient is important for establishing different tissues along the D-V axis.

Although the *bcd* and *dl* proteins are both sequence-specific transcription factors, they are completely unrelated. *bcd* contains a homeobox (Frigerio et al., 1986; Berleth et al., 1988), while *dl* contains the REL domain and is related to the mammalian regulatory factor NF-kB and the oncoprotein *rel* (Steward, 1987; Ghosh et al., 1990; Kieran et al., 1990). In addition, the *bcd* and *dl* gradients are formed through totally different processes. In the case of *bcd*, the mRNA is trapped at its site of entry at the anterior pole of growing oocytes (Frigerio et al., 1986; Macdonald and Struhl, 1988). This mRNA serves as the source of the morphogen, which is translated in anterior regions and diffuses into more posterior regions (Driever and Nusslein-Volhard, 1988). In contrast to this process of mRNA localization, the *dl* gradient is formed by regulated nuclear transport (Roth et al., 1989; Rushlow et al., 1989; Steward, 1989). The *dl* protein is initially detected throughout the cytoplasm of unfertilized eggs and early embryos. About 90 minutes after fertilization, protein in ventral regions enters nuclei while that present in dorsal regions remains in the cytoplasm. Despite these considerable differences, we have obtained evidence that the two morphogens use a similar strategy to produce sharp patterns of gene expression, particularly stripes, in the early embryo.

To understand how the *bcd* and *dl* morphogens work, we have analyzed the expression of target genes that they directly regulate. We have examined several such genes, but here we will focus on two: the *bcd* target *even-skipped* (*eve*) and the *dl* target *rhomboid* (*rho*).

The expression patterns of target genes

eve encodes a homeodomain protein that plays an important role in segmentation: mutant embryos that lack *eve* protein fail to form segment borders (Nusslein-Volhard et al., 1985; Harding et al., 1986; Macdonald et al., 1986). The *eve* protein is initially expressed throughout all embryonic nuclei at about 2 h after fertilization. During a short period,

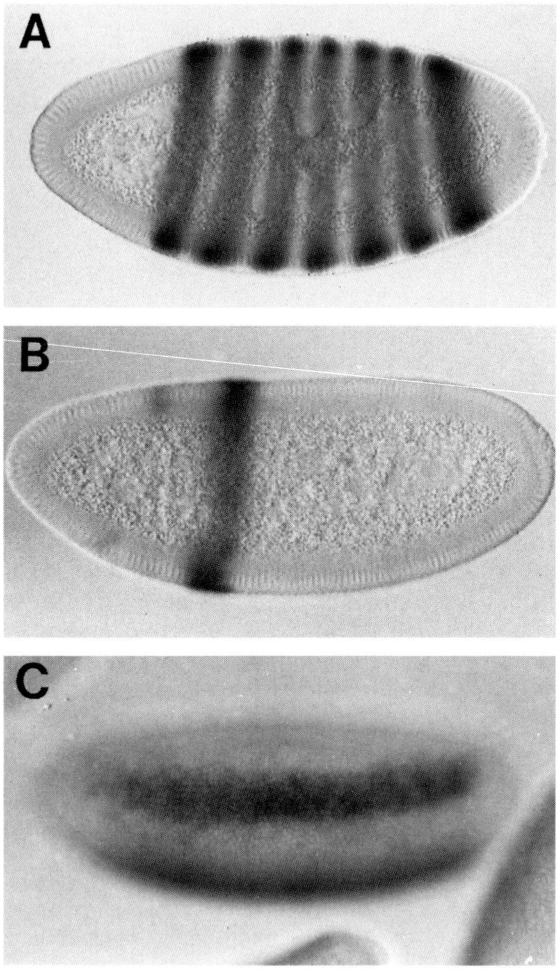

Fig. 1. Expression patterns of *bcd* and *dl* target genes. (A) An optical sagittal section of a wild-type *Drosophila* embryo stained with an anti-*eve* antibody. Anterior is to the left and dorsal is up. *eve* is expressed in a series of 7 transverse stripes at the cellular blastoderm stage. (B) A cellularizing embryo from a line transformed with a *LacZ* reporter gene under the control of a 500-bp promoter element spanning −1.6 kb to −1.1 kb upstream of the *eve* transcription start site. Expression (here and in C) was detected by in situ hybridization with an anti-sense *LacZ* RNA probe. This construct directs a stripe of expression at the position of the 2nd endogenous *eve* stripe. (C) A surface view of an embryo from a line transformed with a *LacZ* fusion gene driven by a 600-bp *rho* promoter element spanning −2.2 kb to −1.6 kb. This construct directs the expression of two ventral lateral stripes.

just 20 to 30 min, this general staining pattern gives way to a series of 7 transverse stripes (see Fig. 1A; Frasch et al., 1987). When they are first formed, the stripes are broad and symmetric, with each including about 5 to 6 nuclei. During another 30 to 60 min, leading to the onset of gastrulation, the stripes are refined to include just 2 or 3 cells. In addition, by this time each stripe shows clear polarity with peak levels of protein at the anterior margin and lower levels in more posterior cells. Genetic studies suggested that the formation of the initial broad stripes depends on the gap class of segmentation genes (Frasch and Levine, 1987; Goto et al., 1989). The subsequent refinement of the stripes involves *eve* autoregulation (Jiang et al., 1991b). We have analyzed the regulation of several of the *eve* stripes, but the one that has been examined in the most detail is stripe #2. It is expressed in an anterior region of the embryo, at the future boundary between the posterior head and anterior thorax.

rho encodes a putative transmembrane receptor protein that plays an important role in the differentiation of the ventral epidermis that arises from the neuroectoderm (Mayer and Nusslein-Volhard, 1988). The gene is expressed in a highly dynamic pattern that includes a number of different tissues during embryogenesis (Bier et al., 1990). The initial pattern includes two ventral-lateral stripes that extend along the length of the embryo. Each of these stripes is about 8 to 10 cells wide and covers the presumptive neuroectoderm, which gives rise to the neurons of the CNS and the ventral epidermis.

First we will consider *eve* stripe 2 and then describe the regulation of the *rho* lateral stripes. We will then summarize the very striking parallels in the regulation of these two patterns.

eve *stripe 2*

The first step in the regulation of stripe 2 by *bcd* is the activation of the gap genes. Two of the best characterized gap genes, *hunchback* (*hb*) and *Kruppel* (*Kr*) (Tautz et al., 1987; Knipple et al., 1985; Preiss et al., 1985), have been shown to encode proteins that are expressed in broad, overlapping domains in precellular embryos (Stanojevic et al., 1989). *hb* is initially expressed in the anterior half of the embryo, while *Kr* is expressed in central regions. Both proteins contain zinc fingers (Tautz et al., 1987; Knipple et al., 1985), but with regard to stripe 2 regulation (see below), *hb* functions as a transcriptional activator while *Kr* is a repressor. *hb* and *Kr* represent just two of the five gap genes, and together they encode a series of transcriptional activators and repressors that span the length of the embyro. We wish to understand how these different permutations of gap gene expression generate on/off stripes of *eve* expression.

To address this issue we have examined the interaction of gap proteins with defined regions of the *eve* promoter. These studies have been facilitated by the demonstration that different regions of the *eve* promoter control the expression of individual stripes (Goto et al., 1989; Harding et al., 1989). For example, a 500-bp fragment located between about −1.6 kb and −1.1 kb is sufficient to direct expression of stripe 2 (Fig. 1B), while a different 500-bp fragment between −3.8 kb and −3.3 kb is responsible for stripe 3. We have investigated the genetic regulation of stripe 2 by examining the expression of the *eve* stripe 2-*LacZ* fusion gene in a variety of segmentation mutants. These experiments, along with previous studies, have identified four key regulatory genes that control the expression of the stripe. Two of the genes function as activators while two are repressors. The genetic regulation of stripe 2 is summarized in Fig. 2A. The *bcd* morphogen activates the expression of *hb*. Together, *bcd* and *hb* define a broad domain in the anterior half of the embryo where stripe 2 can be initiated. The borders of the stripe depend on selective repression: *giant* (*gt*) in anterior regions and *Kr* in pos-

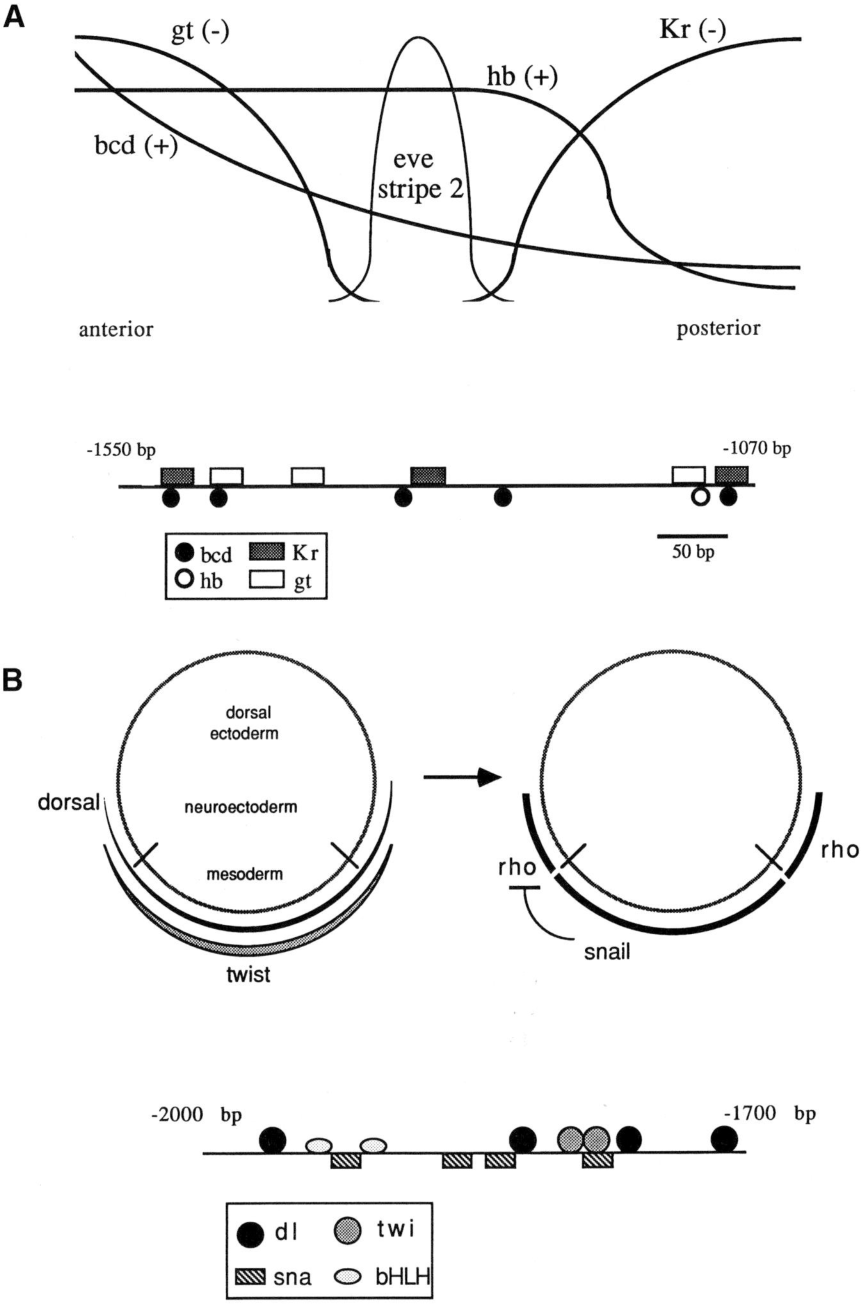

Fig. 2. *Cis*- and *trans*- control of stripes in the *Drosophila* blastoderm. (A) Regulation of *eve* stripe 2. The top schematic shows the relation of stripe 2 to its genetically defined regulators. Gradients of *bcd* and *hb* activate stripe 2 in a broad anterior domain. The anterior and posterior borders of the stripe are formed through selective repression by the gap genes *gt* and *Kr*, respectively. The protein products of all four putative regulators bind to high affinity sites within the 500-bp *eve* stripe 2 promoter element (bottom). (B) Regulation of *rho* lateral stripes. The top diagrams represent transverse sections of blastoderm-stage embryos showing the relation of *rho* stripes to their genetically defined regulators. The *dl* and *twi* gradients cooperate to activate *rho* expression in the ventral half of the embryo. The *sna* repressor prevents activation in the presumptive mesoderm and thus restricts expression of *rho* to two lateral stripes in neuroectodermal regions. The 300-bp neuroectoderm element of the *rho* promoter spans the region from –2.0 kb to –1.7 kb, and contains a total of 10 binding sites for *dl*, *twi* or *sna*. There are also two E-box sequences (bHLH). Various bHLH proteins are able to bind to these sites with different affinities in vitro.

terior regions. The key conclusion of the genetic studies is that repression plays the decisive role in defining the stripe borders (Goto et al., 1989; Frasch et al., 1987; Small et al., 1991). The *bcd* and *hb* activators merely serve to define a broad domain where the stripe can be expressed.

The importance of repression is illustrated by examining the expression of stripe 2-*LacZ* fusion genes in early embyros (Small et al., 1992). We have shown that by the onset of cellularization they are expressed in tight stripes. However, at earlier stages (nuclear cleavage cycle 12 and 13), these same fusion genes are expressed in very broad patterns covering most of the anterior half of the embryo. During a period of just 20 to 30 min this broad pattern is refined into stripe 2. First the posterior border is established, and then the anterior border. The kinetics of this refinement process correspond quite well with the known time of appearance of the stripe 2 regulators. The *bcd* and *hb* activators are maternally expressed and present prior to the time when the *Kr* and *gt* repressors first appear.

The first indication that *bcd*, *hb*, *gt* and *Kr* might directly regulate stripe 2 expression came from DNA binding studies. All four genes encode sequence-specific DNA binding proteins which interact with multiple sites in the *eve* stripe 2 promoter element (Stanojevic et al., 1989; Small et al., 1991). There are a total of 12 factor binding sites, 6 activators and 6 repressors (Fig. 2A). Interestingly, there is very close linkage of *bcd*/*hb* activator sites and *gt*/*Kr* repressor sites. This linkage suggests that *gt* and *Kr* define the stripe borders through a short range mechanism of transcriptional repression. Several lines of evidence suggest that the stripe 2 element functions as a genetic on/off switch to produce a stripe of gene transcription. In anterior regions, there are high levels of *bcd* and *hb*, which bind to the six activator sites and turn the stripe 2 element on. In progressively more posterior regions, there are increasing amounts of *Kr*, which binds to the three *Kr* repressor sites. This blocks the binding or activities of the *bcd* and *hb* activators, which leads to repression and the formation of the posterior stripe border. Similarly, in anterior regions containing high levels of *gt*, the *gt* repressor sites are filled, leading to the formation of the anterior border.

The most critical test of this switch model involves disrupting individual binding sites in the context of otherwise normal *eve-LacZ* fusion genes. The first experiments that were done involved the use of a large region of the *eve* promoter, spanning the first 5.2 kb of 5′ flanking sequence (Stanojevic et al., 1991). This not only directs the expression of stripe 2, but also drives equally intense expression of stripes 3 and 7. Thus, these latter stripes serve as internal controls of even mild disruptions in stripe 2 expression. Disruptions in all three *gt* binding sites have no effect on stripes 3 and 7, but there is a pronounced anterior expansion of stripe 2, similar to that observed in *gt*– mutants. This result provides strong evidence that *gt* directly interacts with the stripe 2 element to repress transcription and form the anterior border of the stripe. In contrast, mutations in 2 of the 6 activator sites (2 *bcd* sites) result in a marked reduction in the levels of stripe 2 expression relative to stripes 3 and 7. This suggests that *bcd* directly interacts with the stripe 2 element to activate transcription. Although mutations in the *bcd* sites reduce expression, the residual stripe is expressed within the normal limits, underscoring the importance of the repressors in defining the stripe borders.

Future studies will focus on understanding how relatively small changes in the ratios of activators and repressors lead to an on/off switch in transcription. Consider the anterior border of the stripe. A cell just anterior to the anterior-most cell in stripe 2 contains about the same levels of the *bcd* and *hb* activators as its neighbor, yet does not express *eve*. Co-immunofluorescence staining suggests that there may be no more than a 2-fold difference in the levels of the *gt* repressor present in these two cells. How does this 2-fold reduction in *gt* give an on/off switch in *eve* transcription? Detailed biochemical studies will be required for a complete answer to this question, but recent observations suggest that the key to the on/off switch involves cooperative interactions among the *bcd* and *hb* activators. There are catastrophic reductions in the levels of expression when individual activator binding sites are disrupted in the context of the minimal 500-bp stripe 2 element. There are 5 *bcd* binding sites in this element and mutations in either of the two right-most *bcd* sites cause nearly a complete loss of expression. Mutations in either the middle *bcd* site or the solo *hb* site cause a substantial reduction in expression (Small et al., 1992). These results suggest that cooperative interactions among the activators are required for turning the stripe 2 element on. The binding of the *gt* or *Kr* repressor would have to interfere with just one or two of the activators in order to effectively repress transcription and form the stripe borders.

rhomboid lateral stripes

We are also studying the regulation of stripes of gene expression along the dorsal-ventral (D-V) axis of the early embryo, which involves the *dl* morphogen. To contrast crudely the *bcd* and *dl* morphogens, consider muscle cells stretched along the length of the embryo. *bcd* controls the expression of segmentation genes such as *eve*, which are responsible for making the muscles different in the head versus tail. In contrast, *dl* is responsible for the formation of muscle in the first place. Thus, the regulation of gene expression along the D-V axis is a problem of tissue differentiation.

Genetic studies suggest that the *dl* morphogen gradient is responsible for establishing three basic embryonic tissues: mesoderm, neuroectoderm and dorsal ectoderm (summarized in Fig. 2B). Peak levels of the protein in ventral regions activate regulatory genes, including *twist* (*twi*) and *snail* (*sna*), which are important for the formation of the mesoderm (Simpson, 1983; Thisse et al., 1987). Low levels of *dl* in lateral regions coincide with genes that are expressed within the neuroectoderm. *dl* also functions as a transcriptional repressor, and restricts the expression of certain genes to dorsal regions where they are important for the differentiation of the dorsal ectoderm (Rushlow et al., 1987; St Johnston and Gelbart, 1987). We have become interested in determining how *dl* regulates the expression of neuroectodermal regulatory genes in lateral regions of the early embryo.

rho is expressed in two ventral lateral stripes that extend along the length of the embryo (Fig. 1C). Each *rho* stripe is about 8 to 10 cells in width and encompasses most or all of the presumptive neuroectoderm. A 300-bp region of the *rho* promoter is sufficient to direct expression of a *LacZ* reporter gene within the limits of the ventral-lateral stripes (Ip et al., 1992b). Experimental results suggest that the *dl* morphogen acts in concert with two of its targets, *twi* and *sna*, to establish the *rho* stripes (summarized in Fig. 2B).

twi and *sna* encode regulatory proteins containing a basic helix-loop-helix (bHLH) and zinc fingers, respectively (Thisse et al., 1988; Boulay et al., 1987). Mutations in either gene block the formation of the ventral furrow and prevent differentiation of the mesoderm. The *sna* and *twi* expression patterns overlap, but the dorsal limits of *twi* extend beyond the *sna* pattern, since intermediate levels of *dl* in ventral-lateral regions are sufficient to activate *twi* but not *sna* (Jiang et al., 1991a; Ip et al., 1992a). *twi* functions primarily as a transcriptional activator that is required for the activation of target genes such as *nautilus* (Michelson et al., 1990), which are expressed in specific cell types in the mesoderm. While *twi* is expressed in a ventral-to-lateral gradient, with peak levels present in ventral regions, *sna* is expressed at uniformly intense levels throughout the presumptive mesoderm. Furthermore, the expression of *sna* is tightly restricted, and the dorsal limits of its expression abruptly end at the boundary between the mesoderm and neuroectoderm. These sharp *sna* borders represent a key early step in mesoderm differentiation, and are reminiscent of the sharp anterior margins of the pair-rule genes *eve* and *ftz*. Double staining studies indicate that a number of neuroectodermal regulatory genes, such as T3 (*lethal of scute*) and *rho*, are expressed in patterns that abut, but do not cross, the *sna* borders. These observations prompted the proposal that *sna* is responsible for establishing the boundary between mesoderm and neuroectoderm (Kosman et al., 1991; Leptin, 1991).

The first indication that *dl*, *twi* and *sna* might directly regulate the *rho* lateral stripes came from DNA binding studies. The minimal 300-bp *rho* promoter element that is sufficient for these stripes, the NEE (neuroectoderm element), contains a total of ten *dl*, *twi* and *sna* binding sites (summarized in Fig. 2B). As was seen for the *eve* stripe 2 element, there is clustering of the factor binding sites, and close linkage between activator and repressor sites. For example, none of the four *dl* binding sites maps more than 50 bp away from a *sna* repressor site. In addition to the four *dl* activator sites, there are two *twi* sites, as well as two 'E box' sequences that resemble the binding sites for the prototypic E12/E47 and *daughterless* (*da*) proteins (Murre et al., 1989). We have obtained evidence that this *rho* NEE functions as an on/off switch. In the ventral half of the early embryo, there are high levels of *dl* and *twi*. These proteins fill the activator sites and switch the NEE on. In ventral regions there are high levels of *sna*, which fill the repressor sites and switch off expression by interfering with the binding or activities of nearby activators. Repression might involve a competition mechanism, since *sna* binds to sequences similar to the divergent E box recognized by the bHLH proteins. In fact, at least one of the *twi* binding sites, t2, is also recognized by the *sna* repressor (see Fig. 2B).

Evidence that the NEE functions as a switch element was obtained by mutagenizing individual factor binding sites in the context of otherwise normal *rho-LacZ* fusion genes. Point mutations that disrupt three of the four *dl* sites virtually abolish expression, suggesting that *dl* directly interacts with the *rho* NEE to activate its transcription. Mutations in the two *twi* sequences cause a substantial reduction in expression. This is consistent with the possibility that *twi* directly activates the NEE, although it should be noted that bHLH proteins all recognize related E box sequences in vitro, and it is conceivable that other, as yet unidentified, bHLH proteins might contribute to *rho* expression. The most dramatic result was obtained when the 4 *sna* repressor sites were mutagenized. This experiment was done in a way that does not interfere with the binding of *twi* to overlapping or neighboring activator sites. Disrupting the *sna* binding sites causes a complete derepression of the *rho* pattern, such that staining extends throughout ventral regions that correspond to the presumptive mesoderm. This result suggests that *sna* directly represses *rho* transcription, and provides evidence that it is directly responsible for establishing the boundary between the mesoderm and neuroectoderm (Ip et al., 1992b).

In conclusion, *dl* and *twi* act in concert to define a broad region in the ventral half of the embryo where *rho* can be activated (summarized in Fig. 2B). The ventral limits of the *rho* lateral stripes depend on selective repression by *sna*. This situation is similar to the regulation of *eve* stripe 2, where repressors are crucial for establishing the borders. It is possible that the dorsal limits of the *rho* stripes are formed by limiting amounts of the *dl* and/or *twi* activators.

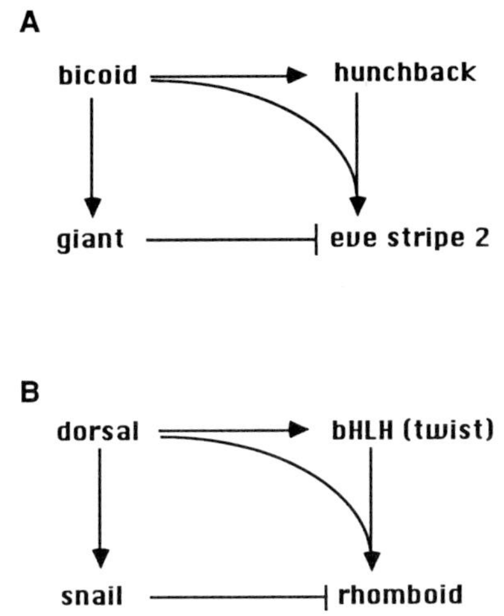

Fig. 3. *bcd* and *dl* generate stripes through similar mechanisms. Diagram of the regulatory circuitry involved in the formation of *eve* stripe 2 (A) and the *rho* lateral stripes (B). The arrows indicate activation and the lines with T-ends represent repression. For example, *bcd* activates *hb* and the two proteins cooperatively activate stripe 2, while *gt* represses the expression.

Discussion

There are remarkable parallels between the regulation of *eve* stripe 2 and the *rho* lateral stripes, as summarized in Fig. 3. In the case of stripe 2, the *bcd* morphogen activates a 'helper', *hb*, and together these genes define a broad region where stripe 2 can be turned on. *bcd* also activates a repressor, *gt*, which is responsible for defining the anterior border of the stripe. Similarly, *dl* activates *twi*, and it appears that these 2 genes are able to initiate the *rho* NEE in the entire ventral half of the embryo. *dl* also activates *sna*, which defines the ventral limit of the *rho* stripes. In addition to the obvious parallels in the two regulatory circuits, in both cases the encoded proteins interact with short (300 to 500 bp) elements which generate stripes. For the most part the two sets of proteins are distinct. *bcd* contains a homeobox and *hb* contains zinc fingers, while the *gt* repressor contains a basic leucine zipper domain. In contrast, *dl* contains a REL domain and probably interacts with bHLH proteins; the *sna* repressor contains zinc fingers. Despite these differences the outcome is the same: stripes.

We propose that the tight clustering of activator and repressor sites observed for the *eve* stripe 2 element and the *rho* NEE reflects weak protein-protein interactions among the regulatory factors active in the early embryo. Studies on gene regulation, primarily in mammalian systems, have identified instances of strong protein-protein interactions, such as jun-fos interactions (Gentz et al., 1989). But we believe that such interactions represent the minority of the cases. Overall, it is likely that the combinatorial control of gene expression depends on weak interactions. Perhaps weak interactions can work to turn promoters on and off coordinately if the interacting factors are bound to closely linked sites. In a sense, the evolution of pattern in the early embryo may involve bringing factor binding sites into close proximity within small promoter elements. This notion can be tested by determining whether random combinations of regulatory proteins present in the early embryo can be made to interact by juxtaposing their binding sites. A strength of the *Drosophila* system is that many of the regulatory factors active in the early embryo have been identified, and their expression patterns have been determined in precellular embryos. By knowing these expression patterns it is possible to predict the novel patterns that should be obtained when two factors are able to interact. For example, if *bcd* activator sites in the *eve* stripe 2 element are replaced by *dl* binding sites, then a hemistripe of expression in the ventral half of the embryo should be observed.

Y.T.I. is a Hoffmann-La Roche fellow of the Life Sciences Research Foundation. S.J.S. is supported by NIH postdoctoral fellowship. This work is funded by NIH grants GM 46638 and GM 34431.

References

Berleth, T., Burri, M., Thoma, G., Bopp, D., Richstein, S., Frigerio, G., Noll, M. and Nusslein-Volhard, C. (1988). The role of localization of *bicoid* RNA in organizing the anterior pattern of the Drosophila embryo. *EMBO J.* **7**, 1749-1756.

Bier, E., Jan, L. Y. and Jan, Y. N. (1990). *rhomboid*, a gene required for dorso-ventral axis establishment and peripheral nervous system development in *Drosophila melanogaster. Genes Dev.* **4**, 190-203.

Boulay, J.-L., Dennefeld, C. and Alberga, A. (1987). The Drosophila developmental gene *snail* encodes a protein with nucleic acid binding fingers. *Nature* **330**, 395-398.

Driever, W. and Nusslein-Volhard, C. (1988). A gradient of *bicoid* protein in Drosophila embryos. *Cell* **54**, 83-93.

Frasch, M., Hoey, T., Rushlow, C., Doyle, H. J. and Levine, M. (1987). Characterization and localization of the *even-skipped* protein of Drosophila. *EMBO J.* **67**, 749-759.

Frasch, M. and Levine, M. (1987). Complementary patterns of *even-skipped* and *fushi-tarazu* expression involve their differential regulation by a common set of segmentation genes in Drosophila. *Genes Dev.* **1**, 981-995.

Frigerio, D., Burri, M., Bopp, D., Baumgartner, S. and Noll, M. (1986). Structure of the segmentation gene *paired* and the Drosophila PRD gene set as part of a gene network. *Cell* **47**, 735-746.

Gentz, R., Rauscher III, F. J., Abate, C. and Curran, T. (1989). Parallel association of Fos and Jun leucine zippers juxtaposes DNA binding domains. *Science* **243**, 1695-1699.

Ghosh, S., Gifford, A. M., Riviere, L. R., Tempst, P., Nolan, G. P. and Baltimore, D. (1990). Cloning of the p50 DNA binding subunit of NF-kB: homology to *rel* and *dorsal*. *Cell* **62**, 1019-1029.

Goto, T., Macdonald, P. and Maniatis, T. (1989). Early and late periodic patterns of *even-skipped* expression are controlled by distinct regulatory elements that respond to different spatial cues. *Cell* **57**, 413-422.

Harding, K., Rushlow, C., Doyle, H. J., Hoey, T. and Levine, M. (1986). Cross-regulatory interactions among pair-rule genes in Drosophila. *Science* **233**, 953-959.

Harding, K., Hoey, T., Warrior, R. and Levine, M. (1989). Autoregulatory and gap gene response elements of the *even-skipped* promoter of Drosophila. *EMBO J.* **8**, 1205-1212.

Ip, Y. T., Park, R. E., Kosman, D., Yasdanbaksh, K. and Levine, M. (1992a). *dorsal-twist* interactions establish *snail* expression in Drosophila embryos. *Genes Dev.* **6**, 1518-1530.

Ip, Y. T., Park, R. E., Kosman, D., Bier, E. and Levine, M. (1992b). The *dorsal* gradient morphogen regulates stripes of *rhomboid* expression in the presumptive neuroectoderm of the Drosophila embryo. *Genes Dev.* **6**, 1728-1739.

Jiang, J., Kosman, D., Ip, Y. T. and Levine, M. (1991a). The *dorsal* morphogen gradient regulates the mesoderm determinant *twist* in early Drosophila embryos. *Genes Dev.* **5**, 1881-1891.

Jiang, J., Hoey, T. and Levine, M. (1991b). Autoregulation of a segmentation gene in Drosophila: combinatorial interaction of the *even-skipped* homeobox protein with a distal enhancer element. *Genes Dev.* **5**, 265-277.

Kieran, M., Blank, V., Logeat, F., Vandekerckhove, J., Lottspeich, F., Le Bail, O., Urban, M. B., Kourilsky, P., Baeuerle, P. A. and Israel, A. (1990). The DNA binding subunit of NF-kB is identical to factor KBF1 and homologous to the *rel* oncogene product. *Cell* **62**, 1007-1018.

Knipple, D. C., Seifert, E., Rosenberg, U. B., Preiss, A. and Jackle, H. (1985). Spatial and temporal patterns of *Kruppel* gene expression in early Drosophila embryos. *Nature* **317**, 40-44.

Kosman, D., Ip, Y. T., Levine, M. and Arora, K. (1991). Establishment of the mesoderm-neuroectoderm boundary in the Drosophila embryo. *Science* **254**, 118-122.

Leptin, M. (1991). *twist* and *snail* as positive and negative regulators during Drosophila mesoderm development. *Genes Dev.* **5**, 1568-1576.

Macdonald, P. M., Ingham, P. and Struhl, G. (1986). Isolation, structure, and expression of *even-skipped*: a second pair-rule gene of Drosophila containing a homeo box. *Cell* **47**, 721-734.

Macdonald, P. M. and Struhl, G. (1988). *Cis*-acting sequences responsible for anterior localization of *bicoid* mRNA in Drosophila embryos. *Nature* **336**, 595-598.

Mayer, U. and Nusslein-Volhard, C. (1988). A group of genes required for pattern formation in the ventral ectoderm of the Drosophila embryo. *Genes Dev.* **2**, 1496-1511.

Michelson, A. M., Abmayr, S. M., Bate, M., Martinez Arias, A. and Maniatis, T. (1990). Expression of a MyoD family member prefigures muscle pattern in Drosophila embryos. *Genes Dev.* **4**, 2086-2097.

Murre, C., McCaw, P. S. and Baltimore, D. (1989). A new DNA binding and dimerization motif in immunoglobulin enhancer binding, *daughterless*, *MyoD*, and *Myc* proteins. *Cell* **56**, 777-783.

Nusslein-Volhard, C., Kluding, H. and Jurgens, G. (1985). Genes affecting the segmental subdivision of the Drosophila embryo. *Cold Spring Harbor Symp. Quant. Biol.* **50**, 145-154.

Preiss, A., Rosenberg, U. B., Kienlin, A., Seifert, E. and Jackle, H. (1985). Molecular genetics of *Kruppel*, a gene required for segmentation of the Drosophila embryo. *Nature* **313**, 27-32.

Roth, S., Stein, D. and Nusslein-Volhard, C. (1989). A gradient of nuclear localization of the *dorsal* protein determines dorsoventral pattern in the Drosophila embryo. *Cell* **59**, 1189-1202.

Rushlow, C. A., Frasch, M., Doyle, H. and Levine, M. (1987). Maternal regulation of *zerknullt*: a homeobox gene controlling differentiation of dorsal tissues in Drosophila. *Nature* **330**, 583-586.

Rushlow, C. A., Han, K., Manley, J. L. and Levine, M. (1989). The graded distribution of the *dorsal* morphogen is initiated by selective nuclear transport in Drosophila. *Cell* **59**, 1165-1177.

Simpson, P. (1983). Maternal-zygotic gene interactions during formation of the dorsoventral pattern in Drosophila embryos. *Genetics* **105**, 615-632.

Small, S., Kraut, R., Hoey, T., Warrior, R. and Levine, M. (1991). Transcriptional regulation of a pair-rule stripe in Drosophila. *Genes Dev.* **5**, 827-839.

Small, S., Blair, A. and Levine, M. (1992). Regulation of *even-skipped* stripe 2 in the Drosophila embryo. *EMBO J* **11**, 4047-4057.

Stanojevic, D., Hoey, T. and Levine, M. (1989). Sequence-specific DNA-binding activities of the gap proteins encoded by *hunchback* and *Kruppel* in Drosophila. *Nature* **341**, 331-335.

Stanojevic, D., Small, S. and Levine, M. (1991). Regulation of a segmentation stripe by overlapping activators and repressors in the Drosophila embryo. *Science* **254**, 1385-1387.

Steward, R. (1987). *Dorsal*, an embryonic polarity gene in Drosophila is homologous to the vertebrate proto-oncogene, *c-rel*. *Science* **238**, 692-694.

Steward, R., Zusman, S. B., Huang, L. H. and Schedl, P. (1988). The *dorsal* protein is distributed in a gradient in early Drosophila embryos. *Cell* **55**, 487-495.

Steward, R. (1989). Relocalization of the *dorsal* protein from the cytoplasm to the nucleus correlates with its function. *Cell* **59**, 1179-1188.

St Johnston, R. D. and Gelbart, W. M. (1987). *decapentaplegic* transcripts are localized along the dorsal-ventral axis of the Drosophila embryo. *EMBO J.* **6**, 2785-2791.

St Johnston, D. and Nusslein-Volhard, C. (1992). The origin of pattern and polarity in the Drosophila embryo. *Cell* **68**, 201-219.

Tautz, D., Lehmann, R., Schnurch, H., Schuh, R., Seifert, E., Kienlin, K. and Jackle, H. (1987). Finger protein of novel structure encoded by *hunchback*, a second member of the gap class of Drosophila segmentation genes. *Nature* **327**, 383-389.

Thisse, B., Stoetzel, C., El Messal, M. and Perrin-Schmitt, F. (1987). Genes of the Drosophila maternal dorsal group control the specific expression of the zygotic gene *twist* in presumptive mesodermal cells. *Genes Dev.* **1**, 709-715.

Thisse, B., Stoetzel, C., Gorostiza-Thisse, C. and Perrin-Schmitt, F. (1988). Sequence of the *twist* gene and nuclear localization of its protein in endomesodermal cells of early Drosophila embryos. *EMBO J.* **7**, 2175-2183.

Journal of Cell Science, Supplement 16, 39-51 (1992)
Printed in Great Britain © The Company of Biologists Limited 1992

Transcriptional control by *Drosophila* gap genes

HERBERT JÄCKLE, MICHAEL HOCH, MICHAEL J. PANKRATZ, NICOLE GERWIN, FRANK SAUER and GÜNTER BRÖNNER

Abt. Molekulare Entwicklungsbiologie, Max-Planck-Institut für biophysikalische Chemie, Am Faßberg, Postfach 2841, W-3400 Göttingen, Germany

Summary

The segmented body pattern along the longitudinal axis of the *Drosophila* embryo is established by a cascade of specific transcription factor activities. This cascade is initiated by maternal gene products that are localized at the polar regions of the egg. The initial long-range positional information of the maternal factors, which are transcription factors (or are factors which activate or localize transcription factors), is transferred through the activity of the zygotic segmentation genes. The gap genes act at the top of this regulatory hierarchy. Expression of the gap genes occurs in discrete domains along the longitudinal axis of the preblastoderm and defines specific, overlapping sets of segment primordia. Their protein products, which are DNA-binding transcription factors mostly of the zinc finger type, form broad and overlapping concentration gradients which are controlled by maternal factors and by mutual interactions between the gap genes themselves. Once established, these overlapping gap protein gradients provide spatial cues which generate the repeated pattern of the subordinate pair-rule gene expression, thereby blueprinting the pattern of segmental units in the blastoderm embryo. Our results show different strategies by which maternal gene products, in combination with various gap gene proteins, provide position-dependent sets of transcriptional activator/repressor systems which regulate the spatial pattern of specific gap gene expression. Region-specific combinations of different transcription factors that derive from localized gap gene expression eventually generate the periodic pattern of pair-rule gene expression by the direct interaction with individual *cis*-acting "stripe elements" of particular pair-rule gene promotors. Thus, the developmental fate of blastoderm cells is programmed according to their position within the anterior-posterior axis of the embryo: maternal transcription factors regulate the region-specific expression of first zygotic transcription factors which, by their specific and unique combinations, control subordinate zygotic transcription factors, thereby subdividing the embryo into increasingly smaller units later seen in the larva.

Key words: *Drosophila* embryogenesis, transcriptional regulation, protein-DNA interactions.

Introduction

The initial establishment of polarity along the longitudinal axis of the *Drosophila* embryo is carried out by three fairly independent genetic pathways of maternal origin. They are required for the development of the anterior, the posterior and the terminal regions of the embryo (Nüsslein-Volhard et al., 1987; Nüsslein-Volhard, 1991; St. Johnston and Nüsslein-Volhard, 1992; for reviews). The first of the zygotic genes to interpret the positional information provided by the maternal genes are members of the gap class of segmentation genes (Nüsslein-Volhard and Wieschaus, 1980). They are responsible for the development of specific, overlapping sets of segment primordia. For example, *hunchback* (*hb*) is needed to establish the gnathal and thoracic segments (Lehmann and Nüsslein-Volhard, 1987), *Krüppel* (*Kr*) for thoracic and anterior-abdominal segments (Gloor, 1954; Nüsslein-Volhard and Wieschaus, 1980; Wieschaus et al., 1984; Preiss et al., 1985), *knirps* (*kni*) and *giant* (*gt*) for abdominal segments (Nüsslein-Volhard and Wieschaus, 1980; Petschek et al., 1987) and *tailless* (*tll*) and *huckebein* (*hkb*) for pattern elements in the head and tail regions (Strecker et al., 1988; Weigel et al., 1990). The correct spatial regulation of the gap genes is critical for normal development since ectopic presence of the gap gene products results in severe alterations of the embryonic body pattern (Jäckle et al., 1986; Gaul et al., 1987; Klingler et al., 1988; Eldon and Pirrotta, 1991; Kraut and Levine, 1991a; Steingrimsson et al., 1991). All of the gap genes studied to date encode proteins containing DNA-binding motifs (Pankratz et al., 1990b; for review). With the exception of the *gt* protein (GT), the above mentioned gap genes encode multiple zinc finger motifs (Rosenberg et al., 1986; Tautz et al., 1987; Nauber et al., 1988; Pignoni et al., 1990; G. Brönner and H. Jäckle, unpublished results), and they are expressed in regions of the preblastoderm embryo that are affected in the larval segmentation pattern seen later in the corresponding mutants (Knipple et al., 1985; Gaul et al., 1987; Tautz et al., 1987; Mohler et al., 1989; Rothe et al., 1989; Pignoni et al., 1990; Brönner and Jäckle, 1991).

The mechanism by which the gap genes are regulated has been studied in detail for one of the gap genes, *hb*, which is expressed in the anterior half of the early embryo (Tautz et al., 1987; Schröder et al., 1988). These studies indicated that *hb* is directly activated in response to the protein product encoded by the anterior maternal organizer gene *bicoid* (*bcd*). After transcription in the nurse cells, *bcd* mRNA is transferred into the growing oocyte, and it eventually becomes localized in the anterior pole region of the egg. After translation, the *bcd* protein (BCD) forms a concentration gradient along the longitudinal axis of the embryo and acts as the anterior morphogen (Frohnhöfer and Nüsslein-Volhard, 1986; Driever and Nüsslein-Volhard, 1988a,b, 1989; Driever et al., 1989; Struhl et al., 1989). The spatial restriction of zygotic *hb* expression to the anterior region of the embryo is due in most part to the limiting amount of the activator BCD in the posterior half. It has been suggested that other zygotic segmentation genes, in particular in the prospective head region, might be activated through a similar mechanism by BCD (Driever et al., 1989; Struhl et al., 1989). The detailed mechanisms by which the gap genes acting in the central and posterior regions, such as *Kr*, *kni* and *gt*, and the gap genes in the posterior terminal region, *tll* and *hkb*, are spatially regulated are the subject of current studies reported here. These studies also address the question of how the maternal morphogenetic input on gap gene expression is then interpreted by the pair-rule genes by making use of the several overlapping gap protein gradients along the anterior-posterior axis.

Combined genetic and molecular analyses indicate that gap genes most likely carry out segmentation function by directing the expression of the pair-rule genes (Akam, 1987; Ingham, 1988; Pankratz and Jäckle, 1990; for reviews). Members of this class of segmentation genes are expressed in tandemly repeating stripes in the blastoderm embryo. The patterning of all pair-rule genes examined to date is altered in embryos mutant for any of the gap genes which act in the trunk region of the embryo, indicating that those gap genes are involved in generating the stripe pattern of pair-rule gene expression. Because a regulatory hierarchy exists even among the pair-rule genes, i.e. primary pair-rule genes such as *hairy* (*h*) or *even-skipped* (*eve*) are required for the normal expression of secondary pair-rule genes such as *fushi tarazu* (*ftz*), the gap genes act most likely at the level of the primary pair-rule genes. In this paper we summarize results indicating that the molecular mechanism underlying gap gene expression in different regions of the *Drosophila* embryo are fundamentally different from and more complex than the activation of *hb* by BCD. We show that *Kr* and *kni* expression patterns arise from superimposition of distinct transcriptional activators and repressors of both maternal and gap gene origin, and that the combinations of distinct gap gene activities provide the regulatory input to the individual stripe elements of the pair-rule gene promotors. Finally, we show that the *Kr* protein (KR) is capable of acting as a concentration-dependent transcriptional activator and repressor, providing the two opposing activities through different portions of the protein.

Anterior-posterior prepattern formed in response to spatially regulated transcription factors

During the early stages of *Drosophila* embryogenesis, the zygotic nuclei divide through a series of twelve mitotic divisions without forming cells. Towards the end of the syncytial phase of development, the nuclei migrate to the periphery to form monolayers of nuclei which then become surrounded by membranes to form the cellular blastoderm (for a review see Campos-Ortega and Hartenstein, 1985). The basic prepattern of the anterior-posterior axis is established prior to the cellularization by the localized expression of the zygotic gap genes (Fig. 1). As one moves from anterior to posterior, the embryo shows transcripts reflecting the expression domains of *hkb*, *tll*, *hb* (and *gt*), *Kr*, *kni*, *gt*, *tll*

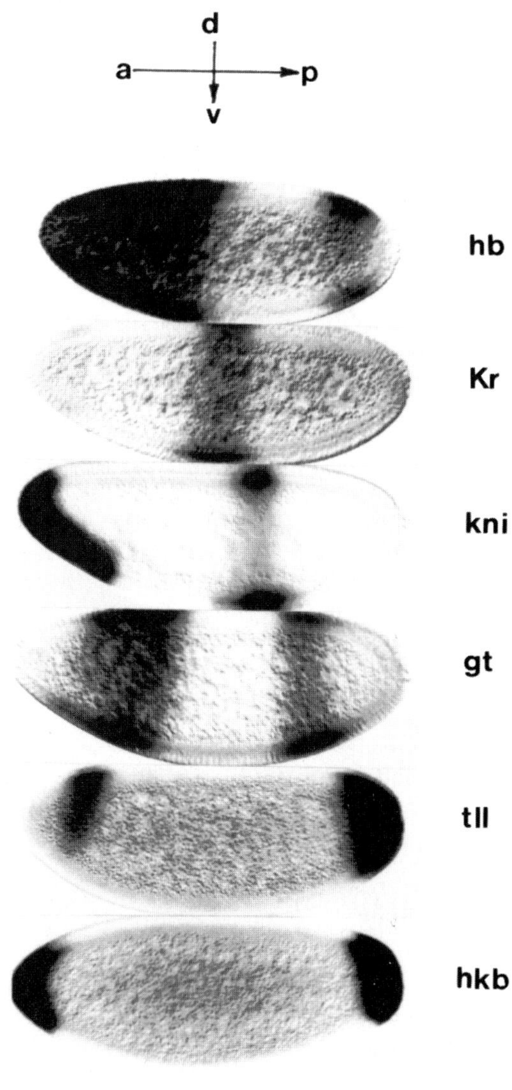

Fig. 1. Expression patterns of gap genes in the blastoderm embryo as visualized by whole mount in situ hybridization. In the text we focus on the anterior domain of *hb* expression (posterior is right; dorsal is up) and the posterior expression domains of the other gap genes. The spatial domains of expression require the genetic input of maternal gene activities and regulatory interactions among the gap genes, as summarized in Fig. 2.

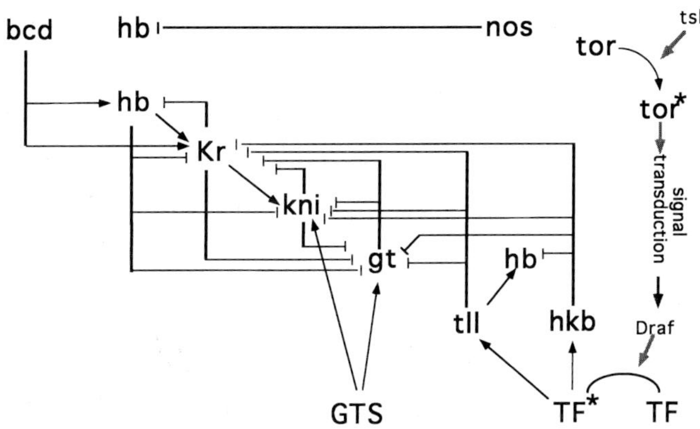

Fig. 2. Genetic circuitry establishing gap gene expression domains in the trunk and tail region of the blastoderm embryo. Arrows indicate positive regulation, lines ending with a vertical bar designate negative inputs. Note that *bcd* activates both *hb* and *Kr*, *hb* activates and represses *Kr* in a (likely) concentration-dependent manner and represses the anterior limits of both *kni* and *gt* expression, *Kr* activates *kni* and represses *gt*, and that the terminal gap genes *tll* and *hkb* act as repressors of *Kr, kni* and *gt*. Note that *tll* is also required to activate posterior *hb* gene expression. For a detailed description of the terminal signal transduction pathway in which a *torso-like (tsl)*-dependent signal (from outside the germ cells) is transmitted through activation (*) of the transmembrane receptor encoded by *torso (tor)* and the *Drosophila raf* oncogene-homologous *Draf* gene, see a recent review by St Johnston and Nüsslein-Volhard (1992). Note that this signal transduction pathway is likely to activate an unknown transcription factor (TF) for reasons described in the text. Note also that *hb* activity is both of maternal and zygotic origin, and that the maternal activity is post-transcriptionally controlled by *nanos* (*nos*) in the posterior half of the embryo (see text). Finally note that *kni* (and possibly *gt*) are activated by a global activator system (GTS) for which no genetic basis is established. A detailed description of the formal interactions depicted here is given in the text. For the position of the expression domains in the embryo see Fig. 1.

and *hkb*. All of these genes encode DNA-binding proteins, the majority of which have been shown to act as transcription factors (see below). The spatial domains of gap gene expression require the input of maternal gene activities and regulatory interactions among themselves, leading to a genetic circuitry required for the formation of trunk and tail segments in the embryo (summarized in Fig. 2).

Expression of hb *in the anterior region*

The early zygotic expression of *hb* is dependent on the BCD gradient as it does not form in *bcd* mutant embryos and it expands posteriorly when the dose of the *bcd* gene is increased in the egg-laying females (Driever and Nüsslein-Volhard, 1988a,b). *bcd* encodes a homeodomain protein (Frigerio et al., 1986; Berleth et al., 1988). Its RNA is localized in the cytoplasm at the anterior pole of the egg (Frigerio et al., 1986; Berleth et al., 1988; St. Johnston et al., 1989). After translation, the BCD transcription factor forms an anterior-to-posterior concentration gradient which determines the domain of zygotic *hb* gene expression extending from the anterior tip to about 50% of the egg length. The *hb cis*-acting region that can direct *bcd*-dependent expression of a reporter gene in the anterior half of the embryo (Schröder et al., 1988) contains six BCD-binding sites; three of them are of high affinity and three of low affinity (Driever and Nüsslein-Volhard, 1989). Four tandem copies of a high-affinity BCD-binding site direct reporter gene expression in the anterior half of the embryo, while the corresponding number of low-affinity sites direct expression in the anterior third only (Driever et al., 1989; Struhl et al., 1989). Addition of extra BCD-binding sites of equal or lower affinity to either construct leads to an increase in the level of expression but it has no significant effect on the spatial expression domains (Driever et al., 1989). Therefore, Driever and Nüsslein-Volhard have concluded that promotors with low-affinity BCD-binding sites require high BCD concentrations to be activated, while the high-affinity binding sites require lower concentrations of BCD to direct expression in more posterior regions of the embryo. There is some doubt whether a threshold concentration of BCD is read solely by the binding site with the highest BCD affinity (Struhl et al., 1989); it appears likely, however, that a given threshold within the BCD gradient is sufficient to activate the *hb* gene above a defined BCD concentration, giving rise to the *hb* anterior domain of expression (Driever et al., 1989; Struhl et al., 1989).

Expression of kni *and* gt *in the abdominal region*

The products of *kni* and *gt* are required for the establishment of the abdominal region of the embryo (Nüsslein-Volhard and Wieschaus, 1980; Petschek et al., 1987). The corresponding expression domains of these "abdominal gap genes" are adjacent to each other, and they are both regulated in response to the maternal posterior organizer gene *nanos* (Rothe et al., 1989; Eldon and Pirrotta, 1991; Kraut and Levine, 1991b). In contrast to *bcd* which is thought to regulate all anteriorly acting gap genes directly, *nanos* provides its activity through regulation of a posterior repressor encoded by maternal *hb* transcripts. This means that *hb* is transcribed not only zygotically (see above), but also during oogenesis, and gives rise to uniformly distributed *hb* transcripts in the mature egg. This maternal transcript is degraded in response to *nanos* activity in the posterior half of the embryo (Wang and Lehmann, 1991), thereby restricting the translation of maternal *hb* mRNA to the anterior half (Tautz, 1988). Thus, *hb* protein (HB) is already present in the anterior half of the embryo before its zygotic complement is expressed in response to the BCD concentration gradient. In *nanos* mutants, both the *hb* transcript and HB are present throughout the embryo (Tautz, 1988). In this case, posterior expression of both *kni* and *gt* are blocked and abdominal segments fail to be established (Nüsslein-Volhard et al., 1987; Rothe et al., 1989; Eldon and Pirrotta, 1991; Kraut and Levine, 1991b). Since HB contains two putative DNA-binding domains, one with four and the other with two zinc finger motifs (Tautz et al., 1987), it is likely that *hb* prevents abdomen formation through repression of the abdominal gap genes *kni* and *gt*. Thus, the role of the posterior maternal organizer system is to prevent posterior expression of the repressor HB (Hülskamp et al., 1989; Irish et al., 1989; Struhl, 1989). In the anterior region, HB serves as a repressor of *kni* and *gt*, setting the anterior borders of their posterior expression domains (Hülskamp et al., 1990).

This interpretation is also consistent with recent genetic studies of Struhl et al. (1992) showing that HB acts as a concentration-dependent morphogen which prevents *kni* and *gt* expression at different concentration values; low levels suppress *kni* expression, and more posteriorly, even lower levels (which do not interfere with *kni* expression) also suppress *gt* expression. Thus, the anterior limits of the *kni* and *gt* expression domains can be set in response to HB. In addition, there is also evidence from genetic studies that the two genes function as weak, mutual repressors (Eldon and Pirrotta, 1991). Thus, KNI is part of the repressor system which delimits the *gt* expression domain (from anterior), and GT may serve similar function on *kni* expression (from posterior; see below).

The posterior limits of the *kni* and *gt* expression domains are mainly set in response to the terminal gap genes, which by genetic means have been shown to be strong repressors of the gap genes *Kr*, *kni* and *gt* (see below). The activator(s) of *kni* and *gt* have not yet been identified, but there is increasing evidence that it might be a factor which is capable of activating KNI throughout the embryo (Pankratz et al., 1992).

Expression of the terminal gap genes in the tail region

Expression of the terminal gap genes *tll* and *hkb* in the posterior pole region in the embryo is almost entirely controlled by the maternal terminal organizer system, and the two genes are likely to encode the zygotic transcription factors which are the functional zygotic response to the activity of a signal transduction pathway (Weigel et al., 1990; Brönner and Jäckle, 1991). In this pathway, the product of the gene *torso* probably acts as a receptor for an extracellular signal that is produced at the two poles of the egg. *torso* encodes a putative transmembrane receptor tyrosine kinase which, in response to a localized extracellular signal (Sprenger et al., 1989; Stevens et al., 1990), activates downstream gene products including the as yet unknown transcription factor which eventually activates *tll* and *hkb*. The activation of *tll* and *hkb* in response to *torso* activity, as well as the effect of ectopic *tll* and *hkb* expression on other gap genes, has been demonstrated in *torso* gain-of-function alleles which encode mutant receptors with constitutive tyrosine kinase activity in the absence of activating ligands. Embryos laid by females containing these gain-of-function alleles develop normal terminal structures but they have defects in the segmented region of the pattern likely to be caused by the correlated ectopic expression of *tll* and *hkb* in these embryos. The segmental pattern defects are suppressed in embryos that are also mutant for *tll* and *hkb*, indicating that the gain-of-function phenotypes are likely due to the ectopic expression of the two terminal gap genes in the central region of the embryos. In fact, their ectopic expression leads to the repression of central gap genes such as *Kr* and *kni* (Brönner and Jäckle, 1991; and references therein).

Ectopic expression of *tll* protein (TLL) under the control of an inducible promotor results in a *torso* gain-of-function, mutant-like phenotype. This is due to repression of *Kr* and *kni* (Klingler et al., 1988; Steingrimsson et al., 1991). Furthermore, in *tll* mutant embryos, the posterior band of *gt* expression (Eldon and Pirrotta, 1991) is broader and extends posteriorly as compared to wild-type expression. In *hkb/tll* double-mutant embryos or embryos which lack the *torso* gene product, *gt* extends to the posterior tip. Similarly, the domain of *kni* expression expands posteriorly in *tll* mutants or in *hkb/tll* double-mutant embryos. These results argue that the activity of the terminal gap genes is capable of repressing the activation of the central gap genes. By contrast, the expression domains of the terminal gap genes are not affected by the absence of central gap gene activity (Brönner and Jäckle, 1991). Since the terminal gap genes do not receive regulatory inputs from the other gap genes in the posterior region of the embryo, the spatial limits of *tll* and *hkb* expression set a stringent posterior boundary in the embryo which separates trunk from tail primordia. In view of the ectopic activation of both *tll* and *hkb* in response to ectopic *torso* activity and the lack of control by other gap genes or by mutual interactions, one must postulate that terminal gap genes are solely dependent on a transcription factor which is locally activated in response to *torso* activity. By analogy to the known long-range morphogen gradient provided by BCD in the anterior region of the embryo, both *hkb* and *tll* may be activated above different concentration values of the active form of this transcription factor, in a manner similar to *hb* activation by BCD.

Expression of Kr *in the central region*

Expression of *Kr* fills the region between the anterior *hb* and posterior *kni* expression domains (Fig. 1). In the absence of zygotic activity of each of these two genes or of *gt*, the *Kr* expression domain expands, to various degrees, towards the wild-type expression domain of the gene whose activity is missing. This suggests that TLL (see above), HB, *kni* protein (KNI) and GT act as potential repressors of *Kr* gene expression, although their ability to repress could be of different strengths (see below). This leaves the question of which factor(s) activate *Kr* in a central region of the wild-type embryo.

In the absence of the anterior morphogen BCD or zygotic HB, *Kr* is expressed in a more anterior region than normal (Gaul and Jäckle, 1987, 1989). In the absence of both BCD and HB (maternal as well as zygotic), however, no *Kr* expression can be observed (Hülskamp et al., 1990). These findings suggest that BCD and HB can each activate *Kr* gene expression. Since *Kr* expression expands anteriorly in the absence of zygotic HB, it appears likely that different HB concentrations would have different effects on *Kr* expression in the anterior region of the embryo: low levels of HB present in the center of the embryo would activate it (Hülskamp et al., 1990) while high levels of HB in a more anterior position would repress *Kr* expression (Gaul and Jäckle, 1987, 1989). This model would explain why *Kr* expression, although dependent on BCD, can expand in the posterior-most region of the embryo: the expansion occurs under the control of maternally supplied HB when posterior repressors such as KNI, GT, TLL and *hkb* protein (HKB) are absent from embryos which lack both *nanos* and *torso* activity (see above).

Once activated, KR influences the expression of its neighbouring genes by acting as a repressor, delimiting secondary zygotic expression of *hb* in the anterior, and *gt* expression in the anterior and posterior region. In addition, it enhances *kni* expression by two means: KR acts directly through interaction with a *kni cis*-acting region and also through repression of *gt* which itself is a repressor of *kni*. Evidence for this proposal comes from reduced levels of *kni* gene expression in *Kr* mutant embryos. In these embryos, *gt* expression extends towards the anterior. However, wild-type levels of *kni* expression are observed in the absence of both *Kr* and *gt* gene activities. The finding of functional KR binding sites in the *Kr*-dependent *kni cis*-acting region (Pankratz et al., 1989; F. Sauer, unpublished results) suggests that *Kr* and *gt* carry antagonizing functions for the control of the level of *kni* gene expression.

The genetic interactions between maternal and gap genes described here (Fig.2) and elsewhere (Gaul and Jäckle, 1990; Nüsslein-Volhard, 1991; Hülskamp and Tautz, 1991; St. Johnston and Nüsslein-Volhard, 1992; for reviews) allow each of the gap genes to be placed in a specific scenario of putative transacting factors required for the normal spatial control of their expression. For comparative reasons and to point out the similarity and differences in the mode of gap-gene spatial regulation with respect to the known regulation of *hb*, we focus here on *Kr* and *kni*. The genetic circuitry leading to *Kr* and *kni* gene expression and the known transacting gene activities capable of spatially controlling the expression of *Kr* and *kni* as deduced from the genetic studies described above are summarized in Fig. 2. In the following we focus on initial studies which, in part, reveal different molecular strategies by which the spatial domains of the gap genes *Kr* and *kni* are set in response to maternal factors and other gap gene products.

Spatial control of *Kr* expression by competitive binding of the BCD activator and repressors to overlapping binding sites

By deletion analysis of the *Kr cis*-acting sequences, we defined a 730 bp minimal *Kr* DNA sequence which is sufficient to mediate gene expression in the authentic early *Kr* expression domain (Hoch et al., 1990, 1991). The corresponding DNA fragment, termed *Kr*730-element, mediates gene expression in *bcd*, *hb*, *kni*, *gt* and *tll* mutant embryos in a manner indistinguishable from the alterations observed for endogenous *Kr* gene expression in these mutants (see above). This suggests that the *Kr*730-element contains the *cis*-acting requirement for the activation of gene expression in response to BCD, and the target sites for the interactions with the gap gene proteins HB, KNI, GT and TLL. The ability of these proteins to bind to *Kr*730-DNA was analysed by DNase I in vitro footprinting using *E. coli*-derived proteins (Hoch et al., 1991, 1992). As summarized in Fig. 3, *Kr*730-DNA contains six binding sites for BCD, seven TLL binding sites, five GT binding sites and one region protected by KNI. In addition, ten HB binding sites were found.

The common feature of BCD and HB binding sites is that they occur in clusters. From 5′ towards 3′ of *Kr*730-DNA, there is a pair of HB binding sites followed by a single site and another single site embedded in a cluster of five BCD binding sites, followed by two HB binding sites, followed by a cluster of four HB binding sites and a single BCD binding site at the end (Fig. 3). One common feature of TLL, GT and KNI in vitro binding sites on *Kr*730-DNA is the overlap with binding sites of BCD, an activator of *Kr* and *Kr*730-mediated gene expression. In order to determine whether the overlap of the BCD binding sites with potential *Kr* repressor protein binding sites might be of functional significance for gene expression in vivo, we placed the *Kr*730-*lacZ* reporter gene into transgenic embryos where KNI, GT and TLL can be expressed under the control of a promotor element responsive to heat-shock.

When TLL was induced throughout the embryo by heat-shock treatment, *lacZ* expression directed by the *Kr*730-element was eliminated (Fig. 4) (Hoch et al., 1992). Similarly, when the reporter gene expression was examined in

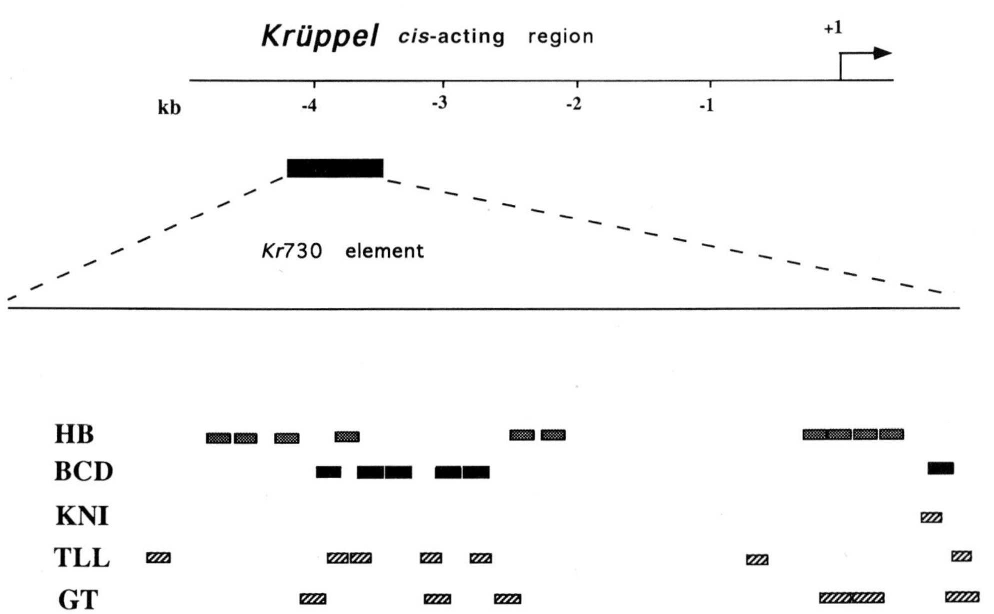

Fig. 3. BCD, HB, KNI, GT and TLL in vitro binding sites in the *Kr*730-element which directs reporter gene expression in place of the *Kr* central domain during blastoderm stage.

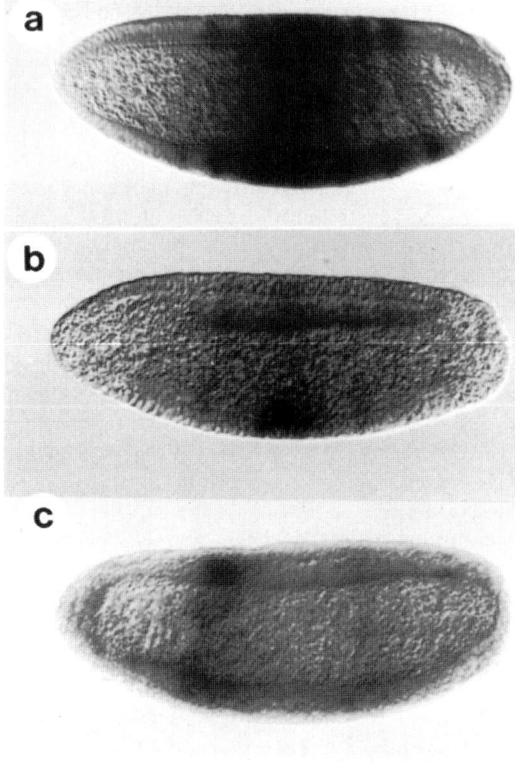

Fig. 4. Expression of *lacZ* directed by the *Kr*730-element in wild-type and experimentally manipulated embryos. (a) *Kr*730-mediated gene expression in early blastoderm embryos. Staining of the *even-skipped* protein (stripe pattern) served as an internal standard for the staining intensity of β-galactosidase expression. (b) *Kr*730-mediated gene expression in transgenic embryos containing the *kni* gene under the control of the heat-inducible hsp70 promotor. After heat shock prior to blastoderm formation resulting in ectopic KNI expression, the amount of *Kr*730-mediated gene expression was reduced in comparison to the amount of *even-skipped* protein staining. The altered stripe pattern is indicative of ectopic *kni* activity. (c) *Kr*730-mediated gene expression in transgenic embryos containing the hsp70*tll* gene; after heat shock, gene expression was absent. The distribution of *even-skipped* protein was abnormal, indicative of ectopic *tll* activity.

embryos where GT was ectopically expressed, *lacZ* expression was greatly reduced or even absent in some of these embryos (Fig. 4). In contrast, KNI ectopic expression reduced *Kr*730 expression only slightly (Hoch et al., 1992). These observations correlate well with genetic data indicating that KNI only weakly represses *Kr* (Gaul and Jäckle, 1987), whereas GT is a strong repressor (Eldon and Pirrotta, 1991; Kraut and Levine, 1991a) and TLL can even abolish *Kr* expression completely (Steingrimsson et al., 1991). Since none of these gap genes interfere with *bicoid* activity or the formation of the BCD gradient, it appeared likely that these repressors prevent BCD-dependent activation of *Kr*730-mediated gene expression by competitive binding to *Kr*730-DNA, thereby preventing the activator from binding.

In order to see whether such a molecular mechanism could account for the results obtained with *Kr*730-DNA, we analysed the single BCD-binding site overlapped by a KNI-binding site in detail. For this, we have chosen the 16-bp sequence (5′)-ACTGAACTAAATCCGG-(3′) at the 3′-end of the *Kr*730-DNA. KNI and BCD competed for binding to this sequence in vitro. Increasing amounts of BCD competed for the binding of KNI and vice versa (Hoch et al., 1992). These experiments demonstrated that each of the two proteins can bind to the 16-bp sequence, but their binding is mutually exclusive. To examine the possible function of interaction of KNI and BCD with the 16-bp element in vivo, we transfected *Drosophila* Schneider tissue culture cells with reporter gene constructs containing the bacterial chloramphenicol acetyltransferase (CAT) gene under the control of this element (Gorman et al., 1982). Cotransfection experiments with plasmid DNA containing the *bcd* or the *kni* gene showed that BCD can activate 16-bp-mediated gene expression. When KNI was introduced along with BCD, BCD-dependent CAT gene activation was suppressed in a manner responsive to the dosage of KNI. KNI did not make the reporter gene inactivable, but it increased the amount of BCD necessary for activation. No repression was observed when the DNA-binding domain of KNI was mutated such that the first Cys in the second zinc finger motif was replaced by a Leu residue, thereby impairing DNA-binding function. These results indicate that KNI can act as an antagonist of BCD-dependent gene activation, suggesting that this function of KNI is provided by competitive binding to the overlapping BCD binding site. In order to see whether KNI could also act as a suppressor of BCD-dependent gene activation in the embryo, we placed the 16-bp element upstream of a *lac*Z reporter gene construct.

The 16-bp *lacZ* reporter gene construct was inserted into the *Drosophila* genome and *lacZ* expression patterns were monitored in the transgenic embryos. No *lacZ* expression was observed in embryos derived from *bcd*-deficient females. Embryos laid by females containing the two normal copies of the *bcd* gene show *lacZ* expression in the anterior third of the embryo, where BCD is present at its highest concentration. When BCD concentration within the embryo was increased by addition of wild-type copies of the *bcd* gene in the females, the extent of *lacZ* expression correspondingly expanded towards the posterior, indicating that the 16-bp element responds to alterations of the BCD concentration along the longitudinal axis. In order to determine whether KNI is also able to interfere with this BCD-dependent gene activation in the embryo, we placed the 16-bp *lacZ* reporter gene into transgenic embryos where KNI was ectopically expressed under the control of the heat-shock responsive promotor element. In cases when *kni* gene expression was induced throughout the embryo, *lacZ* expression directed by the 16-bp element was eliminated (Hoch et al., 1992). Taken together, these findings indicate that the 16-bp element is a responsive site for the action of the activator BCD and the repressor KNI, suggesting a mechanism in which KNI represses by competitive DNA-binding (Ptashne, 1986, 1988; Levine and Manley, 1989) thereby preventing the action of the BCD activator.

In the *Kr*730-element TLL overlaps all BCD-binding sites, and GT overlaps three of them. This observation and the finding that *Kr*730-mediated *lacZ* gene expression is abolished or greatly reduced upon ectopic expression of

TLL and GT, respectively, suggest that gene activation or repression by competition of opposing transcriptional regulators at the same binding sites, as established here for the 16-bp element, may provide a molecular basis for gap gene products to contribute to the spatial control of *Kr* gene expression in response to the morphogen BCD. The concentration of a gap protein at a particular point in its gradient may effectively block gene activation by the amount of BCD found at that location, thus restricting the initial gene activation in reponse to the gradient of the BCD morphogen. This interpretation of our data is consistent with a finding that the initially broad *Kr* expression domain narrows, from both ends, during blastoderm formation when the concentrations of the surrounding gap gene products increase significantly. It leaves open, however, how *Kr* is initially activated in a more posterior position of the embryo than *hb* (see above). In view of the mechanism established for *hb* activation, i.e. the affinity of the BCD-binding sites determines the threshold concentration above which a gene becomes activated in response to BCD, we looked for high-affinity binding sites within *Kr*730-DNA. By direct comparison with the high-affinity binding sites of the *hb* promotor, none of the six BCD-binding sites bound BCD with higher affinity than the sites within the *hb* promotor (M. Hoch, unpublished). This finding suggests that the mechanism by which BCD sets the posterior limit of the expression domain in response to BCD may differ from the one observed with *hb*. In the case of *hb*, the BCD-binding sites are separated by about 140 nucleotides corresponding to one nucleosome equivalent (Driever et al., 1989), whereas the clustered BCD-binding sites within the *Kr*730-DNA are separated by maximum of 40 nucleotides. Thus, BCD-binding to these sites might be susceptible to cooperative interactions leading to a crowding effect and subsequently to positive interaction with the transcription machinery in more posterior regions of the BCD gradient than in the case of *hb*. In this model, each of the repressors provided by TLL, GT and KNI may interfere with the BCD cooperative interactions, and thus may interfere more severely with BCD-dependent activation than by blocking of individual activator sites. However, since we do not know yet how HB activates through the *Kr*730-element, cooperative interactions between BCD and HB cannot be excluded to account for the activation of *Kr* in a more posterior region of the embryo than *hb*.

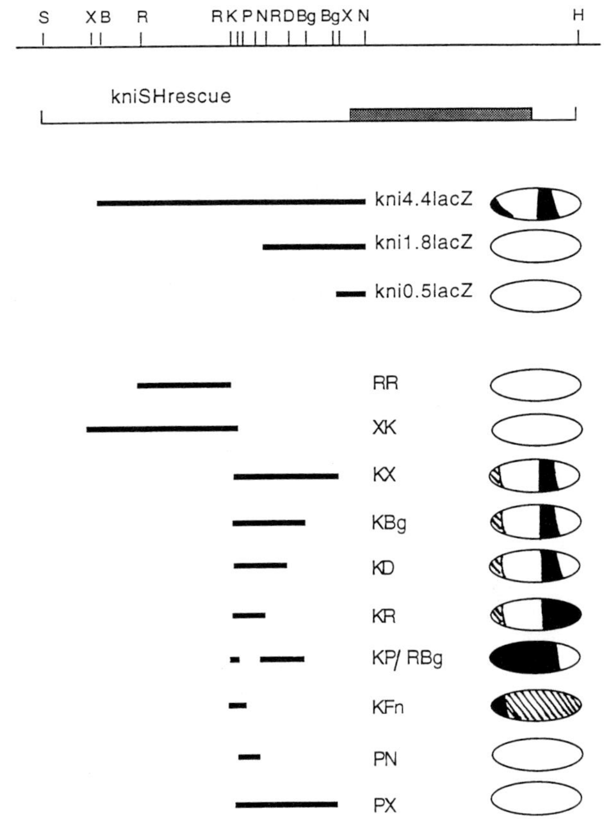

Fig. 5. Transformation constructs derived from the *kni* genomic region. Top line shows the positions of the various restriction sites utilized for making the DNA constructs used for fly transformation experiments (see Pankratz et al., 1992; for details). The hatched bar in the 9-kb kniSH *kni* rescue fragment denotes the transcribed region. This construct, which contains the coding region and approximately 6 kb of flanking sequences, fully rescues the *kni* mutant phenotype The thick lines on the left indicate the fragments used to make the *lacZ* indicator gene fusion constructs. The first three constructs contain the endogenous *kni* promoter. All others are fused to the basal heat shock promoter. The schematized embryos on the right illustrate the ß-gal expression patterns directed by the corresponding fusion constructs. Striped areas denote weaker staining. S, *Sal*I; X, *Xba*I; B, *Bam*HI; K, *Kpn*I; P, *Pst*I; N, *Nru*I; D, *Dra*I; Bg, *Bgl*II; H, *Hin*dIII; T, *Taq*I; C, *Cla*I.

Spatial control of *kni* expression by general activation and local repression through distinct *cis*-acting elements

As with the terminal gap genes there are no genetically identified factors that could activate *kni* and *gt* expression. However, these factors are capable of activating *kni* (and possibly also *gt*) throughout the embryo. The spatial domain of *kni* expression, in the prospective abdominal region, is set by localized repressors which act on top of the activator. HB appears to be responsible for repression in the anterior half of the embryo whereas TLL acts as a repressor in the pole region. Deletion analysis indicates that the *kni cis*-acting region consists of at least three functional elements which respond to these factors (Figs 5, 6). One is needed for activation throughout the embryo, another is required to prevent expression in the anterior half in response to HB, while the third is required to prevent expression in the posterior pole region in response to TLL. The element required for setting the anterior border of *kni* expression contains HB-binding sites while the element required for setting the posterior border contains TLL-binding sites. Deletion of each of the two *kni cis*-acting elements results in an expansion of reporter gene expression towards the site of the repressor gene product which is now unable to interfere with the *cis*-acting element in transgenic wild-type embryos (Pankratz et al., 1992). In the absence of both repressor elements, gene expression can be obtained throughout the embryo. On top of this basic regulatory scheme, additional factors are required to further regulate *kni* wild-type gene

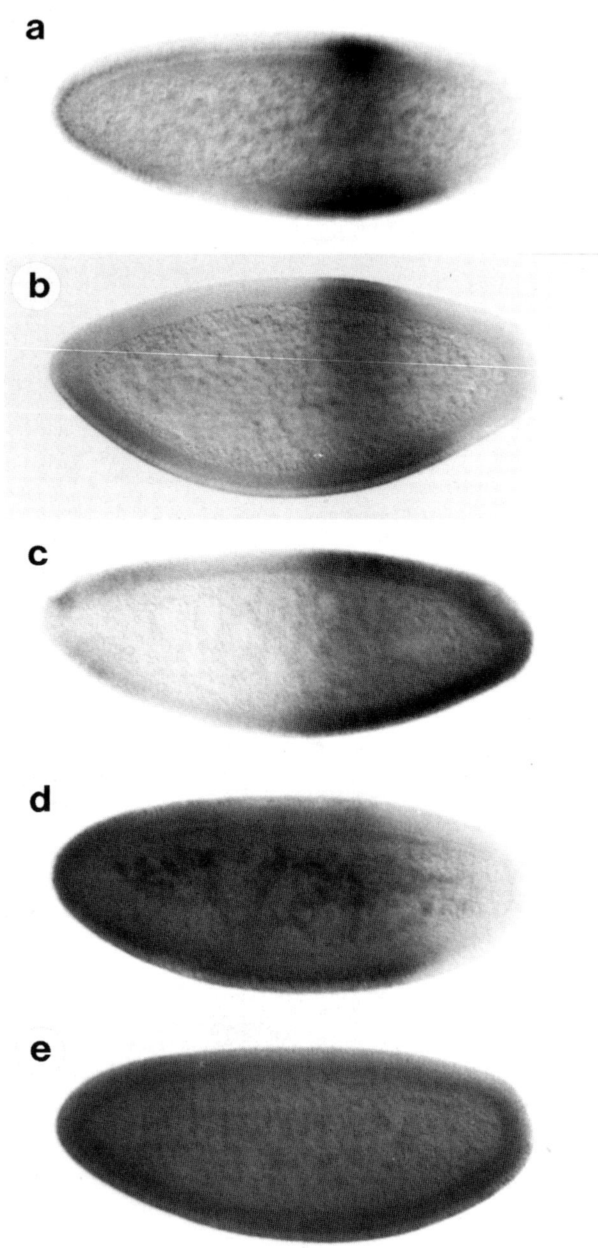

Fig. 6. Reporter gene expression directed by various *kni* upstream sequences in early embryos. Embryos are oriented with anterior to the left and dorsal up, and were stained with anti-ß-gal antibody to visualize reporter gene expression (Pankratz et al.,1992). (a) KBg construct; the spatial pattern of the posterior expression resembles that of the *kni* transcripts. (b) KD construct; the spatial pattern is similar to the above KBg construct. (c) KR construct; the posterior domain is expanded to the posterior tip of the embryo. At late blastoderm (not shown) the posterior domain retracts from the posterior tip of the embryo; this is also seen with endogenous *kni* expression in embryos derived from *torso* females. (d) KP/RBg construct; the posterior domain is expanded to the anterior region of the embryo. As *kni* is also expressed at the anterior tip (see Fig. 1), we do not know to what extent the strong signal in the very anterior region is due to the increased expression of the anterior domain, the anteriorly expanded posterior domain, or both. However, the ectopic expression between approximately 50%-75% egg length (posterior pole is 0%) is due to the anteriorly expanded posterior domain because that expression is still present in embryos derived from *bcd* mutant mothers. The endogenous *kni* expression as well as the reporter gene expression of the kni 4.4 lac Z construct (M. J. Pankratz, unpublished result) in the anterior cap domain are absent in *bcd* embryos. (e) KFn construct; reporter gene is expressed throughout the embryo. As in the KP/RBg construct above, this construct also directs strong expression in the anterior *kni* domain. Constructs PN and PX, which both delete a common 58-bp Kpn/Pst fragment, did not direct any reporter gene expression (not shown).

expression. For example, genetic studies indicate that the genes *Kr* and *gt* which act in the *kni* adjacent domains have opposing effects on *kni* expression: KR may act as a co-activator of *kni* expression (Pankratz et al., 1989), while GT may provide a repressor function. The latter effect is relatively weak since no significant extension occurs in *gt* mutants, and ectopic expression of GT under the control of a heat-shock element does not result in the complete suppression of *kni*. In *Kr;gt* double mutant embryos *kni* expression is expanded and restored to wild-type levels.

As outlined in the formal genetic circuitry that regulates *kni* expression in the prospective abdominal region (see Fig. 2), there are two possibilities as to how KR and GT may work on *kni*: KR may act as a repressor of *gt*, thereby preventing GT repressor function on *kni*. In this case, *Kr* function on *kni* would be indirect and mediated by GT, as proposed by Capovilla et al. (1992). We favour a model in which KR compensates in the wild-type situation the repressive effect of *gt*, and both interactions are of direct nature. In the absence of *gt*, KR will have no detectable regulatory effect on *kni* as the level of *kni* expression is highest due to the absence of its repressor. Furthermore, the level of *kni* expression will be unchanged in double mutant embryos since both the co-activator and the repressor are absent. This model is based on the finding of several KR binding sites in the *cis*-acting region of *kni*. Deletion of these binding sites results in the decrease of reporter gene expression (Pankratz et al.,1989) mediated by the deleted *kni cis*-acting sequences. In tissue culture, the fragment containing the KR sites is able to mediate reporter gene activation in response to co-transfected *Kr* gene DNA (F. Sauer, unpublished). Thus, in the wild-type situation *Kr* expression is likely to compensate for *gt* function on *kni* expression by two means. Firstly, it spatially limits *gt* expression and thereby inhibits its repressor function. Secondly, it restores wild-type expression levels, compensating for GT-dependent repression of *kni*.

Concentration-dependent activation or repression by KR

Based on genetic data, both HB and KR may act as positive and negative regulators of transcription on other genes of the zygotic segmentation hierarchy (see above). Both genes encode zinc finger-type proteins which form steep concentration gradients which extend significantly beyond the domains of transcript accumulation. We have examined the regulatory potential of such gradients by a series of co-transfection experiments in the *Drosophila* Schneider cell-

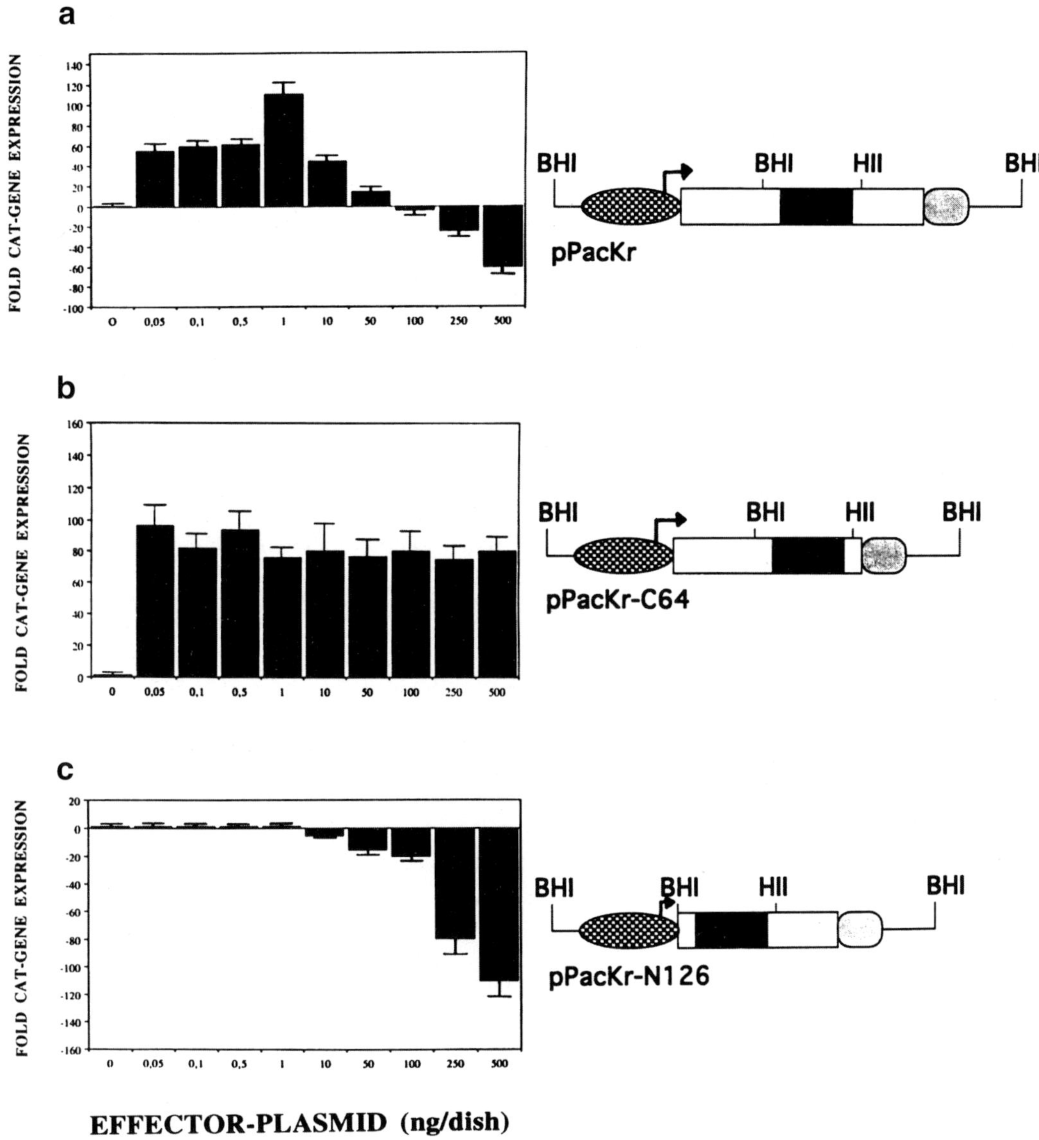

Fig. 7. Concentration-dependent activation and repression by the *Kr* gene require different regions of the protein. Co-transfection experiments were performed using a constant amount of reporter gene plasmid DNA (1 μg) and various concentrations of effector gene plasmid DNA, carrying either the entire *Kr* gene (a) or truncated (b, c) versions of it. The reporter gene DNA contains 5 copies of a KR in vitro binding site in front of a heterologous promoter which drives the expression of the basal level of the reporter gene CAT. (a) Reporter gene activity is dependent on KR: low dosages of *Kr* gene mediate activation while high dosages repress reporter gene expression. (b) Reporter gene expression in response to different dosages of a C-terminal truncated *Kr* protein. Note that this gene only mediates activation. (c) Deletion construct which leads to a N-terminal truncated KR only represses reporter gene expression (Sauer and Jäckle, 1991).

line-system using the *Kr* gene as the source of a specific transcription factor. Different doses of *Kr* expression plasmid were tested for their ability to drive the already mentioned CAT reporter gene expression mediated by a single 11-bp *Kr* in vitro binding site, 5′AAAAGGGTTAA-3′, common to several putative *Kr* target genes (Pankratz et al., 1989; Stanojevic et al., 1989; Treisman and Desplan, 1989) including *kni* (see above). When the KR expression plasmid was used in co-transfection experiments, both activation and repression of the reporter gene were observed (Fig. 7). Low amounts of the KR expression plasmid led to low levels of KR and to transcriptional activation, whereas high amounts of the expression plasmid led to high levels of KR which resulted in reporter gene repression (Licht et al., 1990; Zuo et al., 1990). The basal amount of reporter gene expression was left unchanged by co-transfected *Kr* DNA when the 11-bp-element was absent or replaced in the reporter gene construct, and no activation or repression occurs when plasmids containing the DNA-encoding transcription factors other than KR were used in the co-transfection experiments. Thus, KR is able to provide opposing regulatory effects in a concentration-dependent manner.

In order to determine whether transcriptional activation and repression by KR require specific portions of the protein, we examined truncated versions of the protein in the tissue culture assay (Fig. 7). High and low concentrations of KR lacking the 64 carboxy-terminal amino acids could not repress, but they were able to activate gene expression in a dose-dependent manner. Conversely, KR lacking the 116 amino-terminal aminoacids was not able to activate gene expression, but repression of gene expression was still observed. These results demonstrate that distinct portions of KR other than the DNA-binding domain are essential for gene activation and repression. Preliminary data suggest that KR is able to form a concentration-dependent homodimer, involving the C-terminal region of the protein as a dimerization domain, and that KR can activate transcription as a monomer, but loses this ability when forming homodimers at high concentrations of the protein (F. Sauer, unpublished). If these results obtained with tissue culture and in vitro experiments apply to the wild-type situation in the embryo, *Kr* could only act as a repressor in regions of high protein concentrations (roughly corresponding to the domain of transcript accumulation) and it could activate target genes only in regions which correspond to the domains of adjacent gap gene expression. However, *Kr* does not always act as an activator, as low concentrations of KR can act as transcriptional repressors within the *kni* expression domain (see below).

Repetitive patterns of pair-rule gene expression in response to gap gene-encoded local activators and repressors

The general scheme that emerged from the genetic and molecular analysis of pair-rule genes is that they are required for the establishment of alternating segmental units in the larva, and that they are expressed, apart from a few minor differences, in a basic outward appearance of seven evenly spaced stripes in the blastoderm embryo (Akam, 1987; Ingham, 1988; for reviews). The stripe pattern of expression of pair-rule genes such as *fushi tarazu* (*ftz*) is controlled by a small upstream fragment that directs expression of all the stripes (Hiromi et al., 1985), and their formation depends on the preceding activity of other "primary" pair-rule genes. The work of Howard et al. (1988) suggested that this may not hold true for all the pair-rule genes. They studied mutations in the pair-rule gene *hairy* (*h*) that deleted the expression of specific stripes in the early embryo. Furthermore, studies on the regulatory region of another pair-rule gene, *even-skipped* (*eve*), showed a similar effect: DNA fragments derived from the *eve* upstream region could drive expression of a reporter gene in regions corresponding to particular *eve* stripes (Goto et al., 1989; Harding et al., 1989). These results strongly supported the view that *h* and *eve* are not responding to a preset periodic pattern, but that they respond instead to unique spatial cues that have been established along the axis of the embryo. In addition, proper *ftz* expression requires the preceding activities of *eve* and *h* whereas these two genes do not require *ftz* activity for their normal spatial control. Thus, despite similar expression patterns, the ways in which these patterns emerge are quite different. Both *h* and *eve* are primary pair-rule genes which can be placed at the receiving end of a long range of maternal positional information that is transmitted through localized activity of different gap genes.

We have used the *h cis*-regulatory upstream region (Pankratz et al., 1990a) to ask whether or not individual gap gene proteins are able to interact with *h* DNA. Different gap gene proteins were incubated with small subfragments of the *h* target DNA, and the protein/DNA complexes were then precipitated with antibodies directed against the different proteins. We found that the gap proteins bind, in various combinations and to different degrees, to a number of different fragments. The DNA fragments precipitated by this assay were then used as prospective control elements in front of a *lac*Z reporter gene construct. The precipitated DNA fragments acted as separate control elements giving rise to individual stripes in the transgenic embryos which are localized almost precisely in the position of the stripe of endogeneous *h* gene expression. Only the elements giving rise to stripe 3 and 4 of *h* expression could not be separated. Here we focus on the regulation of stripe number 6.

Genetic studies suggest that the two gap genes *Kr* and *kni* are mainly responsible for the control of *h* stripe 6 expression in the wild-type embryo, which occurs in an area of very low KR and high KNI concentrations (Pankratz et al., 1990a; Riddihough and Ish-Horowicz, 1991). In the absence of *kni*, *h* stripe 6 fails to be expressed. In the absence of *Kr,* reporter gene expression under the control of the *h* stripe 6 element is shifted towards anterior, and the area of expression expands. This suggests that in the wild-type situation, KR would negatively interact with the *h* stripe 6 element to set the anterior border of expression. However, there is a second indirect interaction of KR with the stripe 6 element, provided through KNI. As outlined above, KR co-activates *kni* gene expression. Thus, the absence of KR leads to a lower level of KNI expression, and therefore activation mediated by the *h* stripe 6 element occurs in a more anterior position of the embryo, where the threshold level of KNI would now be sufficient for activation. Due to the concomitant absence of the KR repressor, *h* stripe 6-mediated gene activation would continue throughout the region where KNI is high enough for activation. This mechanism implies that KR can repress *h* stripe 6-mediated gene expression even at low concentrations, and repression overrides activation. Stripe 6 expression would then be observed only in an area with a high enough KNI concentration to activate, but not enough KR to repress transcription. This model is consistent with a finding that ectopic *Kr* expression (via heat-shock control; see above) does not allow for gene expression mediated by the *h* stripe 6 element. At present, we do not know whether activation and repression by KNI and KR, respectively, is provided by competition for overlapping activator/repressor binding sites, as described above for the regulation of *Kr* expression in response to BCD and gap gene proteins, or whether it follows the mode of *kni* gene expression by individual activator/repressor elements. We note, however, that a direct link has been established between binding sites for gap regulatory proteins in the stripe 2 promotor element of *eve* and the expression of this stripe in the embryo (Small et al., 1991; Stanojevic et al., 1991). In this case, regulation of stripe gene expression occurs by overlapping activators (such as the HB and BCD) and repressors (such as GT and KR). Interplay between activators and repressors, as observed for *Kr* expression, is therefore used at the different levels of the segmentation gene hierarchy.

Conclusions

The metameric prepattern in the *Drosophila* blastoderm is generated through a series of DNA-protein interactions that is initiated by at least three transcription factor gradients in the egg. In the terminal regions, one (or several) unknown transcription factor(s) appear to be activated through a signal transduction cascade. Most likely, this results in a local and steep gradient of the activated form of the transcription factor which, at different threshold levels of activity, controls the expression of the terminal gap genes *tll* and *hkb*.

Once activated, their gene products delimit the region of the embryo which gives rise to segmental patterns. This occurs by repression of the other known gap gene activities which cause the periodic pattern of pair-rule gene expression. In the anterior region of the preblastoderm embryo, two morphogen gradients are formed, either by diffusion of the protein from a local source (BCD), or by post-transcriptional regulation of evenly distributed mRNA under the control of *nanos,* to restrict the expression of maternal HB to the anterior half of the embryo. While BCD is likely to provide only an activating function that is limited to the anterior portion of the egg (with the central gap gene *Kr* likely to represent the posterior-most gene controlled by BCD activity), *hb* activity is likely to both activate and repress gene expression and reaches into the prospective abdominal region, as it marks the anterior border of *gt* expression within the posterior region of the embryo.

The mechanisms by which the spatially localized gap gene expression patterns are regulated are fundamentally different. *hb* expression occurs in response to the BCD gradient, likely to be regulated through the affinity of the BCD binding sites within the *hb* promotor: in regions where the BCD concentration is high enough to bind (through the site with the highest affinity within the promotor), *hb* becomes activated. The activation then leads to a pattern which extends from the anterior pole region, where *bcd* concentration is highest, to the position in the gradient where BCD fails to bind. This mechanism may also account for other genes which are expressed in regions of the embryo where the slope of the BCD gradient is steep. In more posterior regions of the embryo, i.e. in the more shallow portions of the BCD gradient, this mechanism for gene activation might not be sufficient to provide sharp boundaries of the expression domain. For this reason, a different mechanism for BCD-dependent gene activation might be used, as in the case *Kr*. In view of the fact that none of the BCD binding sites of the *Kr*730-element shows a higher affinity binding site for BCD than in the *hb* promotor region, we favour a model in which cooperative interactions may lead to a crowding effect to activate *Kr* by BCD. The use of such a mechanism would be consistent with the finding that gap gene proteins such KNI, GT and TLL (which act as repressors to delimit the *Kr* expression domain) have target sequences that overlap the BCD binding sites within *Kr*730 DNA: binding of either TLL, GT or KNI would interfere with BCD binding and thus would lead to repression. This effect would also interfere with cooperative interactions facilitating the filling of the additional BCD binding sites in the *Kr* regulatory region, and the binding of the repressor instead of the activator would shift *Kr* gene expression towards regions of higher BCD concentrations. In the extreme situation, the binding of several repressors should knock out BCD-dependent activation even if high BCD concentrations were present, as for example in the anterior pole position. The way in which HB activates and represses *Kr* gene expression in a concentration-dependent manner and how it may interfere with BCD-dependent activation has not yet been analysed. It appears, however, that the zinc finger-type proteins encoded by the gap genes are capable of providing these opposing functions, as exemplified in the case of KR.

An interplay between activators and repressors as observed for *Kr* expression is also observed for *kni*. However, the mechanism by which *kni* is spatially regulated in the posterior region again differs significantly from the one that regulates *hb* and *Kr*. Our results suggest that *kni* activation occurs throughout the entire embryo through the activity of a general, as yet unidentified activator. The spatial borders are then actively set through repression by factors that are regionally localized: HB, GT and TLL repress through separate modular *cis*-acting elements that are different from the activation element within the *kni* regulatory region. Another feature, common to both *Kr* and *kni* regulation, is interesting to note. Regulatory inputs of the same kind are provided by more than one gene product. For example, *Kr* can be activated by HB when BCD is absent, and vice versa. Furthermore, TLL, HB and GT in the anterior and KNI, GT, TLL and possibly also HKB (Weigel et al.,1990) in the posterior, can provide local repression of *Kr*. Currently, we do not know whether these multiple activation and repressor functions represent a fine tuning system to determine and specify the precise position in the embryo where the target gene is active (or repressed), whether the regulatory inputs are redundant, or whether some of the regulatory pathways represent just evolutionary relics which are not decisive any longer in the *Drosophila* wild-type embryo.

We would like to thank our colleagues in the lab for comments on the manuscript. The work was supported by the Deutsche Forschungsgemeinschaft and the Fonds der Chemischen Industrie.

References

Akam, M. (1987). The molecular basis for metameric pattern in the *Drosophila* embryo. *Development* **101**, 1-22.

Berleth, T., Burri, M., Thoma, G., Bopp, D., Richstein, S., Frigerio, G., Noll, M. and Nüsslein-Volhard, C. (1988). The role of localization of *bicoid* RNA in organizing the anterior pattern of the *Drosophila* embryo. *EMBO J.* **7**, 1749-1756.

Brönner, G. and Jäckle, H. (1991). Control and function of terminal gap gene activity in the posterior pole region of the *Drosophila* embryo. *Mech. Dev.* **35**, 205-211.

Campos-Ortega, J. A. and Hartenstein, V. (1985). *The Embryonic Development of* Drosophila melanogaster. Berlin: Springer.

Capovilla, M., Eldon, E. D. and Pirrotta, V. (1992). The *giant* gene of *Drosophila* encodes a b-ZIP DNA-binding protein that regulates the expression of other segmentation gap genes. *Development* **114**, 99-112.

Driever, W. and Nüsslein-Volhard, C. (1988a). A gradient of *bicoid* protein in *Drosophila* embryos. *Cell* **54**, 83-93.

Driever, W. and Nüsslein-Volhard, C. (1988b). The *bicoid* protein determines postion in the *Drosophila* embryo in a concentration dependent manner. *Cell* **54**, 95-104.

Driever, W. and Nüsslein-Volhard, C. (1989). The *bicoid* protein is a positive regulator of *hunchback* transcription in the early *Drosophila* embryo. *Nature* **337**, 138-143.

Driever, W., Thoma, G. and Nüsslein-Volhard, C. (1989). Determination of zygotic domains of gene expression in the *Drosophila* embryo by the affinity of binding sites for the *bicoid* morphogen. *Nature* **340**, 363-367.

Eldon, E. D. and Pirrotta, V. (1991). Interactions of the *Drosophila* gap gene *giant* with maternal and zygotic pattern forming genes. *Development* **111**, 367-378.

Frigerio, G., Burri, M., Bopp, D., Baumgartner, S. and Noll, M. (1986). Structure of the segmentation gene *paired* and the *Drosophila* PRD gene set as part of a gene network. *Cell* **47**, 735-746.

Frohnhöfer, H. and Nüsslein-Volhard, C. (1986). Organization of anterior pattern in the *Drosophila* embryo by the maternal gene *bicoid*. *Nature* **324**, 120-125.

Gaul, U., Seifert, E., Schuh, R. and Jäckle, H. (1987). Analysis of *Krüppel* protein distibution during early *Drosophila* development reveals post-transcriptional regulation. *Cell* **50**, 639-647.

Gaul, U. and Jäckle, H. (1987). Pole region-dependent repression of the *Drosophila* gap gene *Krüppel* by maternal gene products. *Cell* **51**, 549-555.

Gaul, U. and Jäckle, H. (1989). Analysis of maternal effect mutant combination elucidates regulation and function of the overlap of *Krüppel* and *hunchback* gene expression in the blastoderm *Drosophila* embryo. *Development* **107**, 651-662.

Gaul, U. and Jäckle, H. (1990). Role of gap genes in early *Drosophila* development. *Adv. Genetics* **27**, 239-272.

Gloor, H. (1954). Phänotypen der Heterozygoten bei der unvollständig dominanten, homozygot lethalen Mutante *Kr* (=*Krüppel*) von *Drosophila melanogaster*. *Arch. Jul. Klaus Stiftung* **29**, 277-287.

Gorman, C. M., Moffat, L. F. and Howard, B. H. (1982). Recombinant genomes which express chloramphenicol acetyltransferase in mammalian cells. *Mol. Cell. Biol.* **2**, 1044-1051.

Goto, T., Macdonald, P. and Maniatis, T. (1989). Early and late periodic patterns of *even-skipped* expression are controlled by distinct regulatory elements that respond to different spatial cues. *Cell* **57**, 413-422.

Harding, K., Hoey, T., Warrior, R. and Levine, M. (1989). Auto-regulatory and gap response elements of the *even-skipped* promoter of *Drosophila*. *EMBO J.* **8**, 1205-1212.

Hiromi, Y., Kuroiwa, A. and Gehring, W. J. (1985). Control elements of the *Drosophila* segmentation gene *fushi tarazu*. *Cell* **43**, 603-613.

Hoch, M., Schröder, C., Seifert, E. and Jäckle, H. (1990). Cis-acting control elements for *Krüppel* expression. *EMBO J.* **9**, 2587-2595.

Hoch, M., Seifert, E. and Jäckle, H. (1991). Gene expression mediated by cis-acting sequences of the *Krüppel* gene in response to the *Drosophila* morphogens *bicoid* and *hunchback*. *EMBO J.* **10**, 2267-2278.

Hoch, M., Gerwin, N., Taubert, H. and Jäckle, H. (1992). Competition for overlapping sites in the regulatory region of the *Drosophila* gene *Krüppel*. *Science* **256**, 94-97.

Howard, K., Ingham, P. and Rushlow, C. (1988). Region-specific alleles of the *Drosophila* segmentation gene *hairy*. *Genes Dev.* **2**, 1037-1046.

Hülskamp, M., Schöder, C., Pfeifle, C., Jäckle, H. and Tautz, D. (1989). Posterior segmentation of the *Drosophila* embryo in the absence of a maternal posterior organizer gene. *Nature* **338**, 629-632.

Hülskamp, M., Pfeifle, C. and Tautz, D. (1990). A morphogenetic gradient of *hunchback* protein organizes the expression of the gap genes *Krüppel* and *knirps* in the early *Drosophila* embryo. *Nature* **346**, 577-580.

Hülskamp, M. and Tautz, D. (1991). Gap genes and gradients - the logic behind the gaps. *BioEssays* **13**, 261-268.

Irish, V. F., Lehmann, R. and Akam, M. (1989). The *Drosophila* posterior-group gene *nanos* functions by repressing *hunchback* activity. *Nature* **338**, 646-648.

Ingham, P. W. (1988). The molecular genetics of embryonic pattern formation in *Drosophila*. *Nature* **335**, 25-34.

Jäckle, H., Tautz, D., Schuh, R., Seifert, E. and Lehmann, R. (1986). Cross-regulatory interactions among the gap genes of *Drosophila*. *Nature* **324**, 668-670.

Klingler, M., Erdélyi, M. Szabad, J. and Nüsslein-Volhard, C. (1988). Function of *torso* in determining the terminal anlagen of the *Drosophila* embryo. *Nature* **335**, 275-277.

Knipple, D. C., Seifert, E., Rosenberg, U., Preiss, A. and Jäckle, H. (1985). Spatial and temporal pattern of *Krüppel* gene expression in early *Drosophila* embryos. *Nature* **317**, 40-44.

Kraut, R. and Levine, M. (1991a). Mutually repressive interactions between the gap genes *giant* and *Krüppel* define middle body regions of the *Drosophila* embryo. *Development* **111**, 611-622.

Kraut, R. and Levine, M. (1991b). Spatial regulation of the gap gene *giant* during *Drosophila* development. *Development* **111**, 601-609.

Lehmann, R. and Nüsslein-Volhard, C. (1987). *hunchback*, a gene required for segmentation of an anterior and posterior region of the *Drosophila* embryo. *Dev. Biol.* **119**, 402-417.

Levine, M. and Manley, J. L. (1989). Transcriptional repression of eukaryotic promotors. *Cell* **59**, 405-408.

Licht, J. D., Grossel, M. J., Figge, J. and Hansen, U. (1990). *Krüppel* is a transcriptional repressor. *Nature* **346**, 76-79.

Mohler, J., Eldon, E. D. and Pirrotta, V. (1989). A novel spatial transcription pattern associated with the segmentation gene, *giant*, of *Drosophila*. *EMBO J.* **8**, 1539-1548.

Nauber, U., Pankratz, M., Kienlin, A., Seifert, E., Klemm, U. and Jäckle, H. (1988). Abdominal segmentation of the *Drosophila* embryo requires a hormone receptor-like protein encoded by the gap gene *knirps*. *Nature* **336**, 489-492.

Nüsslein-Volhard, C. and Wieschaus, E. (1980). Mutations affecting segment number and polarity in *Drosophila*. *Nature* **287**, 795-801.

Nüsslein-Volhard, C., Frohnhöfer, H. G. and Lehmann, R. (1987). Determination of anterioposterior polarity in *Drosophila*. *Science* **238**, 1675-1681.

Nüsslein-Volhard, C. (1991). Determination of the embryonic axes of *Drosophila*. *Development Suppl. 1*, 1-10.

Pankratz, M. J., Hoch, M., Seifert, E. and Jäckle, H. (1989). *Krüppel* requirement for *knirps* enhancement reflects overlapping gap gene activities in the *Drosophila* embryo. *Nature* **341**, 337-340.

Pankratz, M. J., Seifert, E., Gerwin, N., Billi, B., Nauber, U. and Jäckle, H. (1990a). Gradients of *Krüppel* and *knirps* gene products direct pair-rule stripe patterning in the posterior region of the *Drosophila* embryo. *Cell* **61**, 309-317.

Pankratz, M. J., Hoch, M., Rothe, M., Nauber, U., Gerwin, N., Brönner, G., Görlich, K. and Jäckle, H. (1990b). DNA-binding proteins in *Drosophila* development. *Nucleic Acids and Molecular Biology*, Vol 4 (ed. F. Eckstein and D. M. J. Lilley), pp. 185-204. Berlin, Heidelberg: Springer-Verlag.

Pankratz, M. J. and Jäckle, H. (1990). Making stripes in the *Drosophila* embryo. *Trends Genet.* **6**, 287-292.

Pankratz, M. J., Busch, M., Hoch, M., Seifert, E. and Jäckle, H. (1992). Spatial control of the gap gene *knirps* in the *Drosophila* embryo by the posterior morphogen system. *Science* **255**, 986-989.

Petschek, J., Perrimon, N. and Mahowald, A. P. (1987). Region specific effects in l(1) *giant* embryos of *Drosophila*. *Dev. Biol.* **119**, 177-189.

Pignoni, F., Baldarelli, R. M., Steingrimsson, E., Diaz, R. J., Patapoutian, A., Merriam, J. R. and Lengyel, J. A. (1990). The *Drosophila* gene *tailless* is expressed at the embryonic termini and is a member of the steroid receptor superfamily. *Cell* **62**, 151-163.

Preiss, A., Rosenberg, U. B., Kienlin, A., Seifert, E. and Jäckle, H. (1985). Molecular genetics of *Krüppel*, a gene required for segmentation of the *Drosophila* embryo. *Nature* **313**, 27-33.

Ptashne, M. (1986). Gene regulation by proteins acting nearby and at a distance. *Nature* **322**, 697-701.

Ptashne, M. (1988). How eukaryotic transcriptional activators work. *Nature* **335**, 683-689.

Riddihough, G. and Ish-Horowicz, D. (1991) Individual stripe regulatory elements in the *Drosophila hairy* promoter respond to maternal, gap and pair-rule genes. *Genes Dev.* **5**, 840-854.

Rosenberg, U., Schröder, C., Preiss, A., Kienlin, A., Coté, S., Riede, I. and Jäckle, H. (1986). Structural homology of the product of the *Drosophila Krüppel* gene with *Xenopus* transcription factor IIIA. *Nature* **319**, 336-339.

Rothe, M., Nauber, U. and Jäckle, H. (1989). Three hormone receptor-like *Drosophila* genes encode an identical DNA-binding finger. *EMBO J.* **8**, 3087-3094.

Sauer, F. and Jäckle, H. (1991). Concentration-dependent transcriptional activation or repression by *Krüppel* from a single binding site. *Nature* **353**, 563-566.

Schröder, C., Tautz, D., Seifert, E. and Jäckle, H. (1988). Differential regulation of the two transcripts from the *Drosophila* gap segmentation gene *hunchback*. *EMBO J.* **7**, 2882-2887.

Small, S., Kraut, R., Hoey, T., Warrior, R. and Levine, M. (1991). Transcriptional regulation of a pair-rule stripe in *Drosophila*. *Genes Dev.* **5**, 827-839.

Sprenger, F., Stevens, L. M. and Nüsslein-Volhard, C. (1989). The *Drosophila* gene *torso* encodes a putative receptor tyrosine kinase. *Nature* **338**, 478-483.

Stanojevic, D., Hoey, T. and Levine M. (1989). Sequence-specific DNA-binding activities of the gap proteins encoded by *hunchback* and *Krüppel* in *Drosophila*. *Nature* **341**, 331-335.

Stanojevic, D., Small, S. and Levine, M. (1991). Regulation of a segmentation stripe by overlapping activators and repressors in the *Drosophila* embryo. *Science* **254**, 1385-1387.

Steingrimsson, E., Pignoni, F., Liaw, G. J. and Lengyel, J. A. (1991). Dual role of the *Drosophila* pattern gene *tailless* in embryonic termini. *Science* **254**, 418-420.

Stevens, L. M., Frohnhöfer, H. G., Klingler, M. and Nüsslein-Volhard, C. (1990). Localized requirement for *torsolike* expression in follicle cells for the development of terminal anlagen of the *Drosophila* embryo. *Nature* **346**, 660-663.

St. Johnston, D., Driever, W., Berleth, T., Richstein, S. and Nüsslein-Volhard, C. (1989). Multiple steps in the localization of *bicoid* RNA to the anterior pole of the *Drosophila* oocyte. *Development (Suppl.)* **107**, 13-19.

St. Johnston, D. and Nüsslein-Volhard, C. (1992). The origin of pattern and polarity in the *Drosophila* embryo. *Cell* **68**, 201-219.

Strecker, T. R., Merriam, J. R. and Lengyel, J. A. (1988). Graded requirement for the zygotic terminal gene, *tailless*, in the brain and tail region of the *Drosophila* embryo. *Development* **102**, 721-734.

Struhl, G., Struhl, K. and Macdonald, P. M. (1989). The gradient morphogen *bicoid* is a concentration-dependent transcriptional activator. *Cell* **57**, 1259-1273.

Struhl, G., Johnston, P. and Lawrence, P. A. (1992). Control of *Drosophila* body pattern by the *hunchback* morphogen gradient. *Cell* **69**, 237-249.

Tautz, D., Lehmann, R., Schnürch, H., Schuh, R., Seifert, E., Kienlin, A., Jones, K. and Jäckle, H. (1987). Finger protein of novel structure encoded by *hunchback*, a second member of the gap class of *Drosophila* segmentation genes. *Nature* **327**, 383-389.

Tautz, D. (1988). Regulation of the *Drosophila* segmentation gene *hunchback* by two maternal morphogenetic centres. *Nature* **332**, 281-284.

Treisman, J. and Desplan, C. (1989). The products of the *Drosophila* gap gene *hunchback* and *Krüppel* bind to the *hunchback* promoters. *Nature* **341**, 335-337.

Wang, C. and Lehmann, R. (1991). Nanos is the localized posterior determinant in *Drosophila*. *Cell* **66**, 637-647.

Weigel, D., Jürgens, G., Klingler, M. and Jäckle, H. (1990). Two gap genes mediate maternal terminal pattern formation in *Drosophila*. *Science* **248**, 495-498.

Wieschaus, E., Nüsslein-Volhard, C. and Kluding, H. (1984). *Krüppel*, a gene whose activity is required early in the zygotic genome for normal embryonic segmentation. *Dev. Biol.* **104**, 172-186.

Zuo, P., Stanojevic, D., Colgan, J., Han, K., Levine, M. and Manley, J. L. (1990). Activation and repression of transcription by the gap proteins *hunchback* and *Krüppel* in cultured *Drosophila* cells. *Genes Dev.* **5**, 254-264.

Journal of Cell Science, Supplement 16, 53-60 (1992)
Printed in Great Britain © The Company of Biologists Limited 1992

Targets of homeotic gene regulation in *Drosophila*

R. A. H. WHITE, J. J. BROOKMAN, A. P. GOULD*, L. A. MEADOWS, L. S. SHASHIDHARA, D. I. STRUTT and T. A. WEAVER

Department of Anatomy, University of Cambridge, Downing Street, Cambridge CB2 3DY, UK

*Present address: National Institute for Medical Research, The Ridgeway, Mill Hill, London NW7 1AA, UK

Summary

We have used a chromatin immunopurification approach to identify target genes regulated by the homeotic gene *Ultrabithorax*. A monoclonal antibody against the *Ultrabithorax* gene product is used to immunopurify in vivo *Ultrabithorax* protein binding sites in embryonic chromatin. The procedure gives an enrichment of sequences with matches to a consensus homeodomain binding site. In one case we have shown that an immunopurified sequence lies within a 4 kb fragment that acts in vivo as a homeotic response element. We anticipate that this approach will enable us to identify further targets, allowing the analysis of their regulation and function. The chromatin immunopurification strategy may be of general application for the identification of direct in vivo targets of DNA-binding proteins.

Key words: homeotic gene, *Ultrabithorax*, DNA-binding protein, *Drosophila* development.

Introduction

A decade or so of intense research on the generation of patterns of gene expression in the early *Drosophila* embryo has provided the elements of a hierarchical system of transcriptional regulation (reviewed by Akam, 1987; Ingham, 1988). In the anteroposterior axis, the gradient of the *bicoid* transcription factor is interpreted by the gap genes to produce overlapping bands of expression of gap gene products and this pattern is, in turn, interpreted by the pair-rule genes to produce stripes. The overlapping stripe patterns of expression of the family of pair-rule genes are then used to establish the more refined stripes of expression of segment polarity genes such as *engrailed*. Transcriptional regulation by the products of gap, pair-rule and segment-polarity genes generates the expression patterns of the next class of genes in the hierarchy, the homeotic genes. The homeotic genes are expressed in overlapping domains in the anteroposterior axis in rather complex patterns such that different segments in the developing fly express specific combinations of homeotic gene products (Fig. 1; Lewis, 1978; reviewed by Akam, 1987). The products of the homeotic genes are transcription factors and thus extend the cascade of transcriptional regulation; however at this point our knowledge of the hierarchy of control breaks down and we are largely ignorant of the nature of the genes at the next level. The analysis of mutant phenotypes that played such an important role in the identification and ordering of genes in the hierarchy thus far, unfortunately provides few clues as to the identity of the genes directly subordinate to the homeotics.

Loss of function of an individual homeotic gene causes a host of morphogenetic consequences for the development of particular segments. For example, mutations in the *Ultrabithorax (Ubx)* gene of the bithorax complex cause the repeat units, parasegments 5 and 6, to develop morphologies similar to the normal parasegment 4 (Lewis, 1978; Sanchez-Herrero et al., 1985). Such mutants die in the larval stages and show transformations not only in the epidermis, as visualised by alterations in the differentiation and patterning of cuticular elements, but also in the peripheral and central nervous systems and in the somatic and visceral musculature (Hartenstein, 1987; Teugels and Ghysen, 1985; Hooper, 1986; Bienz and Tremml, 1988). Analysis of the developmental consequences of homeotic mutations indicates a wide range of cellular processes under homeotic gene control including mitotic rate, cell growth, movement and invagination, cell-cell interactions and cellular differentiation. From this it was proposed, as long ago as 1975, that the homeotic genes functioned to regulate large batteries of "downstream" or target genes (Garcia-Bellido, 1975).

We are interested in the identification of these target genes for several reasons. Firstly, in order to understand how homeotic genes control morphogenesis we need to identify the molecules through which they act. If we can follow the regulatory command chain downstream from the homeotic genes this should lead us to the molecules regulating the cellular processes of morphogenesis. Secondly, the specific combinations of homeotic gene products in different segments presumably produce a unique set of target gene activities in each segment. How they do this raises interesting questions of transcriptional regulation: do the different homeotic genes regulate the same or different sets

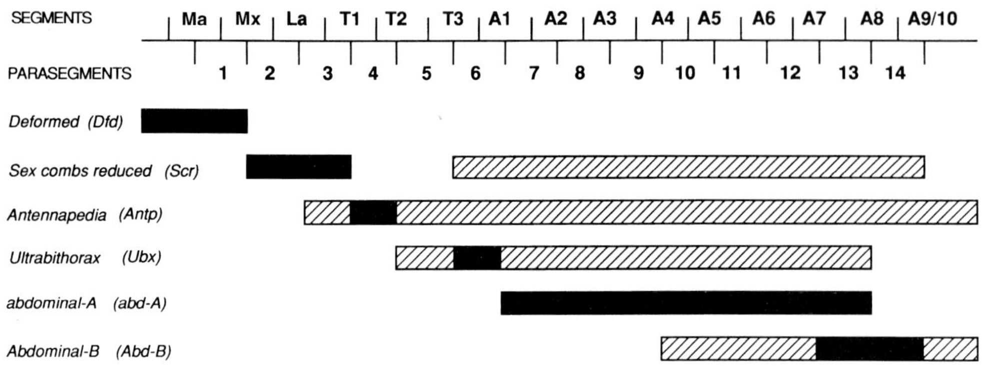

Fig. 1. Domains of expression of homeotic genes in the epidermis of *Drosophila* embryos. The registration along the anteroposterior axis is indicated both in the primordial repeat-units, parasegments (Martinez-Arias and Lawrence, 1985), and also in the later segmental units.

of target genes? Do they compete for binding sites in target gene regulatory sequences? How important are interactions between homeotic gene products or between homeotic gene products and other transcription factors for binding site specificity? How do homeotic genes control the activities of characteristic sets of target genes in different tissues? The solution to all these questions depends on the identification of regulatory sequences of target genes that respond directly to homeotic gene control. The third reason for interest in the identification of direct target genes is that this is not a problem restricted to homeotic genes; in many systems, for example for transcription factors involved in oncogenesis, it is important to identify in vivo targets and we can hope, using *Drosophila* homeotic genes as a model system, to develop strategies for the identification of direct targets of DNA-binding proteins which will be of general application.

Looking for targets

As the homeotic genes control segmental identity, the genes that they control might be expected to show patterns of expression that vary from segment to segment. A number of cloned genes exhibit such patterns. For some, particularly cell type differentiation genes, the basis for this may be trivial; for example the myosin heavy chain gene exhibits a pattern of expression that varies from segment to segment, reflecting the segment-specific arrangments of muscles. Clearly myosin expression may simply be a secondary consequence of earlier processes under homeotic control that specify the patterns of muscle cell development within each segment. Other genes may be more interesting; for example the *decapentaplegic (dpp)* gene is expressed in a number of different tissue types at different stages in development, and is expressed in the visceral mesoderm in a parasegmentally modulated pattern which includes a specific band of expression in parasegment 7. The homeotic gene *Ubx* is normally expressed in parasegment 7 in the visceral mesoderm and *Ubx* function is required both for the formation of a specific morphological feature, the second midgut constriction, and for the expression of *dpp* (Immergluck et al., 1990; Reuter et al., 1990). The *dpp* gene shows homology to the the TGF-β family of vertebrate growth factors and thus provides a clue to at least one class of downstream functions. However, it is not known whether *dpp* is directly or indirectly regulated by *Ubx*.

The technique of P-element mediated enhancer detection makes it possible to screen for genes of interest on the basis of their patterns of expression (O'Kane and Gehring, 1987). Screening for genes whose pattern of expression shows segment-to-segment modulation should produce candidate downstream genes. Wagner-Bernholtz and colleagues (1991) have carried out such a screen, looking for genes whose expression differs between antennal and leg imaginal discs, and have identified the *spalt major* gene as a direct or indirect target of the homeotic gene, *Antennapedia (Antp)*. We have carried out a similar screen on embryos looking for segment-to-segment modulations in expression. Screening several hundred β-galactosidase-expressing lines produced a small number of interesting lines. Two of these are illustrated in Fig. 2. In both cases transcription units lying close to the P-element insertion site exhibited patterns of RNA expression similar to the β-galactosidase expression from the enhancer detector line. The B5 line identified a transcript with a complex expression pattern. In the central nervous system, it is expressed most abundantly in the thoracic neuromeres T1, T2 and T3, and in the epidermis it also shows different patterns from segment to segment including a prominent patch of expression specific to T1. The transcript identified by the line T6 is more restricted to the nervous system where, like B5, it is most prominently expressed in the thoracic neuromeres. The T6 insertion has a wings held-out phenotype when homozygous and fails to complement *eagle* mutations, suggesting that T6 may be allelic to *eagle* (Lindsley and Zimm, 1985). The expression patterns of both B5 and T6 are altered in mutants defective in the genes of the bithorax-complex and thus are directly or indirectly controlled by homeotic genes. Such downstream genes provide useful molecular markers for segment-specific morphogenesis and may provide insights into the gene functions regulated by the homeotics. This analysis could also potentially provide an answer to a major question concerning downstream genes: how many are there? In the largest published P-element mediated enhancer detection screen 7 lines out of about 4,000 showed segment-segment differences not ascribable simply to tissue-type differentiation (Bier et al., 1989). This suggests a rather low number of downstream genes by these criteria but these figures should be treated with extreme caution as it is clear that the P-element mediated enhancer detection technique gives a biased sampling of the genome.

The enhancer detection can usefully provide downstream genes, but it does not immediately lead to direct targets of

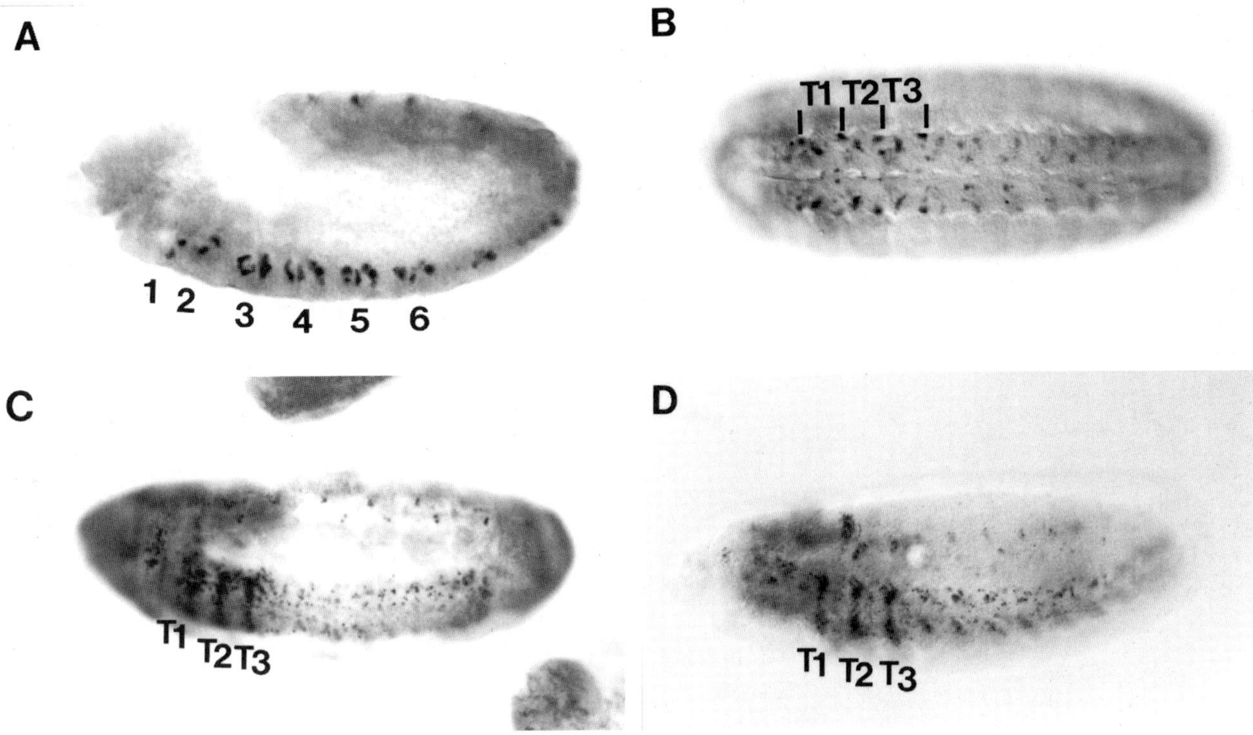

Fig. 2. Expression patterns from the T6 and B5 transcription units. (A,B) Expression of β-galactosidase from the T6 line. (A) Stage-11 embryo, lateral view (staging according to Campos-Ortega and Hartenstein, 1985). Parasegments 1-6 are indicated. The expression is in the central nervous system and extends from parasegment 1-14, but with most prominent expression in parasegments 3-5. (B) Stage-15 embryo, ventral view. Labelling is restricted to the central nervous system with clear segment-to-segment modulation in expression with most prominent expression in thoracic neuromeres. (C) Expression of β-galactosidase from the B5 line. Stage-15 embryo, ventrolateral view. Expression is seen in many tissues but within the central nervous system the expression is most pronounced in the thoracic neuromeres. (D) In situ hybridization on a wild-type embryo using a probe for the B5 transcription unit, showing that the endogenous transcription is very similar to the pattern of β-galactosidase expression from the B5 enhancer-detector line. Stage-15 embryo, ventrolateral view.

homeotic control. A strategy for the identification of such direct targets could be based on the location of in vitro binding sites for homeotic gene products in genomic DNA. However, homeodomain proteins bind rather promiscuously to DNA in vitro, and analysis of the DNA sequence requirements for binding a *Ubx* product in vitro revealed little sequence preference outside a core TAAT motif (Ekker et al., 1991). We have chosen to pursue an in vivo approach designed to isolate DNA sequences occupied by *Ubx* protein in embryonic chromatin (Gould et al., 1990).

The immunopurification strategy

The approach is outlined in Fig. 3. Briefly, embryonic nuclei are prepared according to the low salt method of Wu (1984); the nuclei are digested with *Hae*III and then the soluble chromatin fragments are released by lysis with EDTA. This soluble chromatin is then subjected to an immuno-affinity purification using anti-Ubx antibody in order to enrich for chromatin fragments containing bound Ubx protein. DNA is extracted from the control and experimental immunomatrix beads and cloned into M13 phage. Several experiments gave a small (approximately two-fold) increase in number of recombinant phage in the experimental (+Ab) versus the control (−Ab). To ask whether specific sequences were more highly enriched than bulk DNA we sequenced clones from the +Ab and −Ab pools. We then screened the sequences for matches to the homeodomain consensus binding sites derived from footprint analysis with engrailed protein (Desplan et al., 1988). The results of this screen are shown in Fig. 4. There is a clear enrichment for matches to the homeodomain binding consensus in the +Ab population. In addition, the enrichment is highly sequence-specific; if the preference for A over T in position four of the consensus is reversed then no enrichment is seen (Fig. 4B).

This sequence analysis suggests that the immunopurification successfully enriches for Ubx binding sites and the next question is whether these binding sites lie adjacent to transcriptional units under *Ubx* control. To address this, we took the six clones containing the top scoring matches to the consensus sequence and used these as probes to isolate flanking phage-lambda genomic clones. The lambda clones were then used as probes for in situ hybridization on embryos to detect any transcript expression. In four cases we identified embryonic transcripts within the range of the lambda clones. The distributions of RNA expression from two of these transcription units are shown in Figs 5 and 6. Transcript #48 has a rather complex and dynamic pattern.

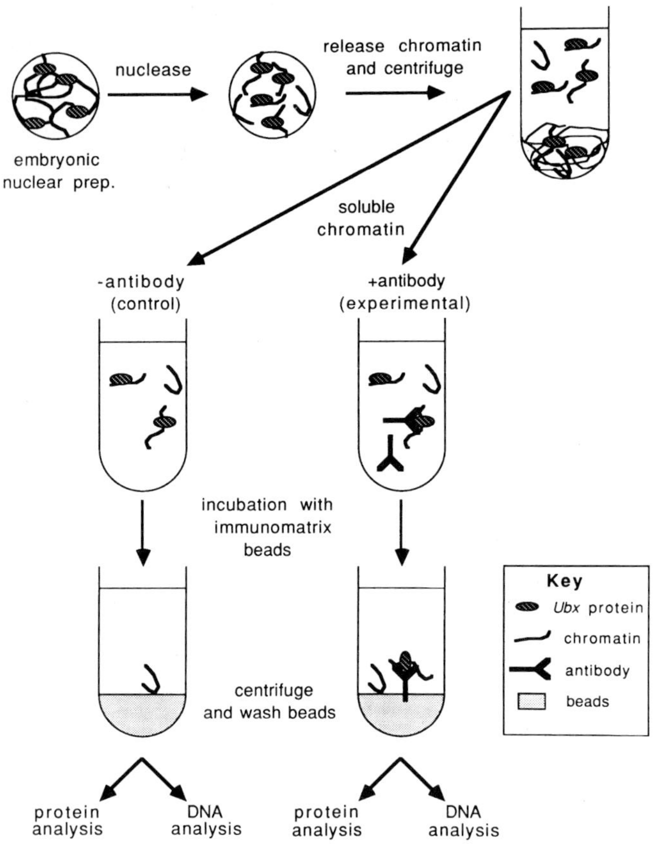

Fig. 3. The chromatin immunopurification procedure. Nuclei from *Drosophila* embryos are digested with *Hae*III and a soluble chromatin fraction is prepared after nuclear lysis in EDTA. Chromatin fragments are immunopurified using biotinylated-FP3.38, a monoclonal antibody against the Ubx protein products (White and Wilcox, 1984) and streptavidin-agarose beads (Gould et al., 1990).

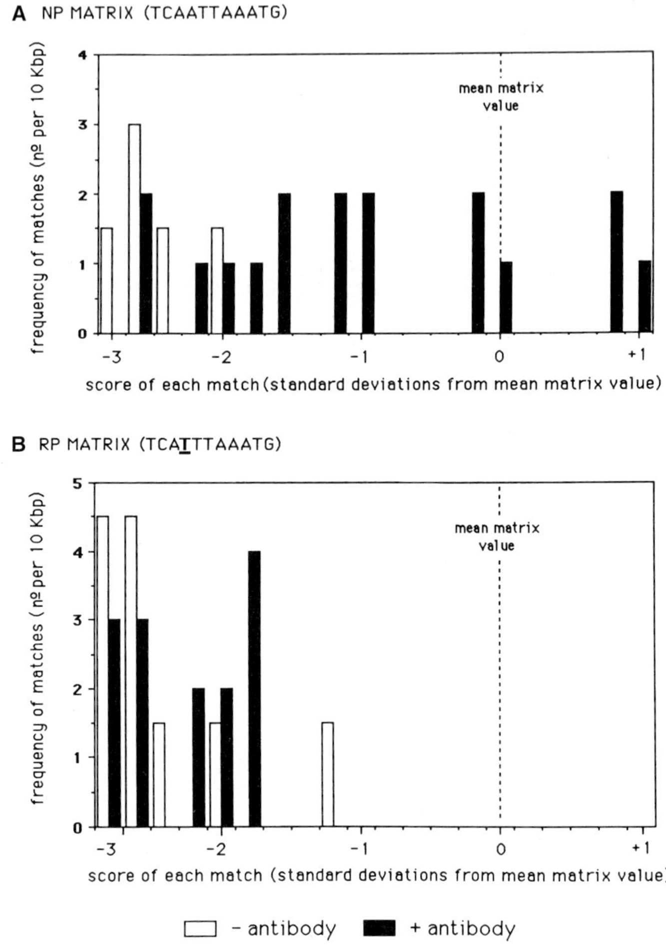

Fig. 4. Sequence enrichment in the chromatin immunopurification. Sequences from the experimental (+Ab) and control (−Ab) DNA clones were concatenated to form two sequence files of lengths 9.8 kb and 6.5 kb respectively. They were screened against a weight matrix (Staden, 1984) constructed from engrailed homeodomain binding sequences (Desplan et al., 1988). (A) The scores and frequency of matches are shown; scores within two standard deviations of the mean matrix value are only seen in the +Ab sequences. (B) Screening against a weight matrix with the preferences for A versus T at the fourth position of the consensus reversed. This RP sequence shows greatly reduced binding to the engrailed homeodomain compared with the NP sequence (Desplan et al., 1988). Screening against the RP matrix reveals no difference between the +Ab and −Ab sequences.

It is initially expressed in a ventral stripe roughly corresponding to the mesodermal anlage at the cellular blastoderm stage. After invagination of the ventral furrow expression shuts off in the mesoderm but expression is activated in the lateral and dorsal ectoderm. In the extended germ band there is a second wave of expression in the mesoderm and it is not until germ band retraction that the first clearly segmentally modulated pattern appears. This expression (Fig. 5) is in epidermal stripes and in the ventrolateral epidermis there is a weak stripe in the first thoracic segment (T1), a very weak stripe in T2, a weak stripe in T3 and then stronger stripes in the abdominal segments. At later stages the #48 transcript is expressed in a segmentally modulated pattern in the visceral mesoderm with predominant expression in parasegment 7, the repeat unit under *Ubx* control in this tissue. Transcript #35 appears initially in segmentally repeated subsets in the visceral mesoderm, then towards the end of the extended germ band stage it is expressed in the somatic mesoderm predominantly in the thoracic segments (Fig. 6). Following the expression in the somatic mesoderm reveals that the labelled cells are precursor cells for specific sets of ventral and pleural muscles. In the central nervous system expression also appears around the time of germ band retraction but it initially shows little segment-to-segment modulation in intensity. However, by stage 16 the most prominent labelling is in the thoracic neuromeres.

Thus, both transcription units #35 and #48 show segmentally modulated patterns of expression consistent with regulation by the homeotic gene *Ubx*. In the case of #35, expression is low in the *Ubx* expression domain suggesting a repressive control, and with #48 expression is high in the *Ubx* domain indicating a positive control. The alterations in the expression patterns of #35 and #48 in *Ubx* mutant embryos are also consistent with a direct regulatory role of the *Ubx* gene on their expression.

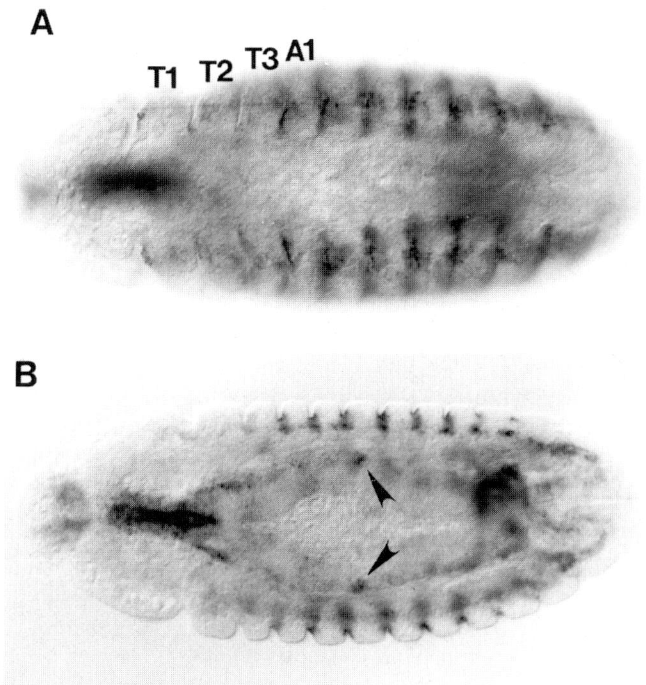

Fig. 5. Expression of the #48 transcription unit. In situ hybridization using a #48 cDNA probe to stage-15 embryos. (A) ventral view, focussing on the ventrolateral epidermis, showing expression predominantly at the anterior margins of the abdominal segments. (B) Ventral view focussing more internally showing the expression in parasegment 7 of the visceral mesoderm (arrowheads).

Regulation of the #35 transcription unit

The immunopurification procedure appears to have successfully led us to transcription units plausibly under the direct regulation of the homeotic gene *Ubx*. However the case would be considerably more compelling if we could demonstrate that the immunopurified DNA fragments, which are putative in vivo binding sites of the *Ubx* protein, are indeed components of regulatory elements under *Ubx* control. We have approached this question for the #35 transcription unit (Gould and White, 1992).

We first asked whether the 110 bp #35 immunopurified fragment had led us to a regulatory region capable of mediating homeotic control. The 110 bp fragment lies some 7 kb to the 3′ side of the #35 transcription unit and thus is not in a region that one would immediately expect to har-

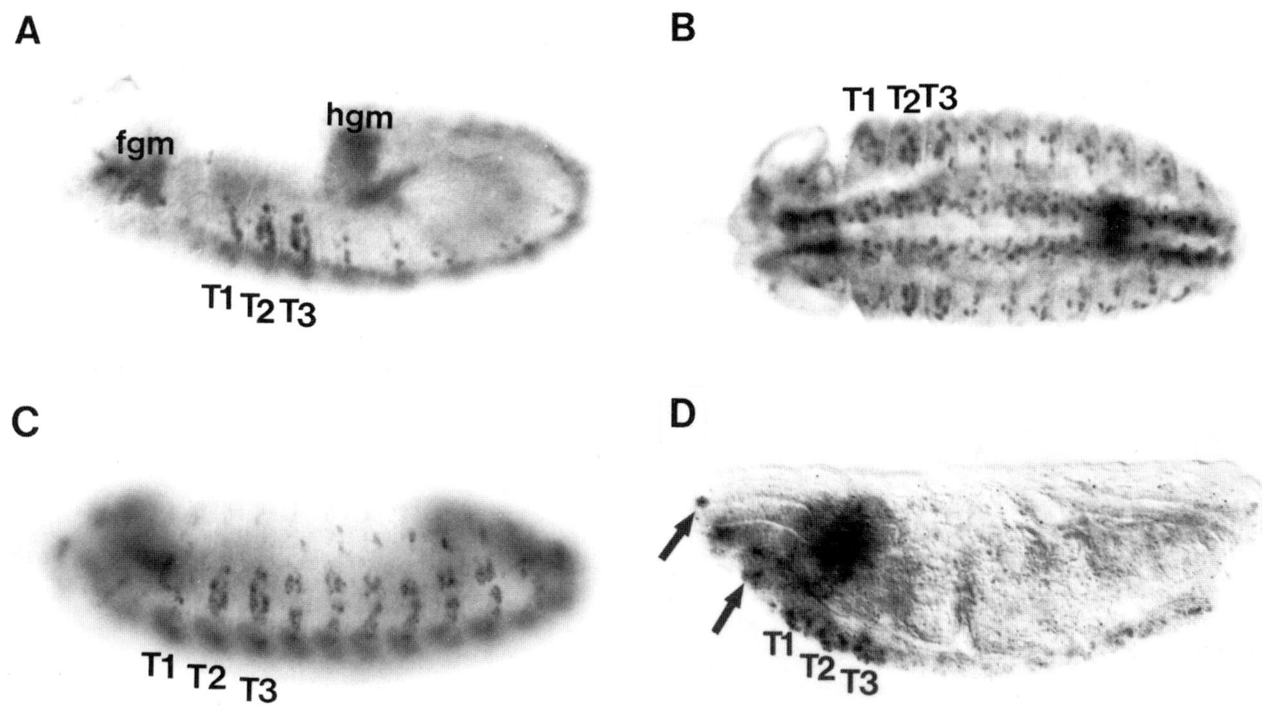

Fig. 6. Expression of the #35 transcription unit (connectin). In situ hybridization to whole embryos. (A) Stage 12. Lateral view focussing on the somatic mesoderm. Labelling is predominantly in specific cell clusters in the thoracic somatic mesoderm. There is also labelling in the visceral mesoderm of the foregut (fgm) and hindgut (hgm). (B) Stage 14. Ventral view. On either side of the midline there is labelling in longitudinal tracts of cells in the central nervous system (cns) with little segment-to-segment modulation in intensity. More laterally, labelling is seen in the somatic mesoderm. (C) Stage 14. Lateral view focussing on the somatic mesoderm. Labelling is seen in the precursors of specific pleural muscles in the thoracic segments and in the precursors of specific ventral and pleural muscles in the abdominal segments. (D) Stage 16. Labelling is predominantly in the brain lobes and in the thoracic neuromeres of the ventral nervous system. The arrows indicate labelling associated with head and gnathal sense organs.

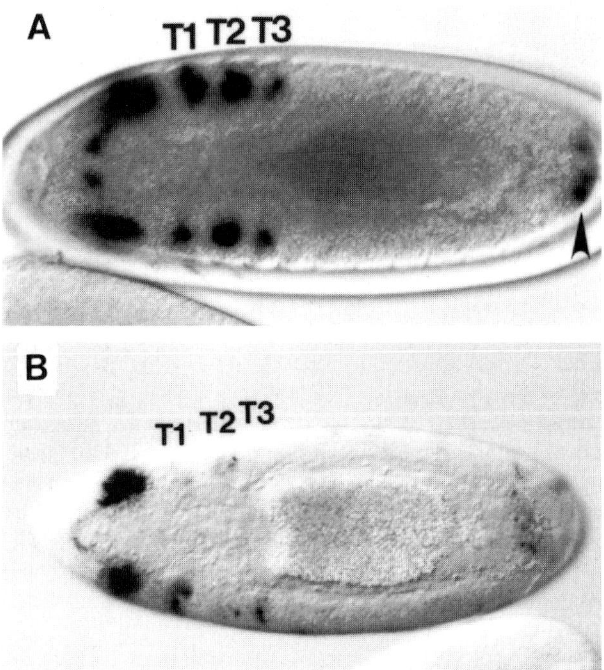

Fig. 7. Expression from #35 regulatory sequences. β-galactosidase expression from reporter constructs containing #35 regulatory sequences in pHZ50 (Hiromi and Gehring, 1987). (A) Expression from line 4.40 transformed with construct containing a 4 kb restriction fragment including the 110 bp #35 immunopurified sequence. Stage 13, ventral view. Labelling is tightly restricted to head and gnathal sensory organs and to the thoracic somatic mesoderm. The pattern represents a subset of the total #35 expression and note that there is no expression in the central nervous system. (B) Expression from line T8.110 transformed with construct containing the 4 kb restriction fragment but with a deletion of the 110 bp #35 immunopurified sequence. In comparison to A, the thoracic labelling in the somatic mesoderm is weak and variable.

bour a control element. However, we took a 4 kb restriction fragment, which included the 110 bp immunopurified sequence, and tested its ability to drive expression of β-galactosidase from a minimal heat-shock promoter in transformed fly lines. The expression pattern produced by this construct is illustrated in Fig. 7. The 4 kb fragment drives β-galactosidase expression in a subset of the #35 expression pattern and, importantly, shows a highly segment-to-segment modulated pattern of expression. The β-galactosidase activity mimics two parts of the endogenous #35 expression pattern with expression in gnathal sense organs and in specific muscle precursor cells in the thoracic segments. No β-galactosidase activity is seen in the central nervous system as presumably the control sequences responsible for expression in this tissue lie outside the 4 kb fragment used here.

The pattern of expression driven by the 4 kb fragment shows that the immunopurification strategy has led us to a sequence that acts as a homeotic response element. Indeed we have shown, by crossing the 4 kb construct into homeotic mutant backgounds, that the segment-to-segment modulation in the expression of the 4 kb construct is dependent on homeotic gene control. Several homeotic genes appear to be involved, with *Antennapedia* (*Antp*) and *Sex combs reduced* (*Scr*) acting to activate expression and *Ubx* and *abdominal-A* (*abd-A*) acting as repressors.

We also wanted to determine whether the 110 bp clone 35 sequence, immunopurified as an in vivo Ubx binding site, was an important component of the homeotic response element within the 4 kb fragment. We constructed a precise deletion of the 110 bp sequence from the 4 kb fragment and transformed this deletion construct (called 4kb-35) into flies. The effect of the deletion was to reduce the intensity of expression predominantly in the somatic mesodermal cells of the thoracic segments; the strong, consistent expression in these cells seen with the 4 kb construct was replaced by weak and variable expression with the 4kb-35 lines. Thus the 110 bp fragment appears to be an important component of the regulatory sequences contained within the 4 kb fragment. It does not, however, define a *Ubx* specific regulatory module. It might have been anticipated, for example, that the removal of the 110 bp sequence, containing a putative in vivo Ubx binding site, would have been equivalent to removal of *Ubx* function. In that case the expression from the deletion construct would have been similar to the expression of the entire 4 kb construct in a *Ubx* mutant backgound. However, in the former, the thoracic expression is reduced and in the latter the expression is derepressed and extends into anterior abdominal segments.

Our interpretation is that the 110 bp is involved in the regulation by several homeotic genes and its deletion affects, not only *Ubx*-dependent repression, but also positive controls by *Antp* and *Scr*. Also, as deletion of the 110 bp sequence weakens rather than abolishes the thoracic expression, it is likely that there are other sites of homeotic control within the 4 kb fragment.

Discussion

The chromatin immunopurification strategy appears to have successfully guided us to a homeotic response element and provides us with a method for the identification of further target genes allowing us to study both their regulation and function. Whilst we have concentrated so far on the validation of the approach as a means for the identification of direct targets, we can ask at this juncture what have we learnt of the nature of target genes and of their regulation.

The sequence of the #35 gene reveals that it is a member of the leucine-rich repeat family and it contains 10 leucine-rich repeats (Gould and White, 1992). It contains a signal sequence but no transmembrane domain. The COOH-terminal has a hydrophobic sequence suggesting that the molecule may be linked to the membrane via a glycolipid anchor (Ferguson and Williams, 1988). We have expressed the #35 product in tissue culture cells and shown that it is cleavable from the membrane by glycosylinositol specific phospholipase C. The presence of leucine-rich repeats does not immediately provide a specific function for the #35 protein: leucine-rich repeats appear to be involved in protein-protein interactions but occur in proteins of a wide variety of functional types (Takahashi et al., 1985; Kataoka et al.,

1985; Titani et al., 1987). However, in *Drosophila*, two cell-surface, leucine-rich repeat molecules, chaoptin and Toll, have been shown to be capable of mediating cell-cell adhesion (Krantz and Zipursky, 1990; Keith and Gay, 1990). Chaoptin is required for the organization of the *Drosophila* eye and like the #35 gene product it is linked to the cell-surface via a glycosyl-phosphatidylinositol anchor. We have shown that the #35 gene product, when expressed in the non-adherent *Drosophila* S2 cell line, is capable of mediating homophilic cell-cell adhesion. This becomes more interesting when taken together with results from Nose et al. (1992), who independently isolated the #35 gene. Using an antibody to the gene product, they have demonstrated that the #35 protein is expressed on a subset of motoneurons and also on their specific target muscles both prior to, and immediately following, innervation. This strongly suggests that the #35 gene product has a role in the formation of specific neuromuscular contacts and the #35 gene has been named connectin. However, at present we have no mutations in the connectin gene to test its requirement for neuromuscular development.

The #48 gene sequence reveals a transmembrane protein with limited sequence similarity to the EGF-receptor family. Its developmental role is unknown but several lethal complementation groups have been mapped in the vicinity of the #48 transcription unit at 97D.

The preliminary analysis of these two target genes allows us to make a couple of points. Firstly, the targets we have identified are not transcription factors, and thus moving downstream of homeotic genes we are moving away from the cascade of transcriptional regulation responsible for the early patterning processes in *Drosophila* development. Secondly, as one of the characterized gene products is a cell surface molecule capable of mediating cell-cell adhesion and the other is a transmembrane protein, it appears that the identification of target genes will provide a useful entry point into processes of cell-cell recognition and communication that must be important in regulating morphogenesis.

What do these genes tell us about how homeotic genes control the activities of target genes? It seems that neither of the genes that we have identified are uniquely controlled by the *Ubx* homeotic gene. These genes were identified on the basis of their proximity to putative Ubx in vivo binding sites; however, their patterns of expression, in wild-type and homeotic mutant backgrounds, indicate that several homeotic genes participate in their regulation. Some homeotic genes activate and others repress; in the case of expression from the 4 kb connectin regulatory element, *Scr* and *Antp* activate transcription whereas *Ubx* and *abd-A* repress. Also, with respect to different target genes, a particular homeotic gene can act as either an activator or a repressor; thus, *Ubx* activates #48 gene expression but represses expression from the (#35) connectin gene. With the identification of DNA sequences that respond to homeotic control we should be able to analyse the molecular basis of these interactions. We want to define the functional homeoprotein binding sites and to ask whether different homeodomain proteins act on the same or different sites. There are currently two complicating factors; firstly the homeotic genes interact and so a mutation in a particular homeotic gene may alter the expression from a homeotic response element indirectly through its affect on another homeotic gene; secondly, the homeotic response element that we have defined is 4 kb long and will need to be dissected into smaller functional units to allow the identification of binding sites and their analysis by mutagenesis.

Will the chromatin immunoaffinity purification approach be generally useful as a method for the isolation of target genes? The method depends on the availability of specific antibody and on a stable interaction between the DNA-binding protein and its binding site in vivo in chromatin and under conditions allowing specific antibody binding. In addition we relied on a second selection based on sequence matches to an in vitro-defined binding consensus. Measurements of the affinity of binding have been made for the *Antp* homeodomain. It binds with relatively high affinity (K_D in the range 10^{-9} to 10^{-10} M) to a consensus homeodomain binding site and the interaction is also relatively stable, with the half-life of the homeodomain-DNA complex estimated to be approximately 90 min (Affolter et al., 1990). However, in comparing Antp with other DNA-binding proteins it should be recognised that it is difficult to extrapolate from the in vitro to the in vivo situation and the binding of many transcription factors in chromatin may be stabilized by protein-protein interactions. The immunopurification approach may also be useful for determining whether a DNA-binding protein interacts directly in vivo with a particular sequence by assaying the enrichment of the sequence on immunopurification. Enrichment could be assayed by representation in a library derived from the immunopurified DNA or by directly assaying the amount of a given sequence in the immunopurification by, for example, PCR.

Whilst the approach may be useful for the identification of target genes of homeotic control in other organisms, and it would be particularly interesting to identify target genes in vertebrates, it may also be possible to use the target genes isolated in *Drosophila* to search for homologues in other organisms. As there is growing evidence that the *Hox* gene clusters in vertebrates serve similar functions to their *Drosophila* homeotic gene counterparts (reviewed by McGinnis and Krumlauf, 1992), it seems likely that the functions of target genes may also be evolutionarily conserved.

This work was supported by the Medical Research Council and the Wellcome Trust. The T6 transformant line was kindly provided by Maria Leptin and Michael Wilcox.

References

Affolter, M., Percival-Smith, A., Muller, M., Leupin, W. and Gehring, W. J. (1990). DNA binding properties of the purified Antennapedia homeodomain. *Proc. Nat. Acad. Sci. USA* **87**, 4093-4097.

Akam, M. (1987). The molecular basis for metameric pattern in the *Drosophila* embryo. *Development* **101**, 1-22.

Bienz, M. and Tremml, G. (1988). Domain of Ultrabithorax expression in Drosophila visceral mesoderm from autoregulation and exclusion. *Nature* **333**, 576-578.

Bier, E., Vaessin, H., Shepherd, S., Lee, K., McCall, K., Barbel, S., Ackerman, A., Carretto, R., Uemura, T., Grell, E., Jan, L. Y. and Jan, Y. N. (1989). Searching for pattern and mutation in the *Drosophila* genome with a P-lacZ vector. *Genes Dev.* **3**, 1273-1287.

Campos-Ortega, J. A. and Hartenstein, V. (1985) *The Embryonic Development of* Drosophila melanogaster. Berlin, Heidelberg: Springer-Verlag.

Desplan, C., Theis, J. and O'Farrell, P. H. (1988). The sequence specificity of homeodomain-DNA interaction. *Cell* **54**, 1081-1090.

Ekker, S. C., Young, K. E., Von Kessler, D. P. and Beachy, P. A. (1991). Optimal DNA sequence recognition by the Ultrabithorax homeodomain of Drosophila. *EMBO J.* **10**, 1179-1186.

Ferguson, M. A. F. and Williams, A. F. (1988). Cell-surface anchoring of proteins via glycosyl-phosphatidylinositol structures. *Annu. Rev. Biochem.* **57**, 285-320.

Garcia-Bellido, A. G. (1975). Genetic control of wing disc development in *Drosophila*. In *Cell Patterning*, Ciba Foundation Symposium **29**, pp. 161-178.

Gould, A. P., Brookman, J. J., Strutt, D. I. and White, R. A. H. (1990). Targets of homeotic gene control in Drosophila. *Nature* **348**, 308-312.

Gould, A. P. and White, R. A. H. (1992). Connectin: a target of homeotic gene control in Drosophila. *Development* in press.

Hartenstein, V. (1987). The influence of segmental compartmentalisation on the development of the larval peripheral nervous system in *Drosophila melanogaster*. *Roux's Arch. Dev. Biol.* **196**, 101-112.

Hiromi, Y. and Gehring, W. J. (1987). Regulation and function of the *Drosophila* segmentation gene *fushi tarazu*. *Cell* **50**, 963-974.

Hooper, J. E. (1986). Homeotic gene function in the muscles of Drosophila larvae. *EMBO J.* **5**, 2321-2329.

Immergluck, K., Lawrence, P. A. and Bienz, M. (1990). Induction across germ layers in Drosophila mediated by a genetic cascade. *Cell* **62**, 261-268.

Ingham, P. W. (1988). The molecular genetics of embryonic pattern formation in Drosophila. *Nature* **335**, 25-34.

Kataoka, T., Broek, D. and Wigler, M. (1985). DNA sequence and characterization of the S. cerevisiae gene encoding adenylate cyclase. *Cell* **43**, 493-505.

Keith, F. J. and Gay, N. J. (1990). The Drosophila membrane receptor Toll can function to promote cellular adhesion. *EMBO J.* **9**, 4299-4306.

Krantz, D. E. and Zipursky, S. L. (1990). Drosophila chaoptin, a member of the leucine-rich repeat family, is a photoreceptor cell-specific adhesion molecule. *EMBO J.* **9**, 1969-1977.

Lewis, E. B. (1978). A gene complex controlling segmentation in *Drosophila*. *Nature* **276**, 565-570.

Lindsley, D. and Zimm, G. (1985). The genome of Drosophila melanogaster. Part 1: genes A-K. *Drosophila Inform. Service* **62**, 104.

Martinez-Arias, A. and Lawrence, P. A. (1985). Parasegments and compartments in the *Drosophila* embryo. *Nature* **313**, 639-642.

McGinnis, W. and Krumlauf, R. (1992). Homeobox genes and axial patterning. *Cell* **68**, 283-302.

Nose, A., Mahajan, V. B. and Goodman, C. S. (1992). Connectin: A homophilic cell adhesion molecule expressed on a subset of muscles and the motoneurons that innervate them in Drosophila. *Cell* **70**, 553-567.

O'Kane, C. J. and Gehring, W. J. (1987). Detection *in situ* of genomic regulatory elements in *Drosophila*. *Proc. Nat. Acad. Sci. USA* **84**, 9123-9127.

Reuter, R., Panganiban, G. E. F., Hoffman, F. M. and Scott, M. P. (1990). Homeotic genes regulate the spatial expression of putative growth factors in the visceral mesoderm of Drosophila embryos. *Development* **110**, 1031-1040.

Sanchez-Herrero, E., Vernos, I., Marco, R. and Morata, G. (1985). Genetic organization of Drosophila bithorax complex. *Nature* **313**, 108-113.

Staden, R. (1984). Computer methods to locate signals in nucleic acid sequences. *Nucl. Acid Res.* **12**, 505-519.

Takahashi, N., Takahashi, Y. and Putnam, F. W. (1985). Periodicity of leucine and tandem repetition of a 24-amino acid segment in the primary structure of leucine-rich a_2-glycoprotein of human serum. *Proc. Nat. Acad. Sci. USA* **82**, 1906-1910.

Teugels, E. and Ghysen, A. (1985). Domains of action of bithorax genes in Drosophila central nervous system. *Nature* **314**, 558-561.

Titani, K., Takio, K., Handa, M. and Ruggeri, Z. M. (1987). Amino-acid sequence of von-Willebrand factor-binding domain of platelet membrane glycoprotein 1b. *Proc. Nat. Acad. Sci. USA* **84**, 5610-5614.

Wagner-Bernholz, J. T., Wilson, C., Gibson, G., Schuh, R. and Gehring, W. J. (1991). Identification of target genes of the homeotic gene Antennapedia by enhancer detection. *Genes Dev.* **5**, 2467-2480.

White, R. A. H. and Wilcox, M. (1984). Protein products of the bithorax complex in *Drosophila*. *Cell* **39**, 163-171.

Wu, C. (1984). Two protein-binding sites in chromatin implicated in the activation of heat-shock genes. *Nature* **309**, 229-234.

Journal of Cell Science, Supplement 16, 61-67 (1992)

Pax genes, mutants and molecular function

GEORGES CHALEPAKIS, PATRICK TREMBLAY and PETER GRUSS

Max-Planck Institute for Biophysical Chemistry, Department of Molecular Cell Biology, D-3400 Göttingen, FRG

Summary

The paired domain is a conserved DNA binding motif which was first found in *Drosophila* segmentation gene products. This paired domain is encoded by a well conserved, paired box DNA sequence, also detected in the genomes of other species. The mouse paired box-containing genes are referred to as *Pax* genes and are expressed in a distinct spatiotemporal pattern during embryogenesis. Pax proteins are able to bind to specific DNA sequences and modulate transcriptional activity. Interestingly, three different *Pax* genes have already been shown to correspond to some mouse and human mutants, emphasizing their role as developmental control genes.

Key words: *Pax*, DNA binding, *Undulated*, *Splotch*, *Waardenburg*, *Small eye*, *Aniridia*.

Introduction

The paired box was found in *Drosophila* segmentation genes (Bopp et al., 1986; Frigerio et al., 1986; Coté et al., 1987; Baumgartner et al., 1987; Morrissey et al., 1991; Dambly-Chaudiere et al., 1992), and also in the genomes of mouse (Deutsch et al., 1988; Dressler et al., 1988; Walther et al., 1991), zebrafish (Krauss et al., 1991a,b,c; Püschel et al., 1992), chicken (M. Goulding and P. Gruss, unpublished), frog, turtle, nematode and man (Burri et al., 1989).

To date, eight murine paired box genes (*Pax*) have been isolated. Comparison of their sequence and genomic organization led to their classification into four classes (Fig. 1). Genes of the same group share similar intron-exon boundaries, similarities in the protein structure and display a related expression pattern during development (Walther et al., 1991; Deutsch and Gruss, 1991; Gruss and Walther, 1992). *Pax* genes are not organized in clusters and, although three *Pax* genes are found on chromosome 2, their loci are far apart (Fig. 1).

The proposed translation products of the *Pax* genes are between 361 (Pax-1) and 479 (Pax-3) amino acids long. The paired domain is 128 amino acids long and is located close to the amino terminus of the protein. Only Pax-3 has an extended amino terminus, containing 33 amino acids amino-terminal to the paired domain. In addition to the paired domain, Pax-3, Pax-7 and Pax-6 have an additional paired type homeodomain of 61 amino acids located carboxy-terminal to the paired domain. Preliminary data also indicate the presence of a paired type, homeobox sequence in Pax-4 (M. Asano and P. Gruss, unpublished data). Pax-2, Pax-5 and Pax-8 have retained the amino-terminal part of the homeo domain including the first α-helix motif.

Computer structure predictions suggest that the paired domain contains three α-helices. The first helix is located at the amino-terminal part of the paired domain while the second and third helices, separated by 8 amino acids, are located in the carboxy-terminal part of the paired domain (Bopp et al., 1989; Burri et al., 1989; Treisman et al., 1991). However, the most conserved residues lie outside of the helices and there is no homology to known helix-turn-helix or the helix-loop-helix motifs.

The paired type homeodomain contains a helix-turn-helix motif and the second helix appears to be responsible for specific DNA recognition. Treisman et al. (1989) have shown that amino acid 9 of this helix plays a critical role in sequence recognition.

Pax-1, Pax-2, Pax-3, Pax-5, Pax-7 and Pax-8, but not Pax-6 (Walther et al., 1991) contain a conserved octapeptide which was initially described for the human HuP1, HuP2 and the *Drosophila* gooseberry proteins (Burri et al., 1989). It is located carboxy-terminal to the paired domain and, in the case of Pax-3 and Pax-7, between the paired domain and the homeodomain. Although this short peptide is well conserved, no function has yet been assigned to it.

Expression pattern of *Pax* genes

The murine *Pax* genes are first expressed between days 8 and 9.5 of embryogenesis and some are also found to be active in adult tissues. They are expressed in axial structures along the entire antero-posterior axis in a tissue-specific manner. *Pax-1*, expressed predominantly in mesodermally-derived tissues, is the only *Pax* gene not expressed in the developing central nervous system.

In general, *Pax* genes containing a homeobox appear to be expressed earlier and to be mainly restricted to mitotically active cells, whereas the others (*Pax-2* and *Pax-8*) are rather restricted to differentiating cells. *Pax-3*, *Pax-7* and *Pax-6*, which encode a homeodomain, in addition to the

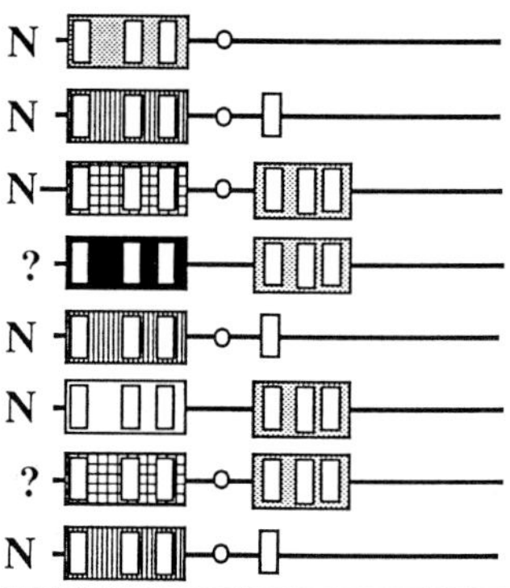

GENE	CHROMOSOME	STRUCTURE PD OP HD	mRNA	PROTEIN	MUTANTS MOUSE	MUTANTS HUMAN
Pax-1	2	N C	3.1 kb	361 aa	*undulated (un)*	
Pax-2	19	N C	4.2 / 4.7 kb	392 / 415 aa		
Pax-3	1	N C	3.3 / 3.6 kb	479 aa	*splotch (Sp)*	*Waardenburg Syndrome I*
Pax-4	6	? ?	?	?		
Pax-5	4	N C	?	391 aa		
Pax-6	2	N C	3.0 kb	422 / 436 aa	*small eye (Sey)*	*Aniridia*
Pax-7	4	? ?	4.9 kb	>300 aa		
Pax-8	2	N C	3.1 kb	457 aa		

Fig. 1. The *Pax* gene family. The chromosomal localization, protein structural features, size of the mRNA and proteins and the correlated mutants are shown. Paired and homeodomains are shown as boxes. The white vertical boxes represent α-helices. The paired domains represented with the same patterns belong to the same paired domain class as judged from protein sequence analysis. aa, amino acids; kb, kilobases; PD, paired domain; OP, octapeptide; HD, homeodomain; ?, not determined.

paired domain, are expressed in the neuroectoderm before neural differentiation starts (day-8.5 p.c.). On the other hand, *Pax-2* and *Pax-8*, which lack a homeobox, are first expressed in the neural tube at day-10 p.c., at the time when neural differentiation begins. Although *Pax* genes, with the exception of *Pax-6*, are expressed in mesodermally-derived segmented structures, they cannot be considered to be primary segmentation genes as they appear after initial segmentation has been established.

Pax-1

Pax-1, unlike the other *Pax* genes, is not expressed in the developing central nervous system (Deutsch et al., 1988). Expression starts at day-9 p.c. in the segmented paraxial mesoderm of the developing vertebral column, specifically in the sclerotome. These *Pax-1*-expressing sclerotomal cells migrate and surround the notochord to form the perichordal condensations. Later on, Pax-*1* expression is confined to the intervertebral disk anlagen. Thus, as suggested by its expression pattern, *Pax-1* could be involved in the specification of the sclerotomal mesoderm.

Moreover, *Pax-1* is expressed in the sternebrae of the segmented sternum and in the thymus. With the exception of *Pax-1* expression in the thymus, all the other tissues expressing *Pax-1* undergo chondrogenesis, indicating that *Pax-1* might also be involved in certain chondrogenic events.

Pax-2, Pax-8 and *Pax-5*

Pax-2 and *Pax-8* are similarly expressed in the nervous system (Nornes et al., 1990; Dressler et al., 1990; Plachov et al., 1990). Firstly, their transcripts have been identified in the neural tube and in the hindbrain. In the neural tube, expression has been shown to be restricted to a specific subset of postmitotic differentiating cells on both sides of the dorso-ventral midline, the sulcus limitans. However, while *Pax-8* transcripts are undetectable after day-13.5 p.c., *Pax-2* continues to be expressed throughout development and in adult tissues. *Pax-2* is also active in the developing ear and eye.

Secondly, around days-9 and -10 p.c., both genes are expressed in the developing excretory system, more specifically in the condensed metanephric mesenchyme and in the resulting epithelial tubules. *Pax-2* expression is also detected in the Wolffian duct and ureter which buds out from the Wolffian duct. Both *Pax-2* and *Pax-8* are also expressed in the Wilms tumor (Dressler and Douglas, 1992; Poleev et al., 1992). Finally, *Pax-8* is expressed in the thyroid, from the earliest developmental stages to the adult stage (Zannini et al., 1992). Pax-8 thus constitutes one of the earliest markers of thyroid development.

Pax-5 is expressed in the developing brain, predominantly at the midbrain-hindbrain boundary, and in the neural tube. The expression pattern of *Pax-5* in the developing brain is different from those of *Pax-2* and *Pax-8*, while the expression pattern in the neural tube is very similar. Unlike *Pax-2* and *Pax-8*, *Pax-5* is not expressed in the developing excretory system or thyroid (Asano and Gruss, 1992; Adams et al., 1992). *Pax-5* is also expressed in pre-B, pro-B and mature B-cells, but not in plasma cells. Transcripts of *Pax-5* are also detected in the testis (Adams et al., 1992).

Pax-3 and *Pax-7*

Pax-3 and *Pax-7* exhibit a similar expression pattern (Goulding et al., 1991; Jostes et al., 1990). Transcripts are detected during early brain and neural tube development before the onset of neural differentiation. At day-8.5 p.c. *Pax-3* is expressed in the dorsal ventricular zone of the neural tube, including the roof plate and neural crest cells and their derivatives. *Pax-7* also presents a similar pattern, excluding, however, the roof plate and neural crest cells. In addition, its onset of expression is slightly later than *Pax-3*: while *Pax-3* transcripts are detected in the neural folds at day-8.5 p.c., *Pax-7* cannot be detected before closure of the neural tube. The expression of both genes in the neural tube and in the brain shows a correlation with mitotically active cells. Moreover, both *Pax-3* and *Pax-7* are expressed on the dorsal aspect of the somites, the dermomyotome, whereas *Pax-3* transcripts are present earlier than *Pax-7*, detected between 8.5 and 11.5 days p.c., in the dorsolateral regions of the somites. While *Pax-3* mesodermal expression is turned off at around day 11.5, *Pax-7* is still expressed until day 14.5 in myotome-derived muscles, indicating that *Pax-7* could be involved in myogenic differentiation. Finally, *Pax-3* transcripts have been also detected in limb bud and in some craniofacial structures.

Pax-6

Pax-6 is the earliest *Pax* gene to be expressed during development (Walther and Gruss, 1991). Its expression is almost exclusively restricted to the developing central nervous system. *Pax-6* transcripts have been detected in the ventral ventricular zone of the neural tube. In the brain, expression starts at day-8.5 p.c. in all vesicles, with a gap of expression in the roof of the mesencephalon and metencephalon. The expression in the brain is maintained until late in development. *Pax-6* mRNA is also detected in the nose and in all developing structures of the eye starting from day-8.5 p.c. It is present in the epithelial layer of the optic vesicle, in the optic stalk and cup, as well as in the surface ectoderm which will later form the lens. At later stages, the transcripts are found in the inner layer of the optic cup, the neural retina and the overlying ectoderm from which the cornea will develop. The expression of *Pax-6* in the developing eye indicates that *Pax-6* could be involved in some inducing events leading to the formation of the eye.

Molecular function of Pax proteins

Many lines of evidence support the idea that Pax proteins act as transcription regulators. Firstly, the *Drosophila* proteins Pox meso and Pox neuro (Bopp et al., 1989) as well as the murine proteins Pax-1 (R. Fritsch and P. Gruss, unpublished) and Pax-2 (Dressler and Douglas, 1992) are localized in the nucleus. Secondly, various paired domain proteins have been shown to bind specific DNA sequences (the *paired* gene product from *Drosophila*, Prd: Treisman et al., 1989, 1991; Pax-3: Goulding et al., 1991; Pax-1: Chalepakis et al., 1991; Pax-2: Dressler and Douglas, 1992; Pax-8: Zannini et al., 1992; Pax-5: Adams et al., 1992; Pax-6: our unpublished data). The nuclear localization and the DNA binding activity of these genes is consistent with their function as transcriptional modulators capable of regulating developmental processes. This property of *Pax* genes has been further characterized.

Treisman et al. (1989, 1991) have shown that Prd can bind to a sequence present in the *even-skipped* promoter (e5). Prd contains both a paired domain and a paired type homeodomain. Each domain, as a separate entity, can bind to different parts of the e5 sequence: while the homeodomain recognizes the upstream half of the e5 site which contains the typical ATTA motif, identified in several other homeodomain recognition sequences (Treisman et al., 1992; Laughon, 1991), the paired domain contacts the downstream half of the e5 site.

Pax-3 also contains a paired domain and a paired type homeodomain and binds, like Prd, to the e5 site. Binding interference experiments with Pax-3 on the e5 site revealed that Pax-3 contacts the upstream ATTA motif and the downstream sequences including the GTTCC motif. This GTTCC sequence was found to be required for efficient binding of all the Pax proteins tested so far. Interestingly, Pax-3 possesses a much lower DNA binding affinity for the isolated GTTCC motif and no binding is detected on the isolated ATTA motif of the e5 site. Thus, the paired domain and homeodomain bind cooperatively on the e5 site which includes both motifs. If the homeodomain is deleted, the resulting truncated protein binds with very high affinity to the GTTCC motif via its paired domain (our unpublished data). These data indicate that the paired domain might, in the native Pax-3 conformation, be masked by intra- or intermolecular interactions and be released when the homeodomain is deleted, thereby increasing its DNA binding affinity. If the homeodomain is also able to bind specific DNA sequences independently from the paired domain, Pax-3 would then exhibit three DNA binding activities: one mediated by the paired domain, one by the homeodomain and the third by both domains.

Pax-1, Pax-2 and Pax-8, which contain only a paired domain without a homeodomain, also bind to a modified e5 sequence. The nucleotide contacts of Pax-1 on the e5 site have been analyzed by binding interference experiments. Although Pax proteins bind to DNA as monomers, the contacted region is unusually large, encompassing 20 to 24 bp, making both minor and major groove contacts. A three-dimensional model of the contacted sites generated by computer graphics revealed that contact points are not confined to only one side but are rather scattered on all sides of the DNA helix (Chalepakis et al., 1991).

Pax-1, Pax-2 and Pax-8 are able to utilize the modified e5 site, subcloned in front of the thymidine kinase (TK) promoter, and to confer transcriptional activity to the TK promoter. The carboxy-terminal part of these proteins is rich in Pro, and also in Ser and Thr, resembling the transactivation domains of Oct-2 (Tanaka and Herr, 1990) and CTF-1 (Mermod et al., 1989). Nevertheless, it remains to be determined whether this region of Pax proteins encodes a functional transcriptional activation domain.

Pax-3 has been analyzed in more detail: it contains an autonomous transactivation domain encompassing the 80 carboxy-terminal amino acids and a transrepression domain located in the first 90 amino-terminal residues of the protein. Pax-3 is also able to utilize the e5 sequence, in gene transfer experiments, and to confer promoter inducibility at low protein concentrations. At higher protein concentrations, the promoter activity is suppressed. This repression activity mediated by Pax-3 appears to be independent from specific DNA binding and to require the intact homeodomain (our unpublished data).

The DNA binding specificity of different paired domains was analyzed using 20 different sequences derived from the e5 site. The paired domains of Pax-1, Pax-3 and Pax-7 exhibit the same DNA binding specificity. Pax-2 and Pax-8 bind DNA with the same specificity and can be distinguished from Pax-1, Pax-3 and Pax-7 by their binding to a modified e5 sequence which is not bound by Pax-2 and Pax-8. On the other hand, the Pax-6 paired domain interacts preferentially with other sequences weakly recognized by other paired domains. Thus, based on the DNA binding specificity of the paired domains one can subgroup the Pax proteins into three families: the first includes Pax-1, Pax-3 and Pax-7, the second includes Pax-2 and Pax-8 and the third Pax-6. This classification closely resembles the previous classification based on protein conservation of the *Pax* genes (Fig. 1 and Walther et al., 1991).

Pax-8 has also been shown to bind a sequence in the proximal promoter region of thyroglobulin and thyroperoxidase and to enhance transcriptional activity of these genes (Zannini et al., 1992), reinforcing the idea that Pax-8 is involved in the differentiating events of the thyroid. Pax-1, Pax-6 and Pax-8 can activate the promoter of the *c-fos* oncogene in transient transfection experiments in NIH 3T3 cells. The Pax-mediated activation of 13 different *c-fos* promoter deletions was tested in serum-deprived cells and resembles the serum-mediated induction of these reporter constructs (our unpublished data). A Pax binding sequence was identified in the proximal promoter region. Further analysis will be required to decipher these molecular interactions. Finally, Pax binding sequences were also identified in the promoter region of the *N-CAM* gene which is, like cytotactin, activated by Pax-6 and Pax-8 in transient transfection experiments in NIH 3T3 cells.

Recently, it has been reported that *Pax-5* encodes for the transcription factor BSAP which is expressed at the early stages of B-cell differentiation (Adams et al., 1992). Pax-5 is involved in the regulation of the CD19 gene which encodes a B-lymphoid specific transmembrane receptor involved in signal transduction.

Pax mutants

One of the most surprising features of the *Pax* family is the identification of an unusually large number of both murine and human mutants which have been assigned to some of the *Pax* family members: *Pax-1*, *Pax-3* and *Pax-6*. The semidominant nature of most of these mutations has facilitated the isolation and characterization of these mutants and has also brought some insight into the molecular action of the genes.

After its initial identification, mapping studies localized the *Pax-1* gene on chromosome 2 between the β2-*microglobulin* and *agouti* loci, closely linked to the naturally occurring mutation *undulated* (*un*). Initially documented by Wright (1947) and Grüneberg (1950, 1954), this recessive mutation is characterized by a kinky-tail phenotype. Animals affected suffer from severe vertebral malformations along the entire vertebral axis: intervertebral disks are larger while the vertebral centers are reduced in size, leading to a distorted vertebral column. The primary defect appears to reside in the intrinsic inability of the sclerotomal cells to form perichordal condensations, resulting in the improper allocation of these cells to intervertebral disks and vertebral bodies.

The pattern of *Pax-1* expression correlates with these defects, being first observed in the ventral mesoderm at day-9 p.c., in the condensing sclerotomal cells at day-12 p.c. and, later on, in developing intervertebral disks (Deutsch et al., 1988). This evidence indicated that *Pax-1* could be the gene altered in the *un* mutant. Indeed, sequencing of the *Pax-1* paired box isolated from *un* mice, led to the identification of a single nucleotide mutation within the paired box, resulting in a glycine-to-serine exchange (Balling et al., 1988). Although these results provide only circumstantial evidence that *Pax-1* and *un* are allelic, recent observations have reinforced this idea. First, two other *undulated* mutants, *un-extensive* (un^{ex}) and *Un-short tail* (Un^{s}), also have alterations within the *Pax-1* gene (Balling et al., 1992, Dietrich et al., unpublished data). With the recessive un^{ex} mutation, phenotypically stronger than *un* (Wallace, 1985), embryos contain fewer of the *Pax-1* transcripts. Interestingly, Un^{s}, the only *un* semidominant allele, harbours a complete deletion of *Pax-1* (Balling et al., 1992). While heterozygous Un^{s} animals present a kinky tail phenotype, homozygosity leads to late embryonal or neonatal death (Blandova and Egerov, 1975).

Secondly, gel shift experiments performed with Pax-1 have defined a synthetic core motif recognized by the Pax-1 protein. The point mutation found within the paired domain has similarly been shown to reduce the affinity of the Pax-1^{un} protein for its target consensus sequence. In addition, this mutation enables the Pax-1 protein to bind to previously unrecognized sequences (Chalepakis et al., 1991). These data provide a molecular basis for the effect observed in the *un* mutant, strongly suggesting that this mutation results in the loss of Pax-1 protein's ability to bind to its natural targets and to modulate their expression. Moreover, the demonstration that the Pax-1^{un} mutant has acquired the ability to bind to new target sequences may indicate that some of the effects observed in the *un* mutant might be mediated by some dominant gain of function. Although the final proof of the identity of *Pax-1* and *un* depends on rescue experiments of the *undulated* phenotype, these results taken together leave little doubt about the identity of the *un* mutation, clearly indicating that *Pax-1* is a crucial determinant for the specification of sclerotomal mesoderm.

The *Pax-3* gene (Goulding et al., 1991) has recently been shown to correspond to the genetic locus *Splotch* (*Sp*, Epstein et al., 1991; Goulding et al., unpublished data). This spontaneous mutant (Russell, 1947) exhibits, with several other alleles (*Sp-retarded*: Sp^{r}, *Sp-delayed:* Sp^{d}, Sp^{1H}, Sp^{2H}), a semidominant phenotype (Dickie, 1964; Beechey and Searle, 1986). In the homozygous state, *Sp* is lethal. Abnormalities are observed in various *Pax-3*-expressing structures (Goulding et al., 1991). Within the central nervous system, exencephaly, spina bifida and meningocele are often associated with excessive neural growth (Auerbach, 1954; Franz, 1989, 1990). Moreover, these animals manifest defects associated with neural crest cell deficiency: absence or reduction of dorsal root ganglia, white spotting, Schwann cell and heart defects. Heterozygous animals present only pigmentation defects, detectable as white spotting mostly on the belly, but also on the back, the tail and limbs. Both *Pax-3* and the various *Sp* alleles have been mapped to the proximal end of chromosome 1 (Goulding et al., 1991; Snell et al., 1954; Skow et al., 1988) and various alleles of *Sp* have now been characterized molecularly.

Firstly, the Sp^r allele has been shown to harbour a large cytogenetically detectable deletion encompassing the Pax-*3* locus (Epstein et al., 1991). Secondly, the Sp^{2H} allele contains a 32-bp deletion within the *Pax-3* homeobox, introducing a nonsense mutation and resulting in premature translation termination (Epstein et al., 1991). This results in a Pax-3 protein deprived of part of its homeodomain and of its entire carboxy-terminal domain potentially involved in transactivation (our unpublished data). Thirdly, the Sp^{4H} allele has been shown to harbour a large deletion encompassing most of the *Pax-3* coding region (Goulding et al., unpublished data). Finally, the *Sp* allele generates, in addition to the normal full size transcript, a shortened version of the *Pax-3* messenger, devoid of an exon which normally encodes the octapeptide (Goulding et al., unpublished data). The molecular defect underlying this splicing aberration has not been characterized.

A human syndrome has also been associated with alterations within the *PAX3* gene. The Waardenburg syndromes (type 1 and type 2) present a combination of various defects such as deafness, pigmentary deficiency (heterochromia irides, white forelock and eyelash) as well as, in *WS* type I, dystopia canthorum (lateral displacement of the inner corner of the eye). The effects of these dominant syndromes most probably arise from cranial neural crest deficiencies. In five different families presenting *WS* syndromes, various alterations have been documented within the *PAX3* gene: one allele with an 18-bp deletion within the paired box (WS.05; Tassabehji et al., 1992), one allele with a single bp mutation affecting an invariant residue preceding the first helix of the paired domain (Brazil; Baldwin et al., 1992), one allele harbouring a single bp deletion causing a frame shift within the paired domain (WS.06; A. Read, personal communication), another allele containing a 2-bp deletion causing a frameshift within the octapeptide motif (WS.11; A. Read, personal communication) and a last allele where a single bp mutation results in a glycine-to-alanine substitution within the paired domain (WS15; A. Read, personal communication).

The deficiencies associated with the various *Sp* and *WS* mutants together with the semidominant nature of these mutations indicate that *Pax-3* constitutes a crucial factor in the development of the brain, the neural tube and neural crest derived structures.

Pax-6 is the last member of the *Pax* gene family shown to be mutated in mouse mutants. The semidominant mutant, *Small eye* (*Sey*) which was isolated by breeding of small body weight mice, is lethal at birth in the homozygous state. This condition is characterized by the complete absence of eyes and nose, two structures already shown to express *Pax-6*. In the less dramatic heterozygous state, animals present underdevelopment of the eye to various degrees.

Analysis of different alleles of *small eye* (*Sey*, Sey^H, Sey^{neu}) has led to the identification of various alterations within the *Pax-6* gene. First, analysis of the *Pax-6* gene of *Sey* embryos identified a single bp mutation which introduces a stop codon, truncating the protein between the paired domain and the homeodomain (Hill et al., 1991). Secondly, Sey^{neu}, obtained by chemical mutagenesis (Favor et al., 1988) and presenting a phenotype similar to *Sey*, harbours a 116-nucleotide insert within the *Pax-6* cDNA sequence (Hill et al., 1991). This insert originates from an unspliced intron, consequence of a point mutation within the splice acceptor consensus sequence. It introduces a nonsense mutation, resulting in truncation of the protein downstream from its homeobox. The resulting protein lacks its 115 carboxy-terminal residues, a domain which contains a high serine/threonine content (25%), possibly involved in transcription activation. The Sey^H and Sey^{Dey} allele appears to have the entire *Pax-6* gene deleted (Hill et al., 1991; Ton et al., 1991).

The human condition known as anridia (*AN*) has already been proposed to be the counterpart of the murine *Sey* mutant (Glaser et al., 1990; van der Meer et al., 1990). This dominant inherited disorder presents phenotypic variability: partial or complete absence of iris is often accompanied by impaired vision, cataracts, foveal and optic nerve hypoplasia, corneal opacification and glaucoma (Shaw et al., 1960; Nelson et al., 1984). A putative cDNA candidate for the *AN* locus was recently cloned by chromosome walking and shown to be partially or totally deleted in two patients with aniridia (Ton et al., 1991). The corresponding 2.7-kb message is present in various tissues, most importantly the brain, the eyes and the olfactory bulb. The protein encoded by this cDNA contains a paired domain and a paired-type homeodomain and corresponds to the human counterpart of the murine *Pax-6* cDNA. The drastic phenotype as well as the semidominant nature of the *Sey* and *AN* alleles, underlines the importance of *Pax-6* in the development of the eye and nasal placodes.

The phenotypes observed with these various mutants most probably arise from the lost ability of the Pax gene products to modulate the expression of their normal target genes. However, one cannot exclude the possibility that, for some of these mutants, some particular activities of these proteins remain or that some gain-of-function might result from some of these mutations. This is well exemplified by some of the binding studies performed with Pax-1 and Pax-1^{un} proteins in vitro (see above, Chalepakis et al., 1991).

The identification of five different mutants (*un*, *Sey*, Sp, *AN*, *WS*) of murine or human origin corresponding to members of the *Pax* gene family emphasizes the importance of these genes as critical determinants of developmental processes. On the other hand, the nature of the phenotypes observed also raises important questions about the mechanism of action of these genes.

Firstly, an intriguing aspect of some of the *Pax* mutants resides in the absence of apparent phenotype in some structures expressing these genes during development. For instance, *Pax-1* is present at high levels in the thymus and *Pax-6* is present in the spinal cord, where no phenotypes have been yet documented. Although careful examination of these mutants using various markers might lead to the identification of subtle alterations within these compartments, this phenomenon might reflect some functional redundancies masking the effect of *Pax* gene deficiency. Such arguments have already been put forward for gene knock-out experiments where a phenotype was observed only in a subset of tissues normally expressing the knocked-out protein (*Hox-1.5:* Chisaka and Capecchi, 1991; *Hox-*

1.6: Lufkin et al., 1991; *Hox-3.1*: Le Mouellic et al., 1992; *Wnt-1*: McMahon et al., 1992; *En-2*: Joyner et al., 1989).

Secondly, the semidominant nature of most of these mutants indicates that, according to the tissue, some considerable gene dosage effect may intervene. The gene product dosage may be, in certain instances, too low to allow proper tissue development to occur, while complete absence of the gene product completely annihilates the development of this same tissue. This is the case for the *Sey* mutation which results in the small eye phenotype in heterozygous animals and in the total absence of eye structures in homozygotes. The apparent normal development of the olfactory placode in the heterozygous animals, and its absence in homozygotes, also signifies that this tissue might require a lower threshold of Pax-6 protein (as compared to the eye) for proper development, or that alternatively it possesses a higher level of expression of *Pax-6*. Such an argument could also be made with *Sp* mutants where only pigmentation is affected in heterozygous animals. These observations underline the importance of the precise regulatory mechanisms controlling the expression of these genes during development as well as the precise balancing of the subordinate transcription processes regulated by the Pax proteins.

References

Adams, B., Dörfler, P., Aguzzi, A., Kozmik, Z., Urbanek, P., Maurer-Fogy, I. and Busslinger, M. (1992). *Pax-5* encodes the transcription factor BSAP and is expressed in B lymphocytes, the developing CNS, and adult testis. *Genes Dev.* **6**, 1589-1607.

Asano, M. and Gruss, P. (1992). *Pax-5* is expressed at the midbrain-hindbrain boundary during mouse development. *Mechanisms Development, in press.*

Auerbach, R. (1954). Analysis of the developmental effects of a lethal mutation in the house mouse. *J. Exp. Zool.* **127**, 305-329.

Baldwin, C. T., Hoth, C. F., Amos, J. A., da-Silva, E. O. and Milunsky, A. (1992). An exonic mutation in the *HuP2* paired domain gene causes Waardenburg's syndrome. *Nature* **355**, 637-638.

Balling, R., Deutsch, U. and Gruss, P. (1988). *Undulated*, a mutation affecting the development of the mouse skeleton, has a point mutation in the paired box of *Pax-1*. *Cell* **55**, 531-535.

Balling, R., Lau, C. F., Dietrich, S., Wallin, J. and Gruss, P. (1992). Development of the skeletal system. *Ciba Foundation Symposium* **165**, *132-140.*

Baumgartner, S., Bopp, D., Burri, M. and Noll, M. (1987). Structure of two genes at the gooseberry locus related to the *paired* gene and their spatial expression during embryogenesis. *Genes Dev.* **1**, 1247-1267.

Beechey, C. V. and Searle, A. G. (1986). Mutations at the *Sp* locus. *Mouse News Lett.* **75,** 28.

Blandova, Z. R. and Egorov, I. U. (1975). Sut allelic with un. *Mouse News Lett.* **52**, 43.

Bopp, D., Burri, M., Baumgartner, S., Frigerio, G. and Noll, M. (1986). Conservation of a large protein domain in the segmentation gene *paired* and in functionally related genes of *Drosophila*. *Cell* **47**, 1033-1040.

Bopp, D., Jamet, E., Baumgartner, S., Burri, M. and Noll, M. (1989). Isolation of two tissue-specific *Drosophila* paired box genes, *pox meso* and *pox neuro*. *EMBO J.* **8**, 3447-3457.

Burri, M., Tromvoukis, Y., Bopp, D., Frigerio, G. and Noll, M. (1989). Conservation of the paired domain in metazoans and its structure in three isolated human genes. *EMBO J.* **8**, 1183-1190.

Chalepakis, G., Fritsch, R., Fickenscher, H., Deutsch, U., Goulding, M. and Gruss, P. (1991). The molecular basis of the *undulated/Pax-1* mutation. *Cell* **66**, 873-884.

Chisaka, O. and Capecchi, M. R. (1991). Regionally restricted developmental defects resulting from targeted disruption of the mouse homeobox gene *hox-1. 5*. *Nature* **350**, 473-479.

Coté, S., Preiss, A., Haller, J., Schuh, R., Kienlin, A., Seifert, E. and Jäckle, H. (1987). The *gooseberry-zipper* region of *Drosophila*: five genes encode different spatially restricted transcripts in the embryo. *EMBO J.* **6,** 2793-2801.

Dambly-Chaudiere, C., Jamet, E., Burri, M., Bopp, D., Basler, K., Hafen, E., Dumont, N., Spielman, P., Ghysen, A. and Noll, M. (1992). The paired box gene *pox neuro*: A determinant of poly-innervated sense organs in Drosophila. *Cell* **69,** 159-172.

Deutsch, U., Dressler, G. R. and Gruss, P. (1988). *Pax1*, a member of a paired box homologous murine gene family, is expressed in segmented structures during development. *Cell* **53,** 617-625.

Deutsch, U. and Gruss, P. (1991). Murine paired domain proteins as regulatory factors of embryonic development. *Sem. Dev. Biol.* **2,** 413-424.

Dickie, M. M (1964). New splotch alleles in the mouse. *J. Hered.* **55**, 97-101.

Dressler, G. R., Deutsch, U., Balling, R., Simon, D., Guenet, J-L. and Gruss, P. (1988). Murine genes with homology to *Drosophila* segmentation genes. *Development* **104 Supplement**, 181-186.

Dressler, G. R., Deutsch, U., Chowdhury, K., Nornes, H. O. and Gruss, P. (1990). *Pax2*, a new murine paired box containing gene and its expression in the developing excretory system. *Development* **109,** 787-795.

Dressler, G. R. and Douglas, E. C. (1992). *Pax-2* is a DNA binding protein expressed in embryonic kidney and Wilms tumor. *Proc. Nat. Acad. Sci. USA* **89**, 1179-1183.

Epstein, D. J., Vekemans, M. and Gros P. (1991). *splotch* (Sp^{2H}), a mutation affecting development of the mouse neural tube, shows a deletion within the paired homeodomain of Pax-3. *Cell* **67**, 767-774.

Favor, J., Neuhäuser-Klaus, A. and Ehling, U. H. (1988). The effect of dose fractionation on the frequency of ethynitrosourea-induced dominant cataract and recessive specific locus mutations in germ cells of the mouse. *Mutation Res.* **198**, 269-275.

Franz, T. (1989). Persistent truncus arteriosus in the splotch mutant mouse. *Anat. Embryol.* **180**, 457-464.

Franz, T. (1990). Defective ensheathment of motoric nerves in the splotch mutant mouse. *Acta Anat.* **138**, 246-253.

Frigerio, G., Burri, M., Bopp, D., Baumgartner, S. and Noll, M. (1986). Structure of the segmentation gene *paired* and the *Drosophila* PRD gene sets a part of the gene network. *Cell* **47**, 1033-1040).

Glaser, T., Lane, J. and Housman, D. (1990). A mouse model of the aniridia-Wilms tumor deletion syndrome. *Science* **250**, 823-827.

Goulding, M. D., Chalepakis, G., Deutsch, U., Erselius, J. and Gruss, P. (1991). Pax-3, a novel murine DNA binding protein expressed during early neurogenesis. *EMBO J.* **10**, 1135-1147.

Grüneberg, H. (1950). Genetical studies on the skeleton of the mouse. II. *Undulated* and its modifiers. *J. Genetics* **50**, 142-173.

Grüneberg, H. (1954). Genetical studies on the skeleton of the mouse. XII. The development of *undulated*. *J. Genetics* **52**, 441-455.

Gruss, P. and Walther, C. (1992). Pax in development. *Cell* **69**, 719-722.

Hill, R. E., Favor, J., Hogan, B. L. M., Ton, C. C. T., Saunders, G. F., Hanson, J. M., Prosser, J., Jordan, T., Hastie, N. D. and van Heyningen, V. (1991). Mouse *Small eye* results from mutations in a paired-like homeobox-containing gene. *Nature* **354**, 522-525.

Jostes, B., Walther, C. and Gruss, P. (1990). The murine paired box gene, Pax7, is expressed specifically during the development of the nervous and muscular system. *Mech. Dev.* **33,** 27-38.

Joyner, A. L., Skarnes, W. C. and Rossant, J. (1989). Production of a mutation in the mouse *En-2* gene by homologous recombination in embryonic stem cells. *Nature* **338**, 153-156.

Krauss, S., Johansen, T., Korzh, V. and Fjose, A. (1991a). Expression pattern of zebrafish *pax* genes suggests a role in early brain regionalisation. *Nature* **353,** 267-270.

Krauss, S., Johansen, T., Korzh, V., Moens, U., Ericson, J. U. and Fjose, A. (1991b). Zebrafish *pax[zf-a]*: a paired box-containing gene expressed in the neural tube. *EMBO J.* **10,** 3609-3619.

Krauss, S., Johansen, T., Korzh, V. and Fjose, A. (1991c). Expression of the zebrafish paired box gene *pax[zf-b]* during early embryogenesis. *Development* **113,** 1193-1206.

Laughon, A. (1991). DNA binding specificity of homeodomains. *Biochem.* **30**, 11357-11367.

Le Mouellic, H., Lallemand, Y. and Brület, P. (1992). Homeosis in the mouse induced by a null mutation in the *Hox-3. 1* gene. *Cell* **69**, 251-264.

Lufkin, T., Dierich, A., LeMeur, M., Mark, M. and Chambon, P. (1991). Disruption of the *Hox-1.6* homeobox gene results in defects in a region corresponding to its rostral domain of expression. *Cell* **66**, 1105-1119.

McMahon, A. P., Joyner, A. L., Bradley, A. and McMahon, J. A. (1992). The midbrain-hindbrain phenotype of *Wnt-1⁻/Wnt-1⁻* mice results from stepwise deletion of *engrailed*-expressing cells by 9.5 days postcoitum. *Cell* **69**, 581-595.

Mermod, N., O'Neill, E. A., Kelly, T. J. and Tjian, R. (1989). The proline-rich transcriptional activator of CTF/NF-I is distinct from the replication and DNA binding domain. *Cell* **58**, 741-753.

Morrissey, D., Askew, D., Raj, L. and Weir, M. (1991). Functional dissection of the *paired* segmentation gene in *Drosophila* embryos. *Genes Dev.* **5**, 1684-1696.

Nelson, L. B., Spaeth, G. L., Nowinski, T. S., Margo, C. E. and Jackson, L., (1984). Aniridia. A review. *Surv. Ophthalmol.* **28**, 621-642.

Nornes, H. O., Dressler, G. R., Knapik, E. W., Deutsch, U. and Gruss, P. (1990). Spatial and temporal expression of *Pax2* during murine embryogenesis. *Development* **109**, 797-809.

Plachov, D., Chowdhury, K., Walther, C., Simon, D., Guenet, J.-L. and Gruss, P. (1990). Pax8, a murine paired box gene expressed in the developing excretory system and thyroid gland. *Development* **110**, 643-651.

Poleev, A., Fickenscher, H., Mundlos, S., Winterpacht, A., Zabel, B., Fidler, A., Gruss, P. and Plachov, D. (1992). PAX-8, a human paired box gene: isolation and expression in the developing thyroid, kidney and Wilms tumors. *Development*, in press.

Püschel, A. W., Gruss, P. and Westerfield, M. (1992). Sequence and expression pattern of *pax-6* are highly conserved between zebrafish and mice. *Development* **114**, 643-651.

Russell, W. L. (1947). Splotch, a new mutation in the house mouse *Mus musculus*. *Genetics* **32**, 107.

Shaw, M. W., Falls, H. F. and Neel, J. V. (1960). Congenital aniridia. *Am. J. Hum. Genet.* **12**, 389-415.

Skow, L. C., Donner, M. E., Huang, S.-M., Gardner, J. M., Taylor, B. A., Beamer, W. G. and Lalley, B. A. (1988). Mapping of mouse gamma crystallin genes on chromosome 1. *Biochem. Genet.* **26**, 557-570.

Snell, G. D., Dickie, M. M., Smith, P. and Kelton, D. E. (1954). Linkage of loop-tail, leaden, splotch, and fuzzy in the mouse. *Heredity* **8**, 271-273.

Tanaka, M. and Herr, W. (1990). Differential transcriptional activation by Oct-1 and Oct-2: Independent activation domains include Oct-2 phosphorylation. *Cell* **60**, 375-386.

Tassabehji, M., Read, A. P., Newton, V. E., Harris, R., Balling, R., Gruss, P. and Strachan, T. (1992). Waardenburg's syndrome patients have mutations in the human homologue of the *Pax-3* paired box gene. *Nature* **355**, 635-636.

Ton, C. C. T., Hirvonen, H., Miwa, H., Weil, M. M., Monaghan, P., Jordan, T., van Heyningen, V., Hastie, N. D., Meijers-Heijboer, H., Drechsler, M., Royer-Pokora, B., Collins, F., Swaroop, A., Strong, L. C. and Saunders, G. F. (1991). Positional cloning of a paired box- and homeobox-containing gene from the Aniridia region. *Cell* **67**, 1059-1074.

Treisman, J., Gönczy, P., Vashishita, M., Harris, E. and Desplan, C. (1989). A single amino acid can determine the DNA binding specificity of homeodomain proteins. *Cell* **59**, 553-562.

Treisman, J., Harris, E. and Desplan, C. (1991). The paired box encodes a second DNA-binding domain in the Paired homeodomain protein. *Genes Dev.* **5**, 594-604.

Treisman, J., Harris, E., Wilson, D. and Desplan, C. (1992). The homeodomain: A new face for the helix-turn-helix? *Bioessays* **14**, 145-150.

van der Meer de Jong, R., Dickinson, M. E., Woychik, R. P., Stubbs, L., Hetherington, C. and Hogan, B. L. (1990). Location of the gene involving the *small eye* mutation on mouse chromosome 2 suggests homology with human aniridia 2 (AN2). *Genomics* **7**, 270-275.

Wallace, M. E. (1985). An inherited agent of mutation with chromosome damage in wild mice. *J. Hered.* **76**, 271-278.

Walther, C., Guenet, J.-L., Simon, D., Deutsch, U., Jostes, B., Goulding, M. D., Plachov, D., Balling, R. and Gruss, P. (1991). Pax: A murine multigene family of paired box containing genes. *Genomics,* **11**, 424-434.

Walther, C. and Gruss, P. (1991). *Pax-6*, a murine paired box gene, is expressed in the developing CNS. *Development* **113**, 1435-1449.

Wright, M. E. (1947). *Undulated*: A new genetic factor in Mus musculus affecting the spine and tail. *Heredity* **1**, 137-141.

Zannini, M., Francis-Lang, H., Plachov, D. and Di Lauro, R. (1992). *Pax-8*, a paired domain-containing protein, binds to a sequence overlapping the recognition site of a homeodomain and activates transcription from two thyroid-specific promoters. *Mol. Cell. Biol.* **12**, 4230-4241.

Journal of Cell Science, Supplement 16, 69-76 (1992)
Printed in Great Britain © The Company of Biologists Limited 1992

Retinoid receptors and binding proteins

DAVID LOHNES, ANDRÉE DIERICH, NORBERT GHYSELINCK, PHILLIPE KASTNER, CARMEN LAMPRON, MARIANNE LeMEUR, THOMAS LUFKIN, CATHY MENDELSOHN, HARI NAKSHATRI and PIERRE CHAMBON

Laboratoire de Génétique Moléculaire des Eucaryotes du CNRS, Unité 184 de Biologie Moléculaire et de Génie Génétique de L'INSERM, Institut de Chimie Biologique, Faculté de Médecine, 11, rue Humann-67085 Strasbourg-Cedex, France

Summary

Retinoids, in particular all-*trans* retinoic acid (T-RA), are essential for normal development and homeostasis of vertebrates. Although many effects of retinoids, particularily with regard to teratogenicity, have been described in the literature, the mechanisms by which these simple signalling molecules work has only recently begun to be elucidated. We now recognize at least two classes of retinoid-binding proteins and two families of retinoid receptors. The ultimate interpretation of the retinoid signal within a given cell is probably the result of a complex series of interactions between these proteins, yet little is understood concerning the role each member of this signalling pathway plays. It is therefore imperative to dissect the molecular mechanisms which transduce the effects of these ligands, both in vivo and in isolated systems. One approach we are employing is gene targeting of retinoic acid receptors (RARs) and cellular retinoid-binding proteins to generate mice in which one or more of these genes has been functionally inactivated.

Key words: retinoic acid receptors, retinoid X receptors, cellular retinol-binding protein, cellular retinoic acid-binding protein, gene targeting.

Introduction

The term retinoid refers to any synthetic or natural compound which possesses vitamin A-like activity. At least three naturally occurring, biologically active, retinoids have been found which may regulate growth, differentiation and homeostasis of the developing vertebrate: all-*trans* retinoic acid (T-RA), 9-*cis* retinoic acid (9C-RA; Levin et al., 1992; Heyman et al., 1992) and 3,4-didehydroretinoic acid (dd-RA; Thaller and Eichle, 1990). It has been proposed that retinoids may be natural morphogens (Maden, 1982, 1985; Tickle et al., 1982; Chytil, 1984; Slack, 1987; Thaller and Eichele, 1987; Smith et al., 1989; Eichele, 1989a,b; Brockes, 1989, 1990), although this viewpoint has been challenged recently by a number of reports which question the validity of T-RA as a true morphogen (Brockes, 1991; Wanek et al., 1991; Noji et al., 1991). However, the effects of ectopically administered T-RA, either during development (and in the adult) of several vertebrates or on cells in culture, have often been correlated with alterations in patterns of expression and/or the level of expression of specific genes (eg. Wang and Gudas, 1983; LaRosa and Gudas, 1988; Murphy et al., 1988; Chytil and Haq, 1990; Wang et al., 1990; Nicholson et al., 1990; Simeone et al., 1990; Kessel and Gruss, 1991; Izpisua-Belmonte et al., 1991). These observations strongly suggest that T-RA, in conjunction with its receptor (discussed below), may directly regulate gene expression which would ultimately interpret this putative morphogenetic signal. Alternatively, T-RA may act during development to regulate the expression of other factors which themselves may be true morphogens.

To date, two general classes of retinoid-binding proteins have been implicated in the retinoid signalling pathway. First, the cellular retinoid-binding proteins which include cellular retinol-binding proteins (CRBPs) I and II (see Blomhoff et al., 1991, for review) and cellular retinoic acid-binding proteins (CRABPs) I and II (Blomhoff et al., 1990; Giguère et al., 1990a). These proteins bind their ligands (retinol and T-RA, respectively) with high affinity and specificity. The precise function of these proteins is presently unknown, but it has been speculated that they play a role in retinoid metabolism and in the regulation of free retinoid concentration (see below).

The second class of retinoid-binding proteins consists of the products of two multigene families: the retinoic acid receptors (RARα, β and γ; for review, see Chambon et al., 1991) and the retinoid X receptors (RXR α, β and γ; Hamada et al., 1989; Mangelsdorf et al., 1990, 1992; Rowe et al., 1991; Blumberg et al., 1992; Leid et al., 1992). Both classes of receptors are ligand-inducible, *trans*-regulators belonging to the nuclear receptor multigene family, which includes receptors for steroid hormones, thyroid hormones and vitamin D_3 (Evans, 1988; Green and Chambon, 1988; Beato, 1989; Moore, 1990). RARs can be efficiently activated by either T-RA or 9C-RA at 50 nM concentrations of either ligand. RXRs, on the other hand, are activated efficiently

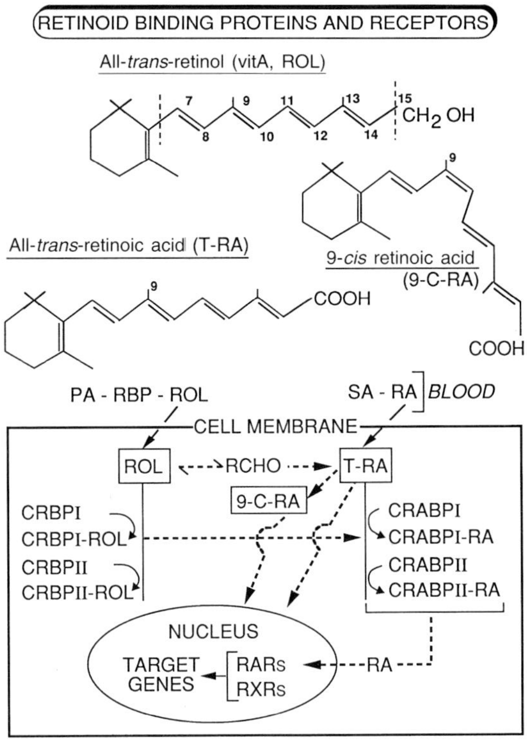

Fig. 1. Schematic representation of the components of the retinoid-signalling pathway. CRBP, cellular retinol-binding protein; CRABP, cellular retinoic acid-binding protein; RBP, retinol-binding protein; RAR, retinoic acid receptor; RXR, retinoid X receptor.

only by 9C-RA in this concentration range. RXRs can, however, be activated by T-RA at high concentrations (1 μM or greater), probably through cellular conversion of T-RA to the 9-*cis* isomer (Heyman et al., 1992; Levin et al., 1992). Thus, with respect to transactivation via RXRs, T-RA could be considered as a prohormone, possibly converted to the 9-*cis* isomer in either a localized or systemic fashion.

The major components of the retinoid-signalling network discussed above are represented schematically in Fig. 1. To investigate how these proteins function ultimately to regulate diverse processes such as those involved in morphogenesis and homeostasis in vertebrates will require efforts along many different experimental avenues. One approach consists of disrupting the corresponding genes in the mouse via gene targeting (see Capecchi, 1989, for a review of this technique). Our objective is to generate animal models in which the genes encoding a specific retinoid-binding protein or retinoid receptor (or a specific isoform of a retinoid receptor - see below) has been functionally inactivated. These mouse models should provide useful tools for investigating the role that each of these proteins plays in transduction of the retinoid signal. Before describing the details of our gene targeting experiments, a review of retinoid-binding proteins and retinoid receptors will first be presented.

Cellular retinoid-binding proteins

The two classes of cellular retinoid-binding proteins (CRBPI and II; CRABPI and II), while not directly implicated in transduction of the retinoid signal, probably play important roles in the ultimate response of a given cell type to retinoids.

CRBPs may play a role in the storage and metabolism of retinol, itself a precursor to T-RA (Blomhoff et al., 1990). CRBPI has been suggested to regulate the formation of the major storage form of retinol, retinly esters (Ong et al., 1987), whereas CRBPII may mediate intracellular transport of retinol during the adsorption of this compound from the intestinal lumen (Ong, 1987; Ong et al., 1987). In addition, transcripts encoding CRBPI are T-RA-inducible (Levin et al., 1987; Smith et al., 1991), which may represent an in vivo feedback mechanism which ultimately regulates the amount of retinol available for conversion to T-RA. The restricted spatio-temporal expression pattern of these two genes during mouse development and in the adult also suggests tissue-specific functions (Dollé et al., 1989; Ruberte et al., 1991).

CRABPI and II bind T-RA with high, although slightly different, affinities (Blomhoff et al., 1990, 1991; Giguère et al., 1990a; Stoner and Gudas, 1989). Again, the precise function that these two proteins fulfil is, at present, unknown. It has been speculated that they may be necessary for transport of T-RA to the nucleus, where it may be transferred to one of the RARs (Takase et al., 1979, 1986). Alternatively, CRABPs have been suggested to control the actual intracellular concentration of free RA within a given cell (Maden et al., 1988; Smith et al., 1989; Ruberte et al., 1991). CRABPII, like CRBPI, is inducible by retinoic acid (Wei et al., 1989; Giguère et al., 1990a; Durand et al., 1992), perhaps providing a feedback mechanism whereby excess ligand could be sequestered, thus preventing inappropriate expression of RA-responsive genes. It is clear that the presence of these proteins is not obligatory for a cellular response to T-RA, as numerous retinoid-responsive cell lines have been documented which contain no detectible CRABPs.

Retinoic acid receptors and retinoid X receptors

Three homologous RARs have been discovered in our laboratory and in others: RARα (Petkovich et al., 1987; Giguère et al., 1987), RARβ (de Thé et al., 1987; Brand et al., 1988; Benbrook et al., 1988) and RARγ (Zelent et al., 1989; Krust et al., 1989; Ishikawa et al., 1990). These genes map to different chromosomes in both mouse (chromosome 11, 14 and 15 for α, β and γ respectively) and human (chromosome 17, 3 and 12 respectively; Mattei et al., 1991). Homologues for all three receptors have also been found in amphibians (Ragsdale et al., 1989; Giguère et al., 1989; Ragsdale and Brockes, 1991) and in chicken (Smith and Eichele, 1991), suggesting that these three genes have been conserved throughout the evolution of higher vertebrates.

More recent work has shown that the RXR family, composed of three subtypes (α, β and γ), is also conserved across species (Leid et al., 1992; Mangelsdorf et al., 1992 and references therein). It is interesting to note that a *Drosophila* homologue, ultraspiracle, has been documented for RXRs (Oro et al., 1990), indicating that this gene family may be more ancestral than the RARs.

Although the discovery of the RXR family with a unique and specific ligand (see above) suggested initially a divergent retinoid-signalling pathway, recent results indicate that the RAR and RXR families may, in fact, represent convergent signalling pathways. This observation comes from several groups which have shown that, at least in vitro, RAR/RXR heterodimers form both in solution and on certain RAREs (retinoic acid response elements), and that this complex transactivates T-RA responsive promoters much more efficiently than homodimers of either receptor (Yu et al., 1991; Kliewer et al., 1992; Leid et al., 1992; Zhang et al., 1992; Durand et al., 1992). Thus, if one argues that these heterodimers function in an analogous manner in vivo, the pleiotropic effects of retinoids may be explained, at least in part, by the multitude of combinatorial possibilities between these two receptor types and the availability of each specific ligand.

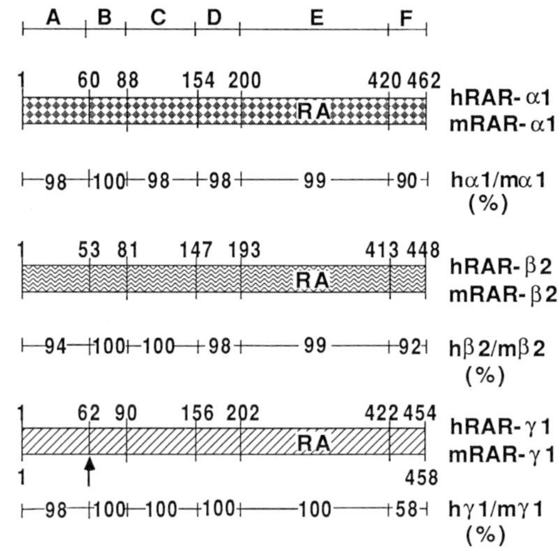

Fig. 2. Comparison of amino acid sequence homology between human and mouse RAR subtypes. The amino acid residue number corresponding to the boundary of each region is given at the top, and percent sequence homology comparison between human and mouse receptors given at the bottom.

Sequence comparison of the retinoic acid receptors

RARs, like all members of the nuclear receptor superfamily, have a modular structure consisting of six regions designated A-F (Green and Chambon, 1988). Previous reseach has shown that region C represents the DNA-binding domain, via folding into a zinc-stabilized structure. This domain also represents the most highly conserved region of this receptor family (Evans, 1988; Green and Chambon, 1988). The E region represents the ligand-binding domain whereas both the A/B and E regions of several steroid hormone receptors have been shown to harbour distinct *trans*-activation domains which may operate in both a cell type and promoter-specific manner (Webster et al., 1988; Bocquel et al., 1989; Tora et al., 1989).

When RARs are compared across species, the amino acid sequences are highly conserved (eg. mouse and human; Krust et al., 1989; Zelent et al., 1989; Fig. 2). However, comparison of the three receptor subtypes within a given species reveals that only regions B, C and E are highly conserved. This observation implies that each RAR may be functionally distinct, possibly regulating a subset of retinoid-responsive genes in a cell-type and/or promotor-dependent fashion. This is further substantiated by the finding of separate spatio-temporal expression patterns for the RARs during embyronic development (see below).

Retinoic acid receptor isoforms

The discovery of isoforms for all mouse RAR (mRAR) types adds an additional level of complexity to RAR expression. To date two main isoforms have been discovered for mRARα (Leroy et al., 1991a), four for mRARβ (Zelent et al., 1991; Nagpal et al., 1992a) and two for mRARγ (Kastner et al., 1990; Giguère et al., 1990b; Fig. 3). The structure of these isoforms shares a common motif; for any RAR type, each isoform diverges only in the 5′-untranslated and A region. The sequence from the B region to the 3′-untranslated sequence remains identical. In the mouse these isoforms are expressed under the control of at least two independent promoters, and result from differential splicing (Kastner et al., 1990; Brand et al., 1990; Leroy et al., 1991a; Zelent et al., 1991; see Fig. 3). It is interesting to note that the isoforms mRARα2 and all mRARβ transcripts are induced by T-RA, suggesting yet another level of complexity in this gene network. Two of these genes, mRARα2 and mRARβ2, contain a similar RARE in their promoter regions which likely mediates the T-RA-induced expression of these genes (de Thé et al., 1990; Sucov et al., 1990; Leroy et al., 1991b). Recently, a divergent RARE has been described for RARγ2 (Lehmann et al., 1992). The significance of the minor isoforms α3 to α7 and γ3 to γ6 is unclear (Kastner et al., 1990; Leroy et al., 1991a; Zelent et al., 1991).

As for RAR types, sequence comparisons of mouse and human RAR isoforms again reveals a striking level of homology across species (Kastner et al., 1990; Leroy et al., 1991a; Zelent et al., 1991). This sequence conservation is exhibited not only in the A region, but also extends to the 5′-untranslated region of each isoform, suggesting the retention of conserved mechanisms that may involve regulation of translation, mRNA stabilization or other events important for proper expression of these proteins.

As discussed above, transcription activating functions (AFs) have been ascribed previously to the A/B and E regions of other members of the steroid hormone superfamily, such as the estrogen receptor. Thus, each mRAR isoform could differ in its transactivation capabilities at a given promoter via the difference in its A region. Indeed, recent work from our laboratory shows that the different RAR subtypes do indeed possess AFs in the E region, as well as different transcription modulation activities in the A/B region that function in a promoter-dependent manner (Nagpal et al., 1992b).

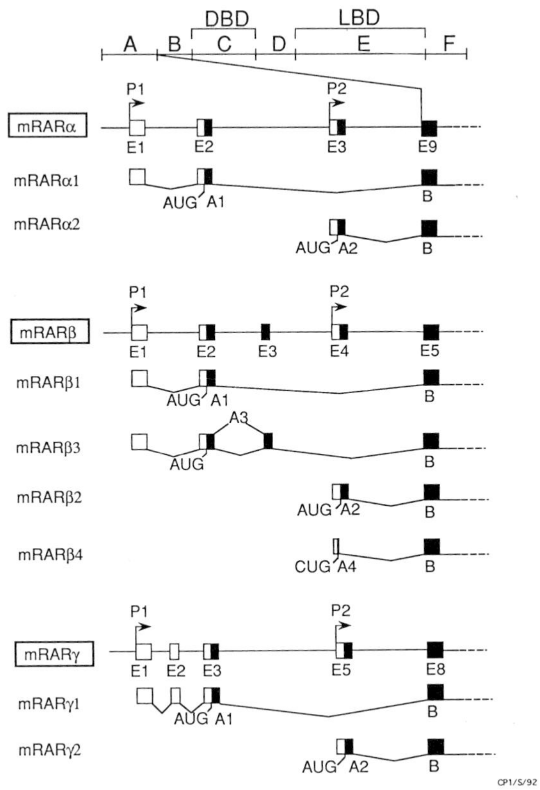

Fig. 3. Schematic representation of the 5′ region of the three mouse RAR genes and of their major isoforms. The exons (E) are indicated by boxes and are numbered as in Kastner et al. (1990), Leroy et al. (1991a) and Zelent et al. (1991). White boxes represent 5′-untranslated sequences and black boxes represent translated sequences. For a given RAR isoform, the numbering A1, A2, A3 and B represents the A1, A2, A3 and B regions of the receptor isoforms. For each receptor, P1 and P2 correspond to the upstream and downstream promoters, respectively.

Mouse retinoic acid receptors and isoforms are differentially expressed during development

Initial northern blot analysis (Petkovich et al., 1987; Krust et al., 1989; Zelent et al., 1989) and in situ analysis (Dollé et al., 1989, 1990; Ruberte et al., 1990, 1991) revealed a wide and differential pattern of expression for all three RARs in both the developing and adult mouse. Further northern blot analysis using probes specific for the major mRAR isoforms has shown additional evidence for isoform-specific expression. In summary, mRARα1 expression has been found to be ubiquitous, which is in keeping with the structure of the promoter for this gene (Brand et al., 1990; Leroy et al., 1991a). The presence of this transcript in essentially all cell types may reflect a general requirement for retinoid-dependent gene regulation. In contrast, mRARα2 has a more restricted pattern of expression, and this transcript is RA-inducible in embryonal carcinoma (EC) cell lines such as P19 and F9, consistent with the presence of a RARE in the promoter of this gene (Leroy et al., 1991b).

mRARβ1 and β3 transcripts are abundant in adult and embryonic brain, which may indicate a requirement for these isoforms in the development of the central nervous system (Zelent et al., 1991). As mentioned above, all three mRARβ isoforms are RA-inducible in EC cells which is consistent with the presence of a functional RARE in the RARβ2 promotor (de Thé et al., 1990; Mendelsohn et al., 1991). A similar motif has not yet been identified in the promoter for the mRARβ1 or β3 genes.

mRARγ1 is expressed mainly in adult skin, whereas mRARγ2 transcripts are more abundant in early embryogenesis and in EC cells (Kastner et al., 1990). Previous in situ analysis indicates that RARγ may also play a role in chondrogenesis, as this gene is highly expressed in condensing precartilage during mouse embryogenesis (Dollé et al., 1989, 1990).

Summary of the system

As discussed above, the response of a given cell to retinoids will depend on a complex series of parameters, only some of which have been discussed here. It is clear that at least three endogenous retinoids may activate gene transcription (T-RA, 9C-RA and ddRA). The synthesis and distribution of these metabolites has not been fully elucidated, although this will surely be of great importance with regard to retinoid-dependent developmental processes. Cellular retinoid-binding proteins (CRBPI and II; CRABPI and II) are likely to play a pivotal role in regulating cellular responses to retinoids, possibly by controlling both the formation and concentration of free ligand capable of binding to nuclear receptors. Finally, at least eight major species of RARs (3 subtypes; 8 isoforms) and three species of RXRs (which may also be expressed as isoforms: Fleischhauer et al., 1992; Leid et al., 1992) have been documented. If one considers the possibility that these receptors may function in vivo as RAR:RXR heterodimers, through activation by two different ligands and possibly in a promoter and cell-type specific fashion, it is clear that the combinatorial possibilities become considerable. This diversity through varied combinations may begin to explain the pleiotropic effects of retinoids in vivo. On the other hand, one may argue that this diversity may represent functional redundancy, a hypothesis currently impossible to refute with any direct evidence.

Gene targeting of components of the RA-signal transduction network

In order to determine which role(s) each of the above proteins may possibly play in development and homeostasis, we have begun gene targeting experiments, with the ultimate goal of establishing mouse lines in which each mRAR (or major mRAR isoform) or cellular retinoid-binding protein has been disrupted. This is being accomplished via homologous recombination in embryonic stem (ES) cells, as reviewed by Capecchi (1989). The results obtained from these experiments should provide valuable insight as to the specific roles these genes play during murine development and homeostasis of the mature animal.

Gene targeting of retinoid receptors and retinoid-binding proteins

The methodology, cell lines and mouse strains used in our gene targeting experiments have been described in detail previously (Lufkin et al., 1991). Briefly, we use the ES cell line D3 (a kind gift from R. Kemler, Frieberg, Germany), cultured under standard conditions. Targeting constructs are generally of the replacement type (Capecchi, 1989), in which a neomycin resistance (*neo*) gene casette has been inserted into exonic sequences of an engineered genomic fragment. The purpose of the *neo* gene is twofold; firstly, it allows selection of integration events via cellular resistance to G418; secondly, it is used to disrupt the coding region of the targeted gene upon homologous recombination with the cognate ES cell allele. These constructs are linearised outside the region of homology to the target allele and used to electroporate D3 ES cells (usually 10^7 cells/electroporation). Selection for the expression of the *neo* gene is performed using 150 μg/ml G418 for 12 days. In some experiments, negative selection against the expression of a HSV *Tk* gene (positive/negative selection; Capecchi, 1989) with 2 μM gancyclovyr is also used. Surviving colonies are isolated, propagated and used for either PCR or genomic Southern blotting analysis to identify clones in which homologous recombination has occured. Cells from positive clones (10-20 cells) are then injected into 3.5-day mouse embryos (blastocysts) and the resulting embryos are reimplanted into pseudo-pregnant recipient females. Resulting male chimeras are then outbred with C57B1/6 females. Germ line transmission (passage of the injected ES cell genotype to the F_1 progeny) can be observed initially via the coat colour of the offspring. Pups carrying the mutation are identified by Southern blot analysis. Heterozygote animals are analyzed for phenotypic alterations and if viable and fertile are then inbred to obtain homozygote mutants for subsequent phenotypic analysis.

Gene targeting of the mRARα1 locus will be used as a paradigm for disruption of other mRARs and retinoid-binding proteins. This isoform is the most ubiquitously expressed mRAR (as discussed above). As described previously, mRAR isoforms differ only in their 5′-untranslated and A regions. The strategy for disrupting a specific isoform, therefore, is to design a construct by which, after homologous recombination, the isoform will undergo premature termination in the A domain specific for that particular mRAR.

For the mRARα locus, a number of genomic clones have been described previously (Leroy et al., 1991a). For homologous recombination, a suitable subclone was obtained which contained exon 2 of the mRARα transcription unit. This exon is specific for mRARα1, and contains a portion of its 5′-untranslated region and the entire A region of this gene. Using this genomic fragment, a construct was prepared in which a *neo* cassette was inserted into the open reading frame of this exon (see Fig. 4). Homologous recombination of this construct with the cognate ES cell allele should result in a null mutation for this isoform via premature termination of the gene product at stop codons located in the 5′ end of the *neo* casette (Fig. 4). In order to avoid possible alteration of expression levels of the mRARα2 gene, which lies downstream of this locus, the *neo* casette used was driven by a minimal promoter, which consists only of the Rous sarcoma virus TATA box and lacks any viral enhancer sequences (derived from the *neo* casette p566 used by Lufkin et al., 1991). As depicted in Fig. 4, the close proximity of the 3′ end of the *neo* casette with respect to the 3′ boundary of the homologous recombination event should allow positive clones to be detected by a PCR reaction via amplification of a 1 kb product specific for the targeted allele.

As expected for a minimal promoter, a low number (22) of G418 resistant colonies were obtained using this targeting construct. PCR analysis of genomic DNA obtained from these clones, however, revealed three positive cell lines (Fig. 5). Thus, at least at this allele, a strong enhancer/pro-

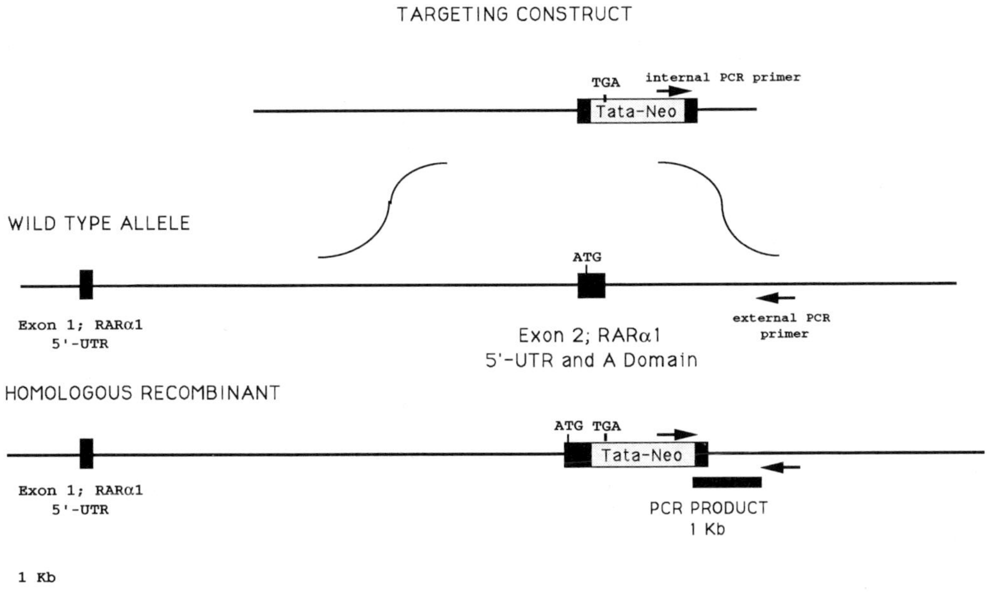

Fig. 4. Homologous recombination strategy for RARα1. The targeting construct, target allele and predicted homologous recombinant are indicated. Exons are denoted as black boxes. The TGA within the Tata-neo cassette represents 5′ stop sites present upstream of the neo coding region. The AUG in exon two of the RARα1 transcription unit represents the initiation codon for the RARα1 protein. The location of PCR primers are noted by horizontal arrows, and the predicted PCR reaction product by a thick horizontal line.

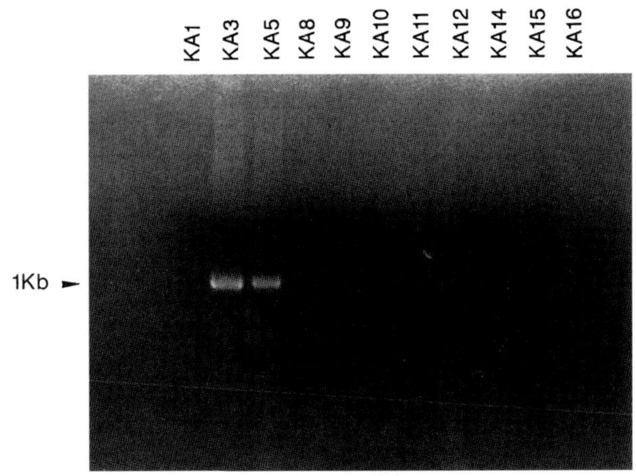

Fig. 5. PCR analysis for homologous recombination at the RARα1 allele. Genomic DNA was isolated from ES cell colonies electroporated with the construct shown in Fig. 4. The pcr was carried out under standard conditions using primers as indicated in Fig. 4. A sample of each reaction was run on an agarose gel and stained with ethidium bromide. Two positive cell lines (KA3 and KA5) are shown in the figure. A third positive line (KA 26) is not indicated.

moter driving the expression of the *neo* gene is not necessary to obtain targeting events. These clones were expanded and injected into blastocysts to generate chimeras. All three clones gave chimeric males which passed the mutated allele to their offspring (Table 1). We have since inbred heterozygote animals to generate homozygote mutant offspring, and are currently examining these animals for any phenotypic changes resulting from the loss of this gene product. It is important to note that recombination of this construct at the mRARα1 locus has no measurable effect on the expression of the mRARα2 gene (data not shown). The strategy of using a minimal promoter to express the *neo* gene may therefore be of general use for targeted disruption of genes which lie in proximity to other transcription units.

Table 1. *Germ line transmission of the RARα1 null mutation*

Chimera	Clone	Offspring	Number of Agouti	Genotype (+/–)
1	KA3	0	0	
2	KA3	20	18	7
3	KA3	0	0	
4	KA3	19	9	3
5	KA3	4	4	1
6	KA3	36	25	18
7	KA5	16	16	11
8	KA5	19	10	7
9	KA5	0	0	
10	KA5	0	0	
11	KA5	31	17	6
12	KA26	13	4	1
13	KA26	17	16	6
14	KA26	55	45	31
15	KA26	20	20	8
16	KA26	37	32	14
17	KA26	10	10	4
18	KA26	4	0	
19	KA26	3	3	1
20	KA26	9	0	
21	KA26	18	16	7
22	KA26	0	0	

The positive ES clones identified by the pcr in Fig. 5 were injected into blastocysts and the resulting male chimeras bred with C57B1/6 females. Germ line transmission was initially scored by the presence of agouti coat colour. Pups carrying the mutated allele were identified by genomic Southern blot analysis of tail tip DNA. (+/–) indicates the number of heterozygote animals present among the agouti offspring.

In analogous experiments, we have successfully targeted a number of mRARs and mRAR isoforms including mRARα1, mRARα (all isoforms), mRARβ2, mRARβ (all isoforms), mRARγ2, mRARγ (all isoforms). In addition, we have successfully disrupted the genes encoding mouse cellular retinoid-binding proteins CRABPI and II.

Conclusions

The ultimate interpretation of retinoid signalling at a cellular level is influenced by a large number of interactive gene products, including: (1) cellular retinoid-binding proteins, which may regulate the formation of active retinoids or control the cellular concentration of free ligand available to associate with retinoid receptors; (2) retinoid receptors which include at least two families, the three RARs and the three RXRs. These receptor families are composed of at least 11 different members (including isoforms), and are under a strict developmental expression program, as evidenced by their different spatio-temporal expression patterns in the developing mouse. An additional level of complexity with regard to control of retinoid-responsive target gene expression may be brought about through heterodimerization of RARs and RXRs to activate differentially, possibly in a cell-type and promoter-specific manner, subsets of these target genes. An additional consideration, not discussed here, is the regulation of retinoid biosynthesis. The enzymes controlling the production, and possibly the catabolism, of the various biologically active retinoids (eg. T-RA, 9C-RA and ddRA) may also play a pivotal role in this network.

In order to further investigate the role(s) that each of these genes play in murine development, we have successfully used gene targeting via homologous recombination in ES cells to create null mutants for a number of the key players in the retinoid signalling pathway. Subsequent analysis of the phenotypic changes brought about by these null mutations should give valuable insight as to the processes that are regulated by each of these proteins, and thus further our understanding of the function(s) performed by each member of this signalling pathway.

The authors would like to thank C. Werlé, B, Boulay, J. M. Lafontaine and the secretarial staff for help in the preparation of the manuscript. We also thank members of the cell culture and animal care facilities for their assistance. The D3 ES cell line was generously provided by Dr R. Kemler. This work was supported by grants from the CNRS, the INSERM, the Centre Hospitalier Universitaire Regionale, the Ministère de la Recherche et de la Technologie, the Fondation pour la Recherche Médicale and the Association pour la Recherche contre le Cancer.

References

Beato, M. (1989). Gene regulation by steroid hormones. *Cell* **56**, 335-344.

Benbrook, D., Lernhardt, E. and Pfahl, M. (1988). A new retinoic acid receptor identified from a hepatocellular carcinoma. *Nature* **333**, 669-672.

Blomhoff, R., Green, M. H., Berg, T. and Norum K. R. (1990). Transport and storage of vitamin A. *Science* **250**, 399-404.

Blomhoff, R., Green, M. H., Berg, T. and Norum, K. R. (1991). Vitamin A metabolism: New perspectives on absorption, transport and storage. *Physiol. Rev.* **71**, 952-990.

Blumberg, B., Mangelsdorf, D. J., Dyck, J. A., Bittner, D. A., Evans, R. M. and De Robertis, E. (1992). Multiple retinoid-responsive receptors in a single cell: families of RXRs and RARs in the *Xenopus* egg. *Proc. Nat. Acad. Sci., USA* **89**, 2321-2325.

Bocquel, M. T., Kumar, V., Stricker, C., Chambon, P. and Gronemeyer, H. (1989). The contribution of the N- and C- terminal regions of steroid receptors to activation of transcription is both receptor and cell-specific. *Nucl. Acids Res.* **17**, 2581-2595.

Brand, N., Petkovich, M., Krust, A., Chambon, P., de Thé, H., Marchio, A., Tiollais, P. and Dejean, A. (1988). Identification of a second human retinoic acid receptor. *Nature* **332**, 850-853.

Brand, N., Petkovich, M. and Chambon, P. (1990). Characterization of a functional promoter for the human retinoic acid receptor-alpha (hRAR-α). *Nucl. Acids Res.* **18**, 6799-6906.

Brockes, J. (1989). Retinoids, homeobox genes and limb morphogenesis. *Neuron* **2**, 1285-1294.

Brockes, J. (1990). Reading the retinoid signal. *Nature* **345**, 766-768.

Brockes, J. (1991). We may not have a morphogen. *Nature* **350**, 15.

Capecchi, M. R. (1989). Altering the genome by homologous recombination. *Science* **244**, 1288-1292.

Chambon, P., Zelent, A., Petkovich, M., Mendelsohn, C., Leroy, P., Krust, A., Kastner, P. and Brand, N. (1991). In *Retinoids: 10 Years On* (ed. J.-H. Saurat), pp. 10-27. Basel, Switzerland: S. Karger.

Chytil, F. (1984). Retinoic acid: biochemistry, pharmacology, toxicology and therapeutic use. *Pharmacol. Rev.* **36**, 93S-100S.

Chytil, F. and Haq, R. (1990). Vitamin A mediated gene expression. *Critical Reviews in Eukaryotic Gene Expression* **1**, 61-73.

de Thé, H., Marchio, A., Tiollais, P. and Dejean, A. (1987). A novel steroid thyroid hormone receptor related gene inappropriately expressed in human hepatocellular carcinoma. *Nature* **330**, 667-670.

de Thé, H., del Mar Vivanco-Ruiz, M., Tiollais, P., Stunnenberg, H., and Dejean, A. (1990). Identification of a retinoic acid responsive element in the retinoic acid receptor β gene. *Nature* **343**, 177-180.

Dollé, P., Ruberte, E., Kastner, P., Petkovich, M., Stoner, C. M., Gudas, L. J. and Chambon, P. (1989). Differential expression of the genes encoding the retinoic acid receptors α, β, γ and CRABP in the developing limbs of the mouse. *Nature* **342**, 702-705.

Dollé, P., Ruberte, E., Leroy, P., Morris-Kay, G. and Chambon, P. (1990). Retinoic acid receptors and cellular retinoid-binding proteins. I. A systematic study of their differential pattern of transcription during mouse organogenesis. *Development* **110**, 1133-1151.

Durand, B., Saunders, M., Leroy, P., Leid, M. and Chambon, P. (1992). All trans and 9-cis retinoic acid induction of mouse CRABP II gene transcription is mediated by RAR/RXR heterodimers. *Cell* **71**, 73-85.

Eichele, G. (1989a). Retinoids and vertebrate limb pattern formation. *Trends Genet.* **5**, 246-251.

Eichele, G. (1989b). Retinoic acid induces a pattern of digits in anterior half wing buds that lack the zone of polarizing activity. *Development* **107**, 863-867.

Evans, R. (1988). The steroid and thyroid hormone receptor family. *Science* **240**, 889-895.

Fleischhauer, K., Park, J. H., DiSanto, J. P., Marks, M., Ozato, K. and Yang, S. Y. (1992). Isolation of a full-length cDNA clone encoding a N-terminally variant form of the human retinoid X receptor β. *Nucl. Acids Res.* **20**, 1801.

Giguère, V., Ong, E. S., Segue, P. and Evans, R. M. (1987). Identification of a receptor for the morphogen retinoic acid. *Nature* **330**, 624-629.

Giguère, V., Ong, E. S., Evans, R. M. and Tabin, C. J. (1989). Spatial and temporal expression of the retinoic acid receptor in the regenerating amphibian limb. *Nature* **337**, 566-569.

Giguère, V., Lyn, S., Yip, P., Siu, C.-H. and Amin, S. (1990a). Molecular cloning of a cDNA encoding a second cellular retinoic acid-binding protein. *Proc. Nat. Acad. Sci., USA* **87**, 6233-6237.

Giguère, V., Shago, M., Zirngibl, R., Tate, P., Rossant, J. and Varmuza, S. (1990b). Identification of a novel isoform of the retinoic acid receptor γ expressed in the mouse embryo. *Mol. Cell. Biol.* **10**, 2335-2340.

Green, S. and Chambon, P. (1988). Nuclear receptors enhance our understanding of transcription regulation. *Trends Genet.* **4**, 309-314.

Hamada, K., Gleason, S. L., Levi, B.-Z., Hirschfeld, S., Apopela, E. and Ozato, K. (1989). H-2RIIBP, a member of the nuclear hormone receptor superfamily that binds to both the regulatory element of major histocompatibility class I genes and the estrogen response element. *Proc. Nat. Acad. Sci., USA* **86**, 8289-8293.

Heyman, R. A., Mangelsdorf, D. J., Dyck, J. A., Stein, R., Eichele, G., Evans, R. M. and Thaller, C. (1992). 9-Cis retinoic acid is a high affinity ligand for the retinoid X receptor. *Cell* **68**, 397-406.

Izpisua-Belmonte, J.-C., Tickle, C., Dollé, P., Wolpert, L. and Duboule, D. (1991). Expression of the homeobox Hox-4 genes and the specification of position in chick wing development. *Nature* **350**, 585-589.

Ishikawa, T., Umesono, K., Mangelsdorf, D. J., Aburatani, H., Stanger, B. Z., Shibasaki, Y., Imawari, M., Evans, R. M. and Takaku, F. (1990). A functional retinoic acid receptor encoded by the gene on the human chromosome 12. *Mol. Endocr.* **4**, 837-844.

Kastner, P., Krust, A., Mendelsohn, C., Garnier, J. M., Zelent, A., Leroy, P., Staub, A. and Chambon, P. (1990). Murine isoforms of retinoic acid receptor gamma with specific patterns of expression. *Proc. Nat. Acad. Sci., USA* **87**, 2700-2704.

Kessel, M. and Gruss, P. (1991). Homeotic transformations of murine vertebrae and concomitant alteration of Hox codes induced by retinoic acid. *Cell* **67**, 89-104.

Kliewer, S. A., Umesono, K., Mangelsdorf, D. J. and Evans, R. M. (1992). The retinoid X receptor interacts directly with nuclear receptors involved in retinoic acid, thyroid hormone and vitamin D_3 signaling. *Nature* **355**, 446-449.

Krust, A., Kastner, P., Petkovich, M., Zelent, A. and Chambon, P. (1989). A third human retinoic acid receptor, hRAR-γ. *Proc. Nat. Acad. Sci., USA* **85**, 329-333.

LaRosa, G. J. and Gudas, L. J. (1988). An early effect of retinoic acid: cloning of an mRNA (Era-I) exhibiting rapid and protein synthesis-independent induction during teratocarcinoma stem cell differentiation. *Proc. Nat. Acad. Sci., USA* **85**, 329-333.

Lehmann, J. M., Zhang, X.-K. and Pfahl, M. (1992). RARγ2 expression is regulated through a retinoic acid response element embedded in Sp1 sites. *Mol. Cell. Biol.* **12**, 2976-2985.

Leid, M., Kastner, P., Lyons, R., Nakshatri, H., Saunders, M., Zacharewski, T., Chen, J.-Y., Staub, A., Garnier, J.-M., Mader, S. and Chambon, P. (1992). Purification, cloning and RXR identity of the HeLa cell factor with which RAR or TR heterodimerizes to bind target sequences efficiently. *Cell* **68**, 377-395.

Leroy, P., Krust, A., Zelent, A., Mendelsohn, C., Garnier, J. M., Kastner, P., Dierich, A. and Chambon, P. (1991a). Multiple isoforms of the mouse retinoic acid receptor alpha are generated by alternative splicing and differential induction by retinoic acid. *EMBO J.* **10**, 59-69.

Leroy, P., Nakshatri, H. and Chambon, P. (1991b). The mouse retinoic acid receptor alpha-2 (mRAR-α2) isoform is transcribed from a promoter that contains a retinoic acid response element. *Proc. Nat. Acad. Sci., USA* **88**, 10138-10142.

Levin, M. S., Li, E., Ong, D. E. and Gordon, J. I. (1987). Comparison of the tissue-specific expression and developmental regulation of two closely linked rodent genes encoding cytosolic retinol-binding proteins. *J. Biol. Chem.* **262**, 7118-7124

Levin, A. A., Sturzenbecker, L. M., Kazmer, S., Bosakowski, T., Huselton, C., Allenby, G., Speck, J., Kratzeisen, C. Rosenberger, M., Lovey, A. and Grippo, J. F. (1992). 9-cis retinoic acid sterioisomer binds and activates the nuclear receptor RXRα, *Nature* **355**, 359-361.

Lufkin, T., Dierich, A., LeMeur, M., Mark, M. and Chambon, P. (1991). Disruption of the Hox-1.6 homeobox gene results in defects in a region corresponding to its rostral domain of expression. *Cell* **66**, 1105-1119.

Maden, M. (1982). Vitamin A and pattern formation in the regenerating limb. *Nature* **295**, 672-675.

Maden, M. (1985). Retinoids and the control of pattern in limb development and regeneration. *Trends Genet.* **1**, 103-107.

Maden, M., Ong, D. E., Summerbell, D. and Chytil, F. (1988). Spatial distribution of cellular proteins binding to retinoic acid in the chick limb bud. *Nature* **335**, 733-735.

Mangelsdorf, D. J., Ong, E. S., Dyck, J. A. and Evans, R. M. (1990).

Nuclear receptor that identifies a novel retinoic acid response pathway. *Nature* **345**, 224-229.

Mangelsdorf, D. J., Umesono, D., Kliewer, S. A., Borgmeyer, U., Ong, E. S. and Evans, R. M. (1991). A direct repeat in the cellular retinol-binding protein type II gene confers differential regulation by RXR and RAR. *Cell* **66**, 555-561.

Mangelsdorf, D. J., Borgmeyer, U., Heyman, R. A., Zhou, J. Y., Ong, E. S., Oro, A. E., Kakizuka, A. and Evans, R. M. (1992). Characterization of the three RXR genes that mediate the action of 9-cis retinoic acid. *Genes Dev.* **6**, 329-344.

Mattei, M.-G., Riviere, M., Krust, A., Ingvarsson, S., Venstrom, B., Islam M. Q., Levan, G., Kastner, P., Zelent, A., Chambon, P., Szpierer, J. and Szpierer, C. (1991). Chromosomal assignment of retinoic acid receptor (RAR) genes in the human, mouse and rat genomes. *Genomics* **10**, 1061-1069.

Mendelsohn, C., Ruberte, E., LeMeur, M., Morriss-Kay, G. and Chambon, P. (1991). Developmental analysis of the retinoic acid-inducible RAR-β2 promoter in transgenic animals. *Development* **113**, 723-734.

Moore, D. D. (1990). Diversity and unity in the nuclear hormone receptors: a terpenoid receptor superfamily. *The New Biologist* **2**, 100-105.

Murphy, S. P., Garbern, J., Odenwald, W. F., Lazzarini, R. A. and Linney, E. (1988). Differential expression of the homeobox gene Hox-1. 3 in F9 embryonal carcinoma cells. *Proc. Nat. Acad. Sci., USA* **85**, 5587-5591.

Nagpal, S., Zelent, A. and Chambon, P. (1992a). RAR-β4, a retinoic acid receptor isoform is generated from RAR-β2 by alternative splicing and usage of a CUG initiator codon. *Proc. Nat. Acad. Sci., USA* **89**, 2718-2722.

Nagpal, S., Saunders, M., Kastner, P., Durand, B., Nakshatri, H. and Chambon, P. (1992b). Differential promoter-context and response element-dependent specificity of the trans acting and modulating functions of the multiple nuclear retinoid receptors (RARs and RXRs) *Cell* **70**, 1007-1013.

Nicholson, R. C., Mader, S., Nagpal, S., Leid, M., Rochette-Egly, C. and Chambon, P. (1990). Negative regulation of the rat stromelysin gene promotor by retinoic acid is mediated by an AP1 binding site. *EMBO J.* **9**, 4443-4454.

Noji, S., Nohno, T., Koyama, E., Muto, K., Ohyama, K., Aoki, Y., Tamura, K., Ohsugi, K., Ide, H., Taniguchi, S. and Saito, T. (1991). Retinoic acid induces polarizing activity but is unlikely to be a morphogen in the chick limb bud. *Nature* **350**, 83-86.

Ong, D. E. (1987). Cellular retinoid binding proteins. *Arch. Dermatol.* **123**, 1693a-1695a.

Ong, D. E., Kakkad, B. and MacDonald, P. N. (1987). Acetyl-CoA-independent esterification of retinol bound to a cellular retinol-binding protein (type II) by microsomes from rat small intestine. *J. Biol. Chem.* **262**, 2729-2736.

Oro, A. E., McKeown, M. and Evans, R. M. (1990). Relationship between the product of the *Drosophila* ultraspiracle locus and the vertebrate retinoid X receptor. *Nature* **347**, 298-301.

Petkovich, M., Brand, N. J., Krust, A. and Chambon, P. (1987). A human retinoic acid receptor which belongs to the family of nuclear receptors. *Nature* **330**, 444-450.

Ragsdale Jr, C. W., Petkovich, M., Gates, P. B., Chambon, P. and Brockes, J. P. (1989). Identification of a novel retinoic acid receptor in regenerative tissues of the newt. *Nature* **341**, 654-657.

Ragsdale Jr, C. W. and Brockes, J. P. (1991). Retinoic acid receptors and vertebrate limb morphogenesis. In *Structure and Function of Hormone Nuclear Receptors*, (ed. M. G. Parker), pp. 269-295. London: Academic Press.

Rowe, A., Eager, N. S. C. and Brickell, P. M. (1991). A member of the RXR nuclear receptor family is expressed in neural-crest-derived cells of the developing peripheral nervous system. *Development* **111**, 771-778.

Ruberte, E., Dollé, P., Krust, A., Zelent, A., Morriss-Kay, G. and Chambon, P. (1990). Specific spatial and temporal distribution of retinoic acid receptor gamma transcripts during mouse embryogenesis. *Development* **108**, 213-222.

Ruberte, E., Dollé, P., Chambon, P. and Morriss-Kay, G. (1991). Retinoic acid receptors and cellular retinoid binding proteins: II. Their differential pattern of transcription during early morphogenesis in mouse embryos. *Development* **111**, 45-60.

Simeone, A., Acampora, D., Arcioni, L., Andrews, P. W., Boncinelli, E. and Mavilio, F. (1990). Sequential activation of HOX2 homeobox genes by retinoic acid in human embryonal carcinoma cells. *Nature* **346**, 763-766.

Slack, J. M. (1987). We have a morphogen. *Nature* **327**, 553-554.

Smith, S. M. and Eichele, G. (1991). Temporal and regional difference in the expression pattern of distinct retinoic acid receptor-β transcripts in the chick embryo. *Development* **111**, 245-252.

Smith, S. M., Pang, K., Sundin, O., Wedden, S. E., Thaller, C. and Eichele, G. (1989). Molecular approaches to vertebrate limb morphogenesis. *Development* **107 Supplement**, 121-131.

Smith, W. C., Nakshatri, H., Leroy, P., Rees, J. and Chambon, P. (1991). A retinoic acid response element is present in the mouse cellular retinol binding protein I (mCRBP I) promoter. *EMBO J.* **10**, 2223-2230.

Stoner, C. M. and Gudas, L. J. (1989). Mouse cellular retinoic acid binding protein: cloning, complementary DNA sequence and messenger RNA expression during the retinoic acid induced differentiation of F9 wild type and RA-3-10 mutant teratocarcinoma cells. *Cancer Res.* **49**, 1497-1504.

Sucov, H. M., Murakame, K. K. and Evans, R. M. (1990). Characterization of an autoregulated response element in the mouse retinoic acid receptor type β gene. *Proc. Nat. Acad. Sci., USA* **87**, 5392-5396.

Takase, S., Ong, D. E. and Chytil, F. (1979). Cellular retinol-binding protein allows specific interaction of retinol with the nucleus in vitro. *Proc. Nat. Acad. Sci., USA* **76**, 2204-2220.

Takase, S., Ong, D. E. and Chytil, F. (1986). Transfer of RA from its complex with cellular retinoic acid binding protein to the nucleus. *Arch. Biochem. Biophys.* **247**, 328.

Thaller, C. and Eichele, G. (1987). Identification and spatial distribution of retinoids in the developing chick limb bud. *Nature* **327**, 625-628.

Thaller, C. and Eichele, G. (1990). Isolation of 3, 4-didehydroretinoic acid, a novel morphogenetic signal in the chick wing bud. *Nature* **345**, 815-819.

Tickle, C., Alberts, B., Wolpert, L. and Lee, J. (1982). Local application of retinoic acid to the limb bud mimics the action of the polarizing region. *Nature* **296**, 564-566.

Tora, L., Gaub, M.-P., Mader, S., Dierich, A., Bellard, M. and Chambon, P. (1989). The human estrogen receptor has two independent nonacidic transcriptional activation functions. *Cell* **59**, 477-487.

Wanek, N., Gardiner, D. M., Muneoka, K. and Bryant, S. V. (1991). Conversion by retinoic acid of anterior cells into ZPA cells in the chick wing bud. *Nature* **350**, 81-83.

Wang, S.-Y. and Gudas, L. J. (1983). Isolation of cDNA clones specific for collagen IV and laminin from mouse teratocarcinoma cells. *Proc. Nat. Acad. Sci., USA* **80**, 5880-5884.

Wang, C., Kelly, J., Bowden-Pope, D. F. and Stiles, C. D. (1990). Retinoic acid promotes transcription of the PDGFα-receptor gene. *Mol. Cell. Biol.* **10**, 6781-6784.

Webster, N. J. G., Green, S., Jin, J. R. and Chambon, P. (1988). The hormone-binding domains of the estrogen and glucocorticoid receptors contain an inducible transcription activation function. *Cell* **54**, 199-207.

Wei, L.-N., Blaner, W. S., Goodman, D. S. and Nguyen-Huu, M. C. (1989). Regulation of the cellular retinoid-binding proteins and their messenger ribonucleic acids during P-19 embryonal carcinoma cell differentiation induced by retinoic acid. *Mol. Endocrinol.* **3**, 454-463.

Yu, V. C., Delsert, C., Andersen, B., Holloway, J. M., Devary, O., Nar, A. M., Kim, S. Y., Boutin, J.-M., Glass, C. K. and Rosenfeld, M. G. (1991). RXR-β: a coregulator that enhances binding of retinoic acid, thyroid hormone and vitamin D receptors to their cognate response elements. *Cell* **67**, 1251-1266.

Zelent, A., Krust, A., Petkovich, M., Kastner, P. and Chambon, P. (1989). Cloning of murine α and β retinoic acid receptors and a novel receptor γ predominantly expressed in skin. *Nature* **339**, 714-717.

Zelent, A., Mendelsohn, C., Kastner, P., Garnier, J. M., Ruffenach, F., Leroy, P. and Chambon, P. (1991). Differentially expressed isoforms of the mouse retinoic acid receptor beta are generated by usage of two promoters and alternative splicing. *EMBO J.* **10**, 71-81.

Zhang, X.-K., Hoffmann, B., Tran, P. B.-V., Graupner, G. and Pfahl, M. (1992). Retinoic X receptor is an auxiliary protein for thyroid hormone and retinoic acid receptors. *Nature* **355**, 441-446.

Journal of Cell Science, Supplement 16, 77-85 (1992)
Printed in Great Britain © The Company of Biologists Limited 1992

Cyclin A recruits p33^{cdk2} to the cellular transcription factor DRTF1

LASANTHA R. BANDARA[1], JÖRG P. ADAMCZEWSKI[2], MARYAM ZAMANIAN[1], RANDY Y. C. POON[2], TIM HUNT[2] and NICHOLAS B. LA THANGUE[1,*]

[1]*Laboratory of Eukaryotic Molecular Genetics, MRC National Institute for Medical Research, The Ridgeway, Mill Hill, London, NW7 1AA, UK*
[2]*Imperial Cancer Research Fund, Clare Hall Laboratories, Blanche Lane, South Mimms, Potters Bar, Herts., EN6 3LD, UK*

*Corresponding author

Summary

Cyclins are regulatory molecules that undergo periodic accumulation and destruction during each cell cycle. By activating p34^{cdc2} and related kinase subunits they control important events required for normal cell cycle progression. Cyclin A, for example, regulates at least two distinct kinase subunits, the mitotic kinase subunit p34^{cdc2} and related subunit p33^{cdk2}, and is widely believed to be necessary for progression through S phase. However, cyclin A also forms a stable complex with the cellular transcription factor DRTF1 and thus may perform other functions during S phase. DRTF1, in addition, associates with the tumour suppressor retinoblastoma (Rb) gene product and the Rb-related protein p107. We now show, using biologically active fusion proteins, that cyclin A can direct the binding of the cdc2-like kinase subunit, p33^{cdk2}, to complexed DRTF1, containing either Rb or p107, as well as activate its histone H1 kinase activity. Cyclin A cannot, however, direct p34^{cdc2} to the DRTF1 complex and we present evidence suggesting that the stability of the cyclin A-p33^{cdk2} complex is influenced by DRTF1 or an associated protein. Cyclin A, therefore, serves as an activating and targeting subunit of p33^{cdk2}. The ability of cyclin A to activate and recruit p33^{cdk2} to DRTF1 may play an important role in regulating cell cycle progression and moreover defines a mechanism for coupling cell-cycle events to transcriptional initiation.

Key words: cyclin, transcription, retinoblastoma, CDK, tumour suppressor.

Introduction

Cyclins are an evolutionarily conserved group of regulatory molecules that undergo periodic accumulation and destruction during the cell cycle and are required for normal cell cycle progression (Hunt, 1989). They were initially identified in marine invertebrates (Evans et al., 1983), and act by binding to and regulating the activity of the mitotic kinase catalytic subunit p34^{cdc2} or related kinase subunits (Pines and Hunter, 1990a). Cyclins, which have been isolated and characterised from a variety of sources, are grouped into several classes. Cyclins of the B class, which function at mitosis, peak towards M phase and are responsible for activating the mitotic kinase subunit p34^{cdc2} (Draetta et al., 1989; Minshull et al., 1990). Cyclin A, however, accumulates somewhat earlier (Giordano et al., 1989; Minshull et al., 1990; Pines and Hunter, 1990b) and in addition to p34^{cdc2} activates the related kinase subunit p33^{cdk2} (Pines and Hunter, 1990b; Tsai et al., 1991). Other cyclins have recently been defined which are thought to regulate progression through earlier phases of the cell cycle (Matsushime et al., 1991; Xiong et al., 1991; Lew et al., 1991; Koff et al., 1991).

Several reports have suggested that cyclin A regulates progression through S phase. In *Xenopus* egg extracts, p33^{cdk2} or a closely related kinase subunit is required for DNA replication (Fang and Newport, 1991) and is believed to fulfill a similar role in other types of cells (Pines and Hunter, 1991). Moreover, inactivation of cyclin A, either through antibody or antisense approaches, prevents cells from completing S phase (Girard et al., 1991). It is likely, therefore, that cyclin A regulates the kinase activity of p33^{cdk2} which in turn modifies protein substrates that are necessary for cell-cycle progression.

The retinoblastoma (Rb) gene product is a negative regulator of the cell cycle that acts by controlling progression through G_1 (Goodrich et al., 1991). This effect is widely believed to result from the activity of the un- or under-phosphorylated Rb protein, because this form of the protein predominates during G_1 (Buchkovich et al., 1989; DeCaprio et al., 1989). Certain viral oncoproteins, such as adenovirus E1a, SV40 large T antigen and HPV E7 bind the Rb protein (Whyte et al., 1988; DeCaprio et al., 1988; Dyson et al., 1989) and the Rb gene is frequently mutated in tumour cells (Hu et al., 1990; Huang et al., 1990). Both of these processes are thought to inactivate the growth suppressing properties of the Rb protein by preventing it from regulating cellular targets. Moreover, the Rb-related protein, p107, is also sequestered by adenovirus E1a and SV40 large T antigen (Whyte et al., 1988; DeCaprio et al., 1988) and, although there is no detailed information available yet on its regulation during the cell cycle, is likely to have an important role in cell-cycle control (Ewen et al., 1991).

It is thought that the Rb protein mediates its growth regulating properties by modulating the activity of molecules that regulate progression through the cell cycle, but only recently have potential candidates been identified (Bandara and La Thangue, 1991; Bagchi et al., 1991; Chellapan et al., 1991; Chittenden et al., 1991; Defeo-Jones et al., 1991). A particularly good candidate is the cellular transcription factor DRTF1 which is found in complexes that contain the Rb protein (Bandara and La Thangue, 1991). DRTF1, first defined in F9 embryonal carcinoma (EC) stem cells as a differentiation-regulated transcription factor (La Thangue and Rigby, 1987), is mostly uncomplexed in F9 EC cells, but in other cell-types forms a stable complex with the Rb protein (Bandara and La Thangue, 1991). The Rb-DRTF1 complex is dissociated by E1a, large T antigen and E7 proteins (Bandara and La Thangue, 1991; Bandara and La Thangue, data not shown), which releases the transcriptionally active protein, the form that predominates in stem cells. Moreover, the protein products of all naturally occurring mutant Rb alleles so far studied fail to bind stably to DRTF1 (Bandara et al., 1991; Bandara and La Thangue, data not shown), underscoring the potential importance of DRTF1 in controlling cellular proliferation.

Recent experiments have shown that the Rb protein can specifically repress the transcription of promoters driven by DRTF1, whereas proteins encoded by some naturally-occurring mutant Rb alleles, which fail to bind DRTF1 in vitro, cannot. Furthermore, Rb-mediated transcriptional repression is relieved by the adenovirus E1a protein (Zamanian and La Thangue, 1992). The Rb protein, therefore, represses the transcriptional activity of DRTF1 which thus provides a potential mechanistic explanation for how the Rb protein exerts its growth regulating properties, given that DRTF1 binding sites occur in the promoters of genes that are necessary for cell cycle progression (Blake and Azizkhan, 1989; Pearson et al., 1991).

DRTF1 binds to a similar DNA sequence to that of the HeLa cell transcription factor E2F, and although the exact relationship between these two transcription factors has yet to be established, they are likely to be related because similar interactions have been defined for E2F (Bagchi et al., 1990; Chellappan et al., 1991).

Cyclin A also binds to DRTF1 (Bandara et al., 1991) and previously we have suggested that this allows cell cycle events to be coupled to transcription. We have developed an in vitro assay in which cyclin A efficiently binds to DRTF1 and established that cyclin A can direct the cdc2-like kinase subunit, p33^{cdk2}, but not the mitotic kinase subunit, p34^{cdc2}, to DRTF1. We suggest that such a kinase will play an important role in regulating the activity of other non DNA-binding proteins, such as the Rb protein and p107, in the DRTF1 transcription factor complex and hence in regulating its transcriptional activity. We also present evidence that the stability of the cyclin A-p33^{cdk2} complex is influenced by DRTF1 or a component within this transcription factor complex.

Results

Cyclin A targets the cdc2-related kinase subunit p33^{cdk2} to DRTF1 complexes that contain either the Rb protein or p107

Cyclin A is part of the DRTF1 transcription factor complex (Bandara et al., 1991). In order to investigate the significance of this interaction we developed an in vitro association assay for cyclin A and DRTF1. Initially, we performed the assay in JM cell extracts where most DRTF1 is bound to the Rb protein (complexed DRTF1 is referred to as DRTF1a) and assessed whether a biologically active cyclin A fusion protein (PA-CA), containing amino acid residues 77 to 432 (Fig.1b), would assemble with JM DRTF1a. Indeed, PA-CA efficiently assembles with DRTF1a, because upon addition of PA-CA to the cell extract the Rb-DRTF1 complex migrates with slower mobility (Fig.1a, compare tracks 2 and 3 with 4 and 5; PA-CA dependent complex indicated by ■). This effect is only apparent with DRTF1a since PA-CA fails to assemble with uncomplexed DRTF1, referred to as DRTF1b (see Fig.2). The binding reaction also requires the integrity of the cyclin box because some derivatives of PA-CA that were mutated in this region fail to assemble with DRTF1a (discussed later). Cyclin A, therefore, contains a domain that enables it to associate with DRTF1a and moreover this association occurs efficiently in vitro.

Since cyclin A regulates the activity of cell cycle kinase subunits, such as p34^{cdc2} (Draetta et al., 1989; Minshull et al., 1990) and p33^{cdk2} (Pines and Hunter, 1990b; Tsai et al., 1991), we reasoned that the cyclin A associated with DRTF1 may provide a targeting function and thus recruit such kinase subunits to a potentially important substrate. We tested this idea directly by determining if an affinity purified fusion protein that contains the entire coding sequence of p33^{cdk2} (amino acid residue 1 to 298; GST-cdk2, Fig.1b) would bind to DRTF1 in JM cell extracts. By itself, GST-cdk2 inefficiently binds to the endogenous DRTF1-Rb complex (Fig.1a, tracks 6 and 7) although in some experiments a weak supershifted cdk2-dependent complex is apparent, presumably caused by a binding reaction with the endogenous cyclin A-DRTF1 complex. Binding is enhanced, however, when additional cyclin A is provided by adding PA-CA at the same time as GST-cdk2, when a novel, slower migrating complex is apparent (Fig.1a, compare tracks 4 and 5 with 8 and 9; indicated by ▲). Thus, GST-cdk2 can assemble into the DRTF1 complex, but does so in a PA-CA-dependent fashion. The association of cdk2 with DRTF1 is thus dependent on cyclin A and therefore cyclin A contains domains that enable it to bind simultaneously to two distinct types of molecule, DRTF1 and cdk2.

The JM DRTF1a complex contains the Rb protein (Fig.1a, compare tracks 14 and 15 with 16 and 17; Rb-supershift indicated by ○). To determine if the cdk2-induced complex also contained the Rb gene product, we tested if the anti-Rb antibody caused a further supershift. Indeed, the cdk2-dependent shift can be further shifted by the Rb monoclonal antibody (Fig.1a, compare tracks 10 and 11 with 12 and 13; cdk2-dependent Rb-supershift indicated by ●). Thus, DRTF1, cyclin A, cdk2 and the Rb gene product can exist in the same molecular complex.

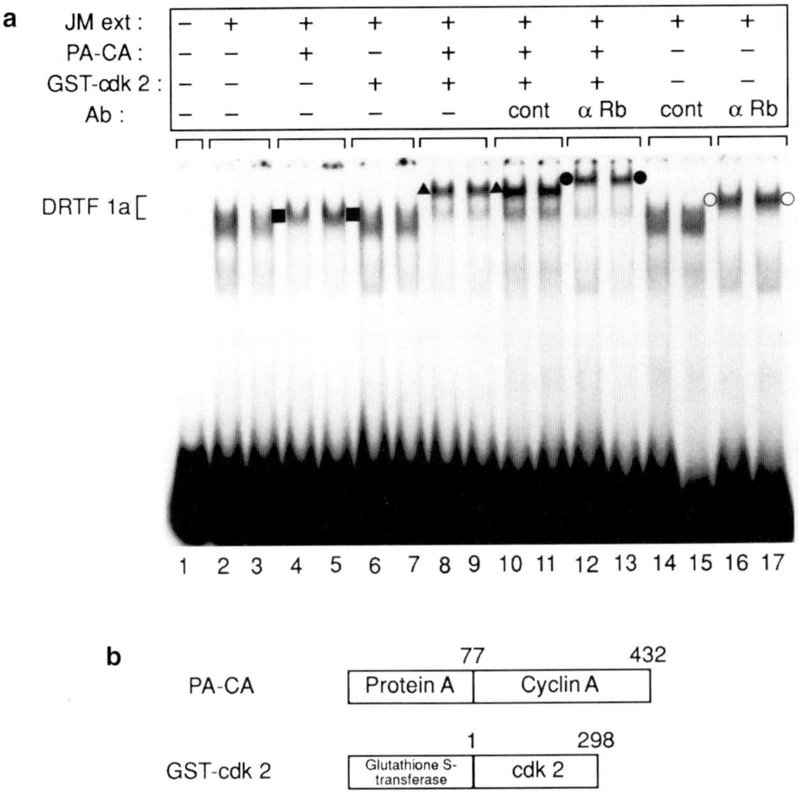

Fig. 1. Cyclin A targets cdk2 to complexed DRTF1. (a) Gel retardation (Shivji and La Thangue, 1991) was performed in a JM whole cell extract (about 2 μg) either before (tracks 2, 3, 14, 15, 16 and 17) or after the addition of PA-CA (about 0.5 μg; tracks 4, 5, 8, 9, 10, 11, 12 and 13) or GST-cdk2 (about 0.5 μg; tracks 6, 7, 8, 9, 10, 11, 12 and 13) in the presence of an anti-Rb, C36 (Whyte et al., 1988), (tracks 12, 13, 16 and 17) or control (tracks 10, 11, 16 or 17) monoclonal antibody; the probe (containing Ad5 E2A promoter sequences from –71 to –50; La Thangue et al, 1990) alone is shown in track 1. DRTF1a is shifted by the anti-Rb monoclonal antibody to ○ (compare tracks 14 and 15 with 16 and 17) and ■ indicates the PA-CA dependent complex (tracks 4 and 5). Note that the cdk-dependent complex, ▲ (tracks 8 and 9), is supershifted to ● by the anti-Rb antibody (compare tracks 10 and 11 with 12 and 13). (b) Structure of fusion proteins used in this experiment. PA-CA was purified as previously described (Bandara et al., 1991). GST-cdk2 contains the entire cdk2 coding sequence (Tsai et al., 1991) fused to glutathione-S-transferase and was purified by affinity chromatography on glutathione Sepharose followed by FPLC on a MonoQ column.

To confirm and extend this result, we tested if cyclin A and cdk2 behaved in a similar fashion in an F9 EC cell extract where most DRTF1 is uncomplexed (Fig.2a, tracks 2 and 3; indicated as DRTF1b). In F9 EC cells, the level of complexed DRTF1 can be increased by adding an Rb fusion protein (containing Rb coding sequence from 379 to 928, GST-Rb) to the cell extract, which efficiently assembles with F9 EC DRTF1b (Bandara et al., 1991 and Fig.2a, compare tracks 2 and 3 with 6 and 7; Rb complex indicated by DRTF1a). When added alone, GST-cdk2 has little effect (Fig.2a, compare tracks 2 and 3 with 10 and 11) and PA-CA a marginal effect (Fig.2a, compare tracks 2 and 3 with 4 and 5). When GST-cdk2 and PA-CA are added together, however, a cdk2-dependent complex with similar mobility to that produced in JM cell extracts is apparent (Fig.2a, compare tracks 2 and 3 with 12 and 13, cdk2-dependent complex indicated by ▲). High levels of uncomplexed DRTF1b still remain, suggesting that GST-cdk2 and PA-CA can only associate with endogenous complexed DRTF1a in F9 EC cell extracts and therefore the amount of cdk2-induced complex reflects the abundance of DRTF1a. This was confirmed by increasing the level of DRTF1a, by adding GST-Rb at the same time as GST-cdk2 and PA-CA when the abundance of the cdk2-dependent complex is enhanced (Fig.2a, compare tracks 12 and 13 with 14 and 15, indicated by ▲); little uncomplexed DRTF1b remains in these conditions. These data indicate that, as in JM extracts, the assembly of p33^{cdk2} with F9 EC DRTF1 is cyclin A-dependent. The combined conclusion of these in vitro assembly assays is thus that the association of GST-cdk2 with DRTF1a is dependent on PA-CA, and that this binding reaction only occurs efficiently with complexed DRTF1a.

The Rb-related protein, p107, which has a degree of similarity with the Rb protein (Ewen et al., 1991), also binds to DRTF1 because a fusion protein containing the p107 coding sequence from amino acid residue 249 to 936 (GST-p107; Fig.2b) efficiently binds to F9 EC DRTF1b (Fig.2a, compare tracks 2 and 3 with 8 and 9). In fact, the characteristics of the p107 and Rb protein binding reactions are very similar indeed both in quality, efficiency of the complex formation and half-life (data not shown).

Since p107 also binds to DRTF1, we assessed whether cyclin A would also bind to the DRTF1-p107 complex and, further, if it is also able to direct p33^{cdk2} to the p107 complex. The p107-induced F9 EC DRTF1a complex is further shifted upon the addition of PA-CA (data not shown), in a similar fashion to the effect that PA-CA has on JM DRTF1a and, once assembled, PA-CA is likewise able to bind GST-cdk2 because the DRTF1-p107 complex is super-shifted upon the addition of GST-cdk2 (Fig.2a, compare 8 and 9 with 16 and 17). Cyclin A can thus bind to and recruit p33^{cdk2} to DRTF1 complexes containing either the Rb gene product or p107.

There are no obvious features of the cdk2-dependent complexes formed with either DRTF1/GST-Rb or GST-p107 that distinguish between them: the mobilities of the complexes are similar (both GST-Rb and GST-p107 cdk2-dependent complexes migrate as a doublet when assayed in F9 EC cell extracts; Fig.3, compare tracks 3 and 4 with 5 and 6) and the stability of these complexes are similar, at least up to 60 min (Fig.3, compare tracks 2 to 6 with 7 to 11, cdk2-dependent complexes indicated by ▲). p33^{cdk2} can therefore, bind via cyclin A to DRTF1 complexes that contain either the Rb protein or p107 and, furthermore, the

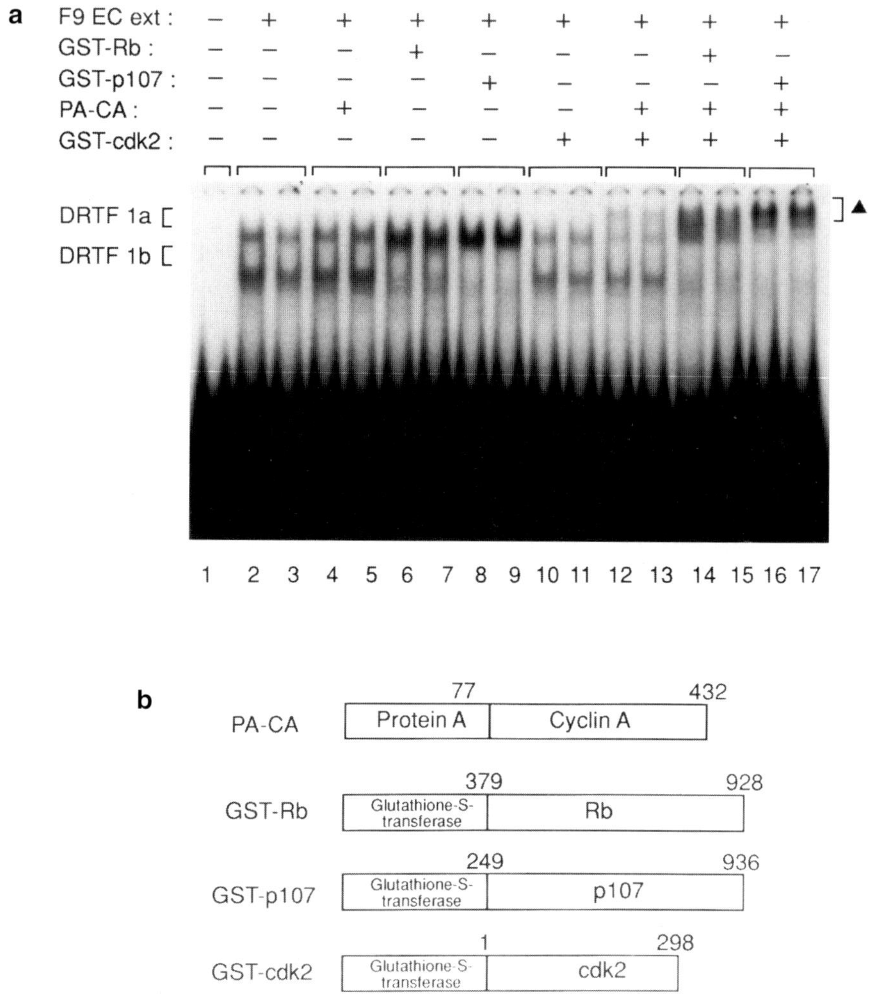

Fig. 2. Cyclin A targets cdk2 to DRTF1 complexes that contain either GST-Rb or GST-p107. (a) Gel retardation was performed in an F9 EC cell extract either before (tracks 2 and 3) or after (about 0.5 μg; tracks 4, 5, 12, 13, 14, 15, 16 and 17) the addition of PA-CA, GST-Rb (about 0.06 μg; tracks 6, 7, 14 and 15), GST-p107 (about 0.02 μg; tracks 8, 9, 16 and 17) or GST-cdk2 (about 0.5 μg; tracks 10, 11, 12, 13, 14, 15, 16 and 17); track 1 shows the probe alone. The cdk2-dependent complex (▲), which migrates as a doublet in the presence of GST-Rb or GST-p107, is apparent in tracks 12 through 17. (b) Structure of fusion proteins used in this experiment. GST-p107 was constructed using a sense oligonucleotide primer for amino acid residues 249-257(5′-CGGGATCCGCAGTC-ATTACTCCTGTTGCATCAGCC) and an anti-sense primer covering amino acid residues 690-732 (5′-GGACAGTGAACTA-AAGTGAATTCAAAAACACG) in a polymerase chain reaction (PCR), using a p107 cDNA (Ewen et al., 1991) as template. In a second PCR reaction, a sense primer for amino acid residues 690-732 (5′-CGTGTTT-TGAATTCACTTTAGTTCACTGTCC) and an antisense primer for amino acid residues 928-936 (5′-CGGGATCCTTAATGATTTG-CTCTTTCACTGAC) were used. Both PCR products were purified, digested with *Bam*HI and *Eco*RI, ligated and cloned into pGEX-2T. GST-p107 and GST-Rb were purified as previously described (Smith and Johnson, 1988).

properties of these binding reactions cannot readily be distinguished in gel retardation assays.

We confirmed that p33^{cdk2} or a closely related subunit is present in the endogenous DRTF1a complexes in F9 EC cell extracts using an antibody raised against the GST-cdk2 fusion protein. The immune but not the preimmune produces a shifted-shift in F9 EC cell extracts (Fig.4, compare tracks 15 and 14); a similar effect is also observed in JM cell extracts (data not shown). We conclude, therefore, that p33^{cdk2} is present in the DRTF1 complex. In this respect, DRTF1 is similar to E2F (Cao et al., 1992; Devoto et al., 1992).

We addressed the specificity of the p33^{cdk2}/DRTF1 interaction by asking if the mitotic kinase subunit, p34^{cdc2}, could also assemble with DRTF1. To test this, the entire coding sequence of a *Xenopus* p34^{cdc2} (Milarski et al., 1991) was expressed as a fusion protein, GST-Xlcdc2 (Fig.4b), affinity purified and then assayed for any binding activity with DRTF1. It was necessary to use *Xenopus* p34^{cdc2} because repeated attempts to express human p34^{cdc2} as a fusion protein have failed. Although GST-cdk2 is able to assemble efficiently with DRTF1a in a cyclin A-dependent manner (Fig.4a, tracks 4 to 11), GST-Xlcdc2 cannot (Fig.4, compare tracks 10 and 11 with tracks 12 and 13). This lack of effect is not simply caused by a fusion protein that lacked any biological activity because GST-Xlcdc2 is able to bind efficiently to PA-CA in the cell extract to produce H1 kinase activity (R. Y. C. Poon and J. P. A., unpublished data). We therefore conclude that p33^{cdk2} is selectively recruited to DRTF1 by cyclin A.

Complex formation between cyclin A and cdk2 fusion proteins produces a biologically active H1 kinase

Since heterodimerisation of an appropriate cyclin molecule with a compatible catalytic subunit is necessary to produce a biologically active protein kinase, we tested if PA-CA would form a heterodimer with GST-cdk2, and furthermore determined if the heterodimer possessed H1 kinase activity. The ability of PA-CA to bind GST-cdk2 was assessed after incubating both fusion proteins together, either in the presence or absence of JM cell extract, and then harvesting GST-cdk2 (and thus any bound PA-CA) on glutathione beads. These complexes were then denatured and assayed by immunoblotting with an anti-GST antibody which would bind to both GST-cdk2 and PA-CA (through the protein A moiety) and therefore indicate whether GST-cdk2 had formed a complex with PA-CA.

In isolation, the ability of GST-cdk2 to bind PA-CA is slight (Fig.5a, track 16, notice weak reactivity with PA-CA) although, we believe, significant, because when several derivatives of PA-CA in which the cyclin box or the C terminus were mutated (Fig.5c, Δ2, Δ18, Δ24 and ΔC-16) are

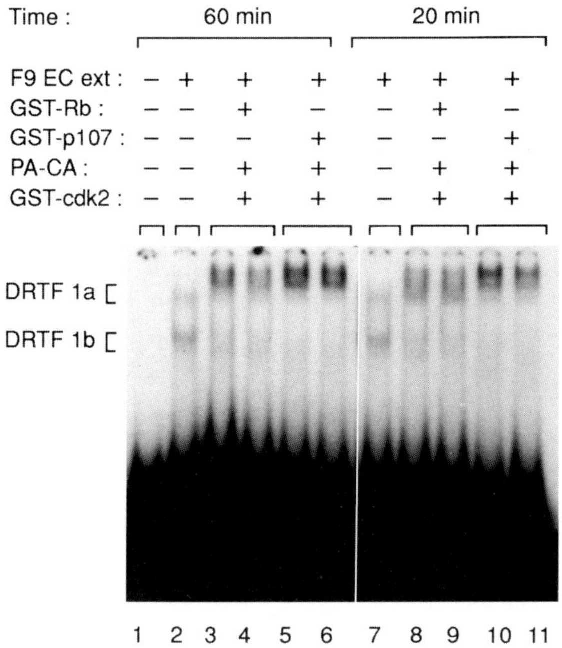

Fig. 3. cdk2-dependent complexes have similar stabilities. Gel retardation was performed in an F9 EC cell extract either alone (tracks 2 and 7) or after the addition of GST-Rb (about 0.06 µg; tracks 3, 4, 8 and 9) or GST-p107 (about 0.02 µg; tracks 5, 6, 10, and 11), together with PA-CA and GST-cdk2 (both at 0.5 µg; tracks 3, 4, 5, 6, 8, 9, 10 and 11), and incubated for either 60 min (tracks 2 to 6) or 20 min (tracks 7 to 11). The cdk2-dependent complex (▲) is indicated. Fusion proteins were as described in Fig. 2.

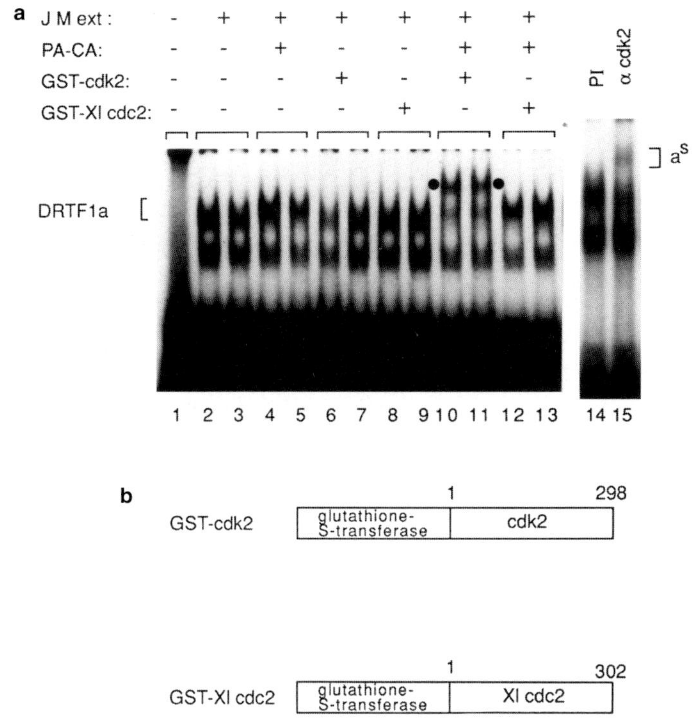

Fig. 4. cdc2 cannot engage with DRTF1. Gel retardation was performed in a JM cell extract either alone (tracks 2 and 3) or after the addition of PA-CA (about 0.5 µg, tracks 4, 5, 10, 11, 12 and 13), GST-cdk2 (about 0.5 µg, tracks 6, 7, 10 and 11) or GST-Xlcdc2 (about 0.2 µg, tracks 8, 9, 12 and 13). GST-Xlcdc2 contains the entire coding sequence of a *Xenopus* cdc2 cDNA (Milarski et al., 1991) in pGEX-2T and was purified as described above. Note that the cdk2-dependent complex, ●, is present only in tracks 10 and 11. For tracks 14 and 15, an F9 EC cell extract was incubated in the presence of either a preimmune (track 14) or anti-GST-cdk2 (track 15) rabbit antiserum. The shifted complex is indicated by a^s.

tested, no detectable binding with PA-CA is apparent (Fig.5a, compare track 16 with 19, 20, 21 and 22). The binding efficiency of PA-CA with GST-cdk2 is, however, enhanced when they are incubated together in the JM cell extract and then assayed (Fig.5a, compare track 9 with 16 and notice the increased amount of PA-CA compared with track 9). This binding requires the integrity of the cyclin box and the C terminus in PA-CA because there is negligible binding when either of these regions are mutated (Fig.5a, compare track 9 through 12 with 15); it is also absolutely dependent on the presence of cdk2, since GST alone or denatured GST-cdk2 do not bind to PA-CA (Fig.5a, compare track 9 with 10 and 11, respectively).

We next determined if heterodimerisation of PA-CA and GST-cdk2 results in H1 kinase activity. Indeed, the affinity purified heterodimer formed in the presence of the JM extract possesses H1 kinase activity (Fig.5b, track 1) which is, as expected, severely compromised when the PA-CA mutants are similarly assayed (Fig.5b, compare track 1 through 4 with 7), and again is dependent on the presence of cdk2 coding sequences (Fig.5b, compare track 1 with tracks 2 and 3). These data show that the PA-CA/GST-cdk2 heterodimer forms a biologically active protein kinase and that an activity in JM cell extracts is necessary for this to occur. Since PA-CA can direct GST-cdk2 to DRTF1 in JM cell extracts, we believe that this would enable heterodimerisation of PA-CA with GST-cdk2 and thus target an active kinase to DRTF1.

The stability of the cyclin A-p33^{cdk2} complex is influenced by DRTF1

The data in the previous section show that PA-CA can bind to GST-cdk2 and that this results in an active H1 kinase. Regions either within the cyclin box or at the C terminus of PA-CA are necessary for this interaction because derivatives in which either region is mutated cannot bind to GST-cdk2 (Fig.5). We wished to determine if these regions were also required for the interaction of PA-CA with DRTF1 and therefore assessed if these mutants assembled with DRTF1a in JM cell extracts. As shown earlier, PA-CA efficiently assembles with DRTF1a, producing a characteristic slower migrating complex (Fig.1, compare tracks 2 and 3 with 4 and 5; PA-CA-dependent complex indicated by ■). Mutant Δ2 is also able to bind to DRTF1 (Fig.6, compare tracks 4 and 5 with 6 and 7; complex indicated by ■) whereas Δ18 and ΔC-16 cannot because there is no effect on the migration of DRTF1a (Fig.6, compare tracks 2 and 3 with 8, 9, 10 and 11); the efficiency of Δ2 binding is, however, lower than the activity of the wild-type sequence. We conclude from these data that the alteration in Δ2 affects the interaction of PA-CA with GST-

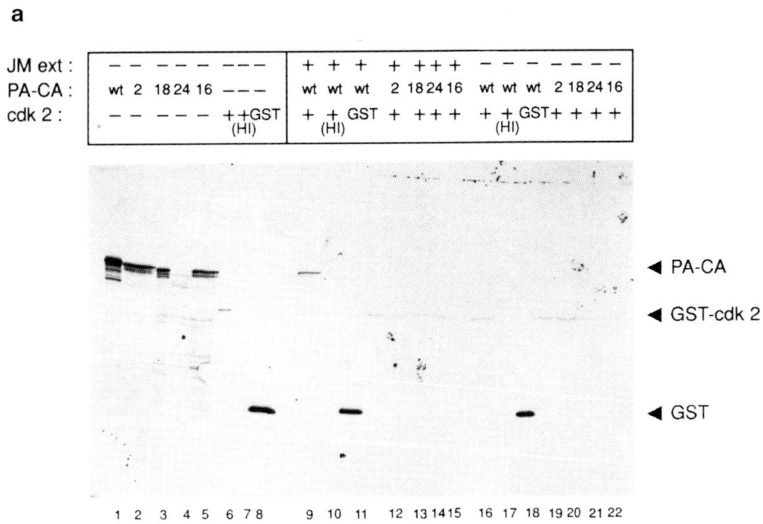

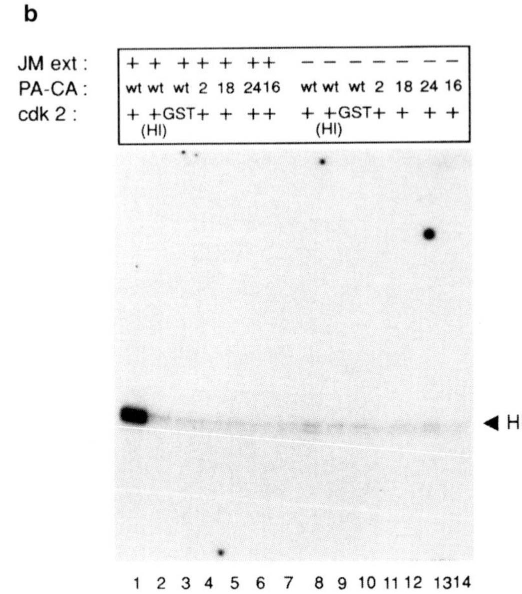

Fig. 5. Heterodimerisation of cyclin A and cdk2 fusion proteins produces an active H1 kinase. (a) PA-CA and GST-cdk2 bind in vitro. Either PA-CA (track 1), PA-CA mutants (Δ2, 18, 24 and C-16; tracks 2 to 5), GST-cdk2 (track 6), heat-denatured GST-cdk2 (track 7) or GST with no fusion (track 8) were assayed after western blotting and probing with an anti-GST antibody. The ability of these fusion proteins to complex in the presence (tracks 9 to 15) or absence (tracks 16 to 22) of the JM extract was assessed for the following combinations: PA-CA and GST-cdk2 (tracks 9 and 16), PA-CA and denatured GST-cdk2 (tracks 10 and 17), PA-CA and GST (tracks 11 and 18), PA-CA ΔAcc2 and GST-cdk2 (tracks 12 and 19), PA-CA Δ18 and GST-cdk2 (tracks 13 and 20), PA-CA Δ24 and GST-cdk2 (tracks 14 and 21) and PA-CA ΔC-16 and GST-cdk2 (tracks 15 and 22). Complexes were affinity purified on glutathione Sepharose, denatured, western blotted and probed with an anti-GST antibody (which binds both to PA-CA and GST). Tracks 1 to 5 show a sample of PA-CA and derivatives loaded directly onto the gel and western blotted. The positions of PA-CA, GST-cdk2 and GST are indicated. (b) Binding of PA-CA to GST-cdk2 produces an H1 kinase. The affinity purified material assayed in Fig.3a, tracks 9 to 22, was assayed in parallel for H1 kinase activity (tracks 1 to 14). The indicated PA-CA fusion proteins (about 300 ng) and GST-cdk2 (about 250 ng) were preincubated either alone or in a JM whole cell extract (about 100 μg) for 15 min at 20°C before the addition of 500 μl of 20 mM Tris, pH 8.0, 100 mM NaCl, 1 mM EDTA, 0.5% NP40 and 1% BSA. Complexes were harvested by incubating with glutathione Sepharose for 2 h and then either assayed for H1 kinase activity (Minshull et al., 1990) or denatured and resolved in a 15% SDS polyacrylamide gel. After transferring onto nitrocellulose, fusion proteins were detected with a polyclonal anti-GST antibody. GST-cdk2 was denatured by heating for 5 min. (c) Description of fusion proteins. PA-CA wt contains the C-terminal 355 amino acid residues of bovine cyclin A, Δ2 has residues 281 and 282 deleted and 283 mutated from serine to threonine, Δ18 lacks residues 273 to 290, Δ24 residues 270 to 293 and ΔC-16 the C-terminal 16 residues; shaded area indicates mutated region.

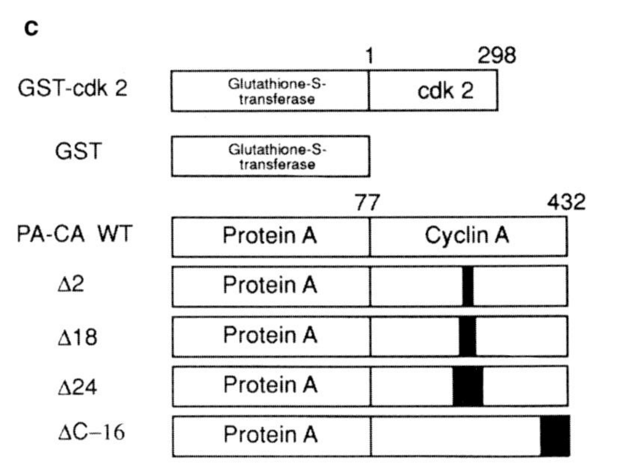

cdk2 more severely than its ability to interact with DRTF1a because the Δ2 mutant cannot bind to GST-cdk2 (Fig.5) but can bind to DRTF1a.

We next assessed if Δ2 can target GST-cdk2 to DRTF1. When both Δ2 and GST-cdk2 are added to the JM extract, the expected cdk2-dependent shift is produced (Fig.6, compare tracks 12 and 13 with 14 and 15, cdk2-dependent shift indicated by ●). Thus, Δ2 can still direct GST-cdk2 to DRTF1. In contrast, but as expected, both Δ18 or ΔC-16 fail to direct GST-cdk2 to DRTF1 (Fig.6, tracks 16 to 19). We conclude from this experiment that mutant Δ2 can still target GST-cdk2 to DRTF1 even though the ability of these two proteins to interact outside the DRTF1 complex is severely compromised (Fig.5). We suggest, therefore, that the stability of the interaction between GST-cdk2 and mutant Δ2 is influenced by DRTF1 or a component within the transcription factor complex.

Discussion

It is known that cyclins regulate the activity of cell cycle kinases and thus participate in controlling the cell cycle (Hunt, 1989). In this study we have defined a new but widely predicted role for cyclin A by demonstrating that it recruits and thus dictates substrate specificity to the catalytic subunit $p33^{cdk2}$. We believe that such a targeting function is likely to be a general principle that enables cyclin-dependent kinase subunits to locate protein substrates during the cell cycle.

Cyclin A directs $p33^{cdk2}$ to DRTF1

The data obtained from the in vitro association assay in which cyclin A targets $p33^{cdk2}$ to DRTF1 clearly lead us to conclude that cyclin A has at least two binding interfaces: one for DRTF1 and the other for $p33^{cdk2}$. The fact that

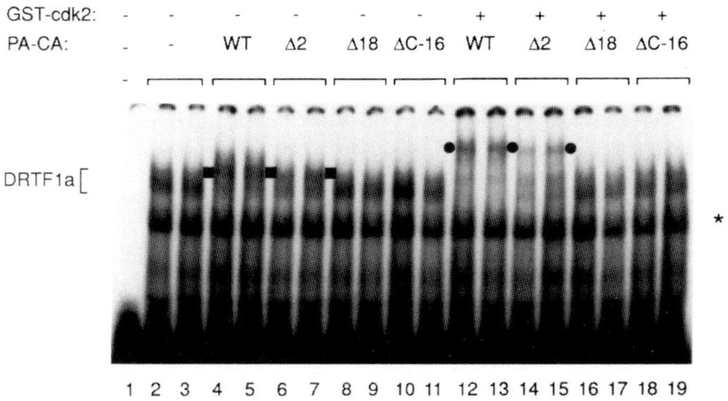

Fig. 6. Evidence that PA-CA directed binding of p33^{cdk2} is influenced by DRTF1. Gel retardation was performed in a JM cell extract either alone (tracks 2 and 3) or after the addition of either wild-type PA-CA (tracks 4, 5, 12 and 13), Δ2 (tracks 6, 7, 14 and 15), Δ18 (tracks 8, 9, 16 and 17) or ΔC-16 (tracks 10, 11, 18 and 19) in the presence (tracks 12 to 19) or absence (tracks 2 to 11) of GST-cdk2. The PA-CA dependent complex is indicated by ■ and note that Δ2 causes the appearence of this complex. Both PA-CA and Δ2 are supershifted upon the addition of GST-cdk2 (indicated by ●). * indicates a non-specific complex.

mutant Δ2, which has amino acid residues 281 and 282 deleted and 283 altered from a serine to a threonine, cannot bind directly to GST-cdk2 (Fig.5a) but still engages with DRTF1a (Fig.6) supports this idea and potentially defines a sequence that distinguishes between these two interaction domains. However, since an increase in the size of the deletion (for example Δ18 which lacks residues 273 to 290) prevents cyclin A from binding to DRTF1, these two binding interfaces may overlap.

Although GST-cdk2 cannot bind directly to Δ2 when it is studied outside the DRTF1 complex it can, nevertheless, be directed and stably bind to the DRTF1 transcription factor complex in an Δ2-dependent fashion (Fig.6). There are at least two possible explanations for this. Perhaps the most likely one is that, in addition to its well characterised binding with cyclin A, p33^{cdk2} interacts with one or more of the components that make up DRTF1 and this serves to stabilise its binding. Alternatively, cyclin A may undergo some modification upon binding to DRTF1 that then enables it to bind more efficiently to p33^{cdk2}. Either of these models would also go someway towards explaining the selective targeting function of cyclin A, namely that cyclin A targets p33^{cdk2} but not p34^{cdc2} to DRTF1. We suggest that this selectivity is also established by an additional interaction that occurs between p33^{cdk2} (but not p34^{cdc2}) and a domain in DRTF1 which may, of course, be the same domain as that involved in enabling the association of p33^{cdk2} with the Δ2-DRTF1 complex.

If indeed such selective targeting and stable association also occurs for p34^{cdc2} substrates (because of analogous interactions between p34^{cdc2} and appropriate substrates), then it would lead us to make the clear prediction that, although both p33^{cdk2} and p34^{cdc2} can bind to cyclin A, they will, nevertheless, stably associate with different, perhaps overlapping, sets of substrates. This is not an altogether surprising suggestion given that p33^{cdk2} and p34^{cdc2} are thought to regulate progression through two distinct parts of the cell cycle, namely S phase and mitosis respectively (Pines and Hunter, 1991b), roles presumably mediated by modulating different substrates. In conclusion, we believe that these data are consistent with a model in which cyclin A provides a general targeting function for catalytic subunits, such as p33^{cdk2} and p34^{cdc2}, that recruits them to a variety of substrates, but that additional interactions between the substrate and catalytic subunit influence whether the catalytic subunit remains stably associated with the substrate.

Cyclin A recruits cdk2 to Rb and p107 complexes

The in vitro assay clearly demonstrates that cyclin A can bind, and recruit p33^{cdk2}, to DRTF1 complexes that contain either the Rb protein or p107. This activity is dependent on the presence of Rb or p107, since cyclin A cannot efficiently bind to free DRTF1. However, in cell extracts, cyclin A is found exclusively in E2F complexes that contain p107 (Cao et al., 1992; Shirodkar et al., 1992). It is clear, however, from other studies that a cdc2-like kinase (containing the conserved PSTAIRE amino acid motif) coprecipitates with the Rb protein in anti-Rb immunoprecipitates (Hu et al., 1992). Furthermore, this Rb-associated kinase contains cyclin A (Hu et al., 1992). We believe, therefore, that the Rb-cyclin A-DRTF1 complex generated in our assays is likely to reflect physiological interactions. The difficulty encountered in detecting this complex in cell extracts by gel retardation may reflect, for example, its short half life in vivo. This seems likely because the active kinase (generated by cyclin A plus p33^{cdk2}) may phosphorylate the Rb protein, and others have reported that the Rb-E2F complex contains exclusively underphosphorylated Rb protein (Chellappan et al., 1991). Our ability to recreate stable complexes in vitro may thus result from physiological interactions that have been ‘frozen’ in vitro. This in vitro assay should prove useful in elucidating the role of this kinase in regulating DRTF1.

A mechanism for coupling cell-cycle events to transcription

The ability of cyclin A to direct p33^{cdk2} to DRTF1 is likely to be an important event in controlling cell-cycle progression because some of the genes that are potentially regulated by DRTF1 are necessary for DNA synthesis (Blake and Azizkhan, 1989; Pearson et al., 1991). Moreover, the Rb protein, which acts in G_1 to prevent cells from entering S phase, represses the transcriptional activity of DRTF1 (Zamanian and La Thangue, 1992) and thus could prevent cells from progressing into S phase by limiting transcription of these gene. Since this is likely to be mediated by un- or under-phosphorylated Rb protein (Buchkovich et al., 1989; DeCaprio et al., 1989) we suggest that the ability of a cyclin to direct a kinase to this complex may regulate the activity of the Rb protein within this transcription factor complex. It is possible, therefore, that the transcriptional activation of DRTF1 is necessary for DNA synthesis to occur. Importantly, we do not wish to imply that p33^{cdk2} is the kinase responsible for regulating the transcriptional activity of DRTF1 but merely, based on the data discussed here, that it is a strong possibility.

Moreover, we cannot rule out that the kinase has a different role from what we have suggested, for example, in modulating the activity of other transcription-related molecules. Whatever the targets for this kinase actually are, it is likely that the ability of cyclin A to target and activate such a kinase is an important event in regulating cell-cycle progression.

We thank Li-Huei Tsai, Matthew Meyerson and Ed Harlow for pGST-cdk2, Bill Kaelin for pGST-Rb, Mark Ewen and David Livingston for a p107 cDNA, Ariel Blocker for construction of cyclin A mutant ΔC-16 and Peter Rigby for many helpful suggestions. J. P. A. acknowledges receipt of a graduate scholarship from the Stiftung Stipendienfonds des VCIeV and L. R. B. is funded by an MRC Collaborative Studentship with Roche Products. N. B. L. T. is a Jenner Fellow of the Lister Institute of Preventive Medicine.

References

Bagchi, S., Raychaudhuri, P. and Nevins, J.R. (1990). Adenovirus E1A proteins can dissociate heteromeric complexes involving the E2F transcription factor: A novel mechanism for E1A trans-activation. *Cell* **62**, 659-669.

Bagchi, S., Weinmann, R. and Raychaudhuri, P. (1991). The retinoblastoma protein copurifies with E2F-1, an E1A-regulated inhibitor of the transcription factor E2F. *Cell* **65**, 1063-1072.

Bandara, L.R. and La Thangue, N.B. (1991). Adenovirus E1a prevents the retinoblastoma gene product from complexing with a cellular transcription factor. *Nature* **351**, 494-497.

Bandara, L.R., Adamczewski, J.P., Hunt, T. and La Thangue, N.B. (1991). Cyclin A and the retinoblastoma gene product complex with a common transcription factor. *Nature* **352**, 249-251.

Blake, M.C. and Azizkhan, J. (1989). Transcription factor E2F is required for efficient expression of hamster dihydrofolate reductase gene in vitro and in vivo. *Mol.Cell.Biol.* **9**, 4994-5002.

Buchkovich, K., Duffy, L.A. and Harlow, E. (1989). The retinoblastoma protein is phosphorylated during specific phases of the cell cycle. *Cell* **58**, 1097-1105.

Cao, L., Faha, B., Dembski, M., Tsai, L.-H., Harlow, E. and Dyson, N. (1992). Independent binding of the retinoblastoma protein and p107 to the transcription factor E2F. *Nature* **355**, 176-179.

Chellappan, S.P., Hiebert, S., Mudryj, M., Horowitz, J.M. and Nevins, J.R. (1991). The E2F transcription factor is a cellular target for the RB protein. *Cell* **65**, 1053-1061.

Chittenden, T., Livingston, D.M. and Kaelin, Jr., W.G. (1991). The T/E1A-binding domain of the retinoblastoma product can interact selectively with a sequence-specific DNA-binding protein. *Cell* **65**, 1073-1082.

DeCaprio, J.A., Ludlow, J.W., Figge, J., Shew, J.-Y., Huang, C.-M., Lee, W.-H., Marsillo, E., Paucha, E. and Livingston, D.M. (1988). SV40 large tumor antigen forms a specific complex with the product of the retinoblastoma susceptibility gene. *Cell* **54**, 275-283.

DeCaprio, J.A., Ludlow, J.W., Lynch, D., Furukawa, Y., Griffin, J., Piwnica-Worms, H., Huang, C-M. and Livingston, D. (1989). The product of the retinoblastoma susceptibility gene has properties of a cell cycle regulatory element. *Cell* **58**, 1085-1095.

Defeo-Jones, D., Huang, P.S., Jones, R.E., Haskell, K.M., Vuocolo, G.A., Hanobik, M.J., Huber, H.E. and Oliff, A. (1991). Cloning of cDNAs for cellular proteins that bind to the retinoblastoma gene product. *Nature* **352**, 152-254.

Devoto, S.H., Mudryj, M., Pines, J., Hunter, T. and Nevins, J.R. (1992). A cyclin A-protein kinase complex possesses sequence-specific DNA binding activity: $p33^{cdk2}$ is a component of the E2F-cyclin A complex. *Cell* **68**, 167-176.

Draetta, G., Luca, F., Westendorf, L., Brizuela, J., Ruderman, J. and Beach, D. (1989). cdc2 protein kinase is complexed with both cyclin A and B: Evidence for proteolytic inactivation of MPF. *Cell* **56**, 829-838.

Dyson, N., Howley, P.M., Munger, K. and Harlow, E. (1989). The human papilloma virus-16 E7 oncoprotein is able to bind to the retinoblastoma gene product. *Science* **243**, 934-936.

Evans, T., Rosenthal, E.T., Youngblom, J., Distel, D. and Hunt, T. (1983). Cyclin: a protein specified by maternal mRNA in sea urchin eggs that is destroyed at each cleavage division. *Cell* **33**, 389-396.

Ewen, M.E., Xing, Y., Lawrence, J.B. and Livingston, D.M. (1991). Molecular cloning, chromosomal mapping and expression of the cDNA for p107, a retinoblastoma gene product-related protein. *Cell* **66**, 1155-1164.

Fang, F. and Newport, J.W. (1991). Evidence that the G_1-S and G_2-M transitions are controlled by different cdc2 proteins in higher eukaryotes. *Cell* **66**, 731-742.

Giordano, A., Whyte, P., Harlow, E., Franza, B.R., Beach, D. and Draetta, G. (1989). A 60Kd cdc2-associated polypeptide complexes with E1A proteins in adenovirus-infected cells. *Cell* **58**, 981-990.

Girard, F., Strausfeld, U., Fernandez, A. and Lamb, N.J.C. (1991). Cyclin A is required for the onset of DNA replication in mammalian fibroblasts. *Cell* **67**, 1169-1179.

Goodrich, D.W., Wang, N.P., Qian, Y-W., Lee, E.Y.-H. and Lee, W-H. (1991). The retinoblastoma gene product regulates progression through the G_1 phase of the cell cycle. *Cell* **67**, 293-302.

Hu, Q., Dyson, N. and Harlow, E. (1990). The regions of the retinoblastoma protein needed for binding to adenovirus E1A or SV40 large T antigen are common sites for mutations. *EMBO J.* **9**, 1147-1155.

Hu, Q., Lees, J.A., Buchkovich, K.J. and Harlow, E. (1992). The retinoblastoma protein physically associates with the human cdc2 kinase. *Mol.Cell.Biol.* **12**, 971-980.

Huang, H.-J.S., Wang, N.-P., Tseng, B.Y., Lee, W.-H. and Lee, E.H.-H.P. (1990). Two distinct and frequently mutated regions of retinoblastoma protein are required for binding to SV40 T antigen. *EMBO J.* **9**, 1815-1822.

Hunt, T. (1989). Maturation promoting factor, cyclin and the control of M-phase. *Curr.Opin.Cell.Biol.* **1**, 268-274.

Koff, A., Cross, F., Fisher, A., Schumacher, J., Leguellec, K., Philippe, M. and Roberts, J.M. (1991). Human cyclin E, a new cyclin that interacts with two members of the cdc2 gene family. *Cell* **66**, 1217-1228.

La Thangue, N.B. and Rigby, P.W.J. (1987). An adenovirus E1A-like transcription factor is regulated during the differentiation of murine embryonal carcinoma stem cells. *Cell* **49**, 507-513.

La Thangue, N.B., Thimmappaya, B. and Rigby, P.W.J. (1990). The embryonal carcinoma stem cell E1a-like activity involves a differentiation-regulated transcription factor. *Nucl. Acids. Res.* **18**, 2920-2938.

Lew, D.J., Dulic, V. and Reed, S. (1991). Isolation of three novel human cyclins by rescue of G_1 cyclin (Cln) function in yeast. *Cell* **66**, 1197-1206.

Matsushime, H., Roussel, M.F., Ashmun, R.A. and Sherr, C.J. (1991). Colony-stimulating factor 1 regulates novel cyclins during the G_1 phase of the cell cycle. *Cell* **65**, 701-713.

Milarski, K.L., Dunphy, W.G., Russell, P., Gould, S.J. and Newport, J.W. (1991). Cloning and expression of *Xenopus* cdc2, a component of MPF. *Cold Spring Harb.Symp.Quant.Biol.* **56**, 377-384.

Minshull, J., Golsteyn, R., Hill, C.S. and Hunt, T. (1990). The A- and B-type cyclin associated cdc2 kinases in *Xenopus* turn on and off at different times in the cell cycle. *EMBO J.* **9**, 2865-2875.

Pearson, B.E., Nasheuer, H.-P. and Wang, T.S.-F. (1991). Human DNA polymerase agene: sequences controlling expression in cycling and serum-stimulated cells. *Mol.Cell.Biol.* **11**, 2081-2095.

Pines, J. and Hunter, T. (1990a). $p34^{cdc2}$. The S and M kinase? *The New Biologist* **2**, 389-401.

Pines, J. and Hunter, T. (1990b). Human cyclin A is adenovirus E1A-associated protein p50 and behaves differently from cyclin B. *Nature* **346**, 760-763.

Pines, J. and Hunter, T. (1991). Cyclin-dependent kinase. A new cell cycle motif? *Trends Cell.Biol.* **1**, 117-121.

Shirodkar, S., Ewen, M., DeCaprio, J.A., Morgan, J., Livingston, D.M. and Chittenden, T. (1992). The transcription factor E2F interacts with the retinoblastoma product and a p107-cyclin A complex in a cell cycle-regulated manner. *Cell* **68**, 157-166.

Shivji, M.K. and La Thangue, N.B. (1991). Multicomponent differentiation-regulated transcription factors in F9 embryonal carcinoma stem cells. *Mol.Cell.Biol.* **11**, 1686-1695.

Smith, D.B. and Johnson, K.S. (1988). Single-step purification of polypeptides expressed in *Escherichia coli* as fusion with glutathione-S-transferase. *Gene* **67**, 31-40.

Tsai, L-H., Harlow, E. and Meyerson, M. (1991). Isolation of the human cdk2 gene that encodes the cyclin A and adenovirus E1A-associated p33 kinase. *Nature* **353**, 174-177.

Whyte, P., Buchovich, K.H., Horowitz, J.M., Friend, S.H., Raybuck, M., Weinberg, R.A. and Harlow, E. (1988). Association between an oncogene and an anti-oncogene: the adenovirus E1A proteins bind to the retinoblastoma gene product. *Nature* **334**, 124-129.

Xiong, Y., Connolly, T., Futcher, B. and Beach, D. (1991). Human D-type cyclin. *Cell* **65**, 691-699.

Zamanian, M. and La Thangue, N.B. (1992). Adenovirus E1A prevents the retinoblastoma gene product from repressing the activity of a cellular transcription factor. *EMBO J.* **11**, 2603-2610.

Journal of Cell Science, Supplement 16, 87-96 (1992)
Printed in Great Britain

SWI6 is a regulatory subunit of two different cell cycle START-dependent transcription factors in *Saccharomyces cerevisiae*

THOMAS MOLL, LÉON DIRICK, HERBERT AUER, JOSH BONKOVSKY and KIM NASMYTH

Institute of Molecular Pathology, Dr. Bohr-Gasse 7, A-1030 Vienna, Austria

Summary

Most genes involved in DNA replication in the yeast *Saccharomyces cerevisiae* are transcribed transiently during late G_1 as cells undergo START. Their promoters all contain one or more versions of an 8-base pair motif (ACGCGTNA) called the *Mlu*I cell cycle box (MCB). MCBs have been shown to be both necessary and sufficient for the late G_1-specific transcription of the *TMP1* thymidylate synthase and *POLI* DNA polymerase genes. A different late G_1-specific transcription element called the SCB (CACGAAAA) is bound by a factor containing the SWI4 and SWI6 proteins. We describe here the formation in vitro of complexes on *TMP1* MCBs that contain the SWI6 protein and, we suggest, a 120 kDa protein that is distinct from SWI4. Transcription due to SCBs and MCBs occurs in the absence of SWI6 but it is no longer correctly cell cycle regulated. We suggest that SWI6 is an essential regulatory subunit of two different START-dependent transcription factors. One factor (SBF) contains SWI4 and binds to SCBs whereas the other (MBF) contains p120 and binds MCBs.

Key words: SWI6, cell cycle, START, transcription, *Saccharomyces cerevisiae*.

Introduction

In many eukaryotic organisms, proliferation is regulated by controls on cells' progression through G_1 phase. In the budding yeast *Saccharomyces cerevisiae*, commitment to the mitotic cell cycle is made at a stage in late G_1 called START (Pringle and Hartwell, 1981). START only takes place if cells have reached a certain size and it is prevented by pheromones that promote conjugation. Gene products required for START include a protein kinase encoded by *CDC28* (Reid and Hartwell, 1977; Lorincz and Reed, 1984) and three G_1-specific cyclin-like proteins encoded by *CLN1*, *CLN2* and *CLN3* (Nash et al., 1988; Richardson et al., 1989). There are indications that an increase in the kinase activity of a complex containing CDC28 and CLN proteins may trigger passage through START (Wittenberg et al., 1990).

Transcriptional controls have an important role at START. *CLN1* and *CLN2* transcripts are absent in early G_1 cells but appear at START, suggesting that CDC28 kinase activity in G_1 may be regulated at least in part by the transcriptional activation of *CLN1* and *CLN2* (Wittenberg et al., 1990). The full activation of *CLN1* and *CLN2* requires the activity of G_1 cyclins and that of the CDC28 protein kinase, implying that it occurs via a positive feedback loop (Cross and Tinkelenberg, 1991; Dirick and Nasmyth, 1991). Transcription of many genes for DNA replication enzymes, such as thymidylate synthase (*TMP1*) and ribonucleotide reductase (*RNR1*), also occurs at START and requires the CDC28 protein kinase (reviewed by Andrews and Herskowitz, 1990).

Two different types of START-dependent *cis*-acting regulatory elements have been identified: one initially identified within the promoter of the cell cycle-regulated *HO* endonuclease (Nasmyth, 1983) whose sequences resemble CACGAAAA (Nasmyth, 1985), and a second found in almost all the cell cycle-regulated genes involved in DNA replication (Pizzagalli et al., 1988) whose consensus ACGCGTNA contains an *Mlu*I restriction site. We refer to these as SCB (swi cell cycle box) and MCB (*Mlu*I cell cycle box) respectively. Both types of element can alone confer START-dependent transcriptional activation on heterologous reporter genes (Breeden and Nasmyth, 1987a; McIntosh et al., 1991; Gordon and Campbell, 1991). A complex containing the SWI4 and SWI6 proteins binds specifically to SCBs found in the *HO* and *CLN2* promoters (Andrews and Herskowitz, 1989; Taba et al., 1991; Nasmyth and Dirick, 1991; Ogas et al., 1991) but it is unclear what factor is responsible for the biological activity of MCBs. Lowndes et al. (1991) have reported the binding of a high molecular weight complex to tandem repeats of ACGCGT (but the same sequence did not give rise to START-dependent transcription in vivo), whereas Verma et al. (1991) have purified a 17 kDa protein that binds to MCBs.

Since both SCBs and MCBs are subject to similar cell cycle control and have some sequence resemblance, we have investigated whether either SWI4 or SWI6 are involved in MCB function. We describe the partial characterization and purification of an activity (named MBF) containing SWI6 and a 120 kDa protein, distinct from SWI4, that binds to both MCBs sites within the *TMP1* promoter.

MBF must have an important role in the biological activity of MCBs in vivo since we find that the transcription of several genes regulated by MCBs is no longer cell cycle regulated in *swi6* mutants. Thus, SWI6 is a common component of distinct complexes that exert START-dependent transcription from MCBs and SCBs.

Results

A factor containing SWI6 binds to MCBs

In the course of studying the role of SWI4 and SWI6 in the activation of G_1 cyclins, we noticed that *swi6* mutations caused subtle but reproducible changes in the pattern of *RNR1* transcription (see below). Since *RNR1* belongs to the group of genes whose promoter contains *Mlu*I sites (MCBs) and is transcribed at START (Elledge and Davis, 1990; Price et al., 1991), we investigated whether SWI4 or SWI6 proteins can bind to MCBs.

The best characterized MCBs are the two required for efficient *TMP1* transcription (McIntosh et al., 1991). We therefore chose as a probe for DNA binding a natural oligonucleotide from the *TMP1* promoter (MCB-TMP1) that contains both these MCBs. Upon mixing crude extracts with MCB-TMP1, we were able to detect the formation of a very slow migrating gel retardation complex that is competed by a fifty-fold molar excess of unlabelled oligonucleotide (*Mlu* 2×) or by oligonucleotides containing multiple SCBs ($CACGA_4$ and sRS2) but not by an unrelated one (Fig. 1A, lanes 7-12). To test whether the formation of this complex requires SWI4 or SWI6, we analysed whether it can also be formed using extracts from *swi4* or *swi6* mutants. We find that the complex can be formed by *swi4* but not by *swi6* mutant extracts (Fig. 1A, lanes 2 and 3). Treatment with SWI6-specific antibodies led to the disappearance of the complex and caused the appearance of a yet slower migrating complex that was barely able to enter the gel (Fig. 1A, lane 5). Neither non-specific nor SWI4-specific antibodies had any effect (Fig. 1A, lanes 4 and 6). We conclude that the SWI6 protein is part of an MCB-binding factor (MBF) but that SWI4 is not.

We have measured the level of MBF activity in a synchronous culture produced by release from α factor-induced G_1 arrest. In this culture (Taba et al., 1991), START-dependent transcription peaks at 75 minutes, bud formation occurs at 80 minutes, and the frequency of anaphase cells peaks at 150 minutes. There do not appear to be any major changes in the level of MBF during this interval (Fig. 1B).

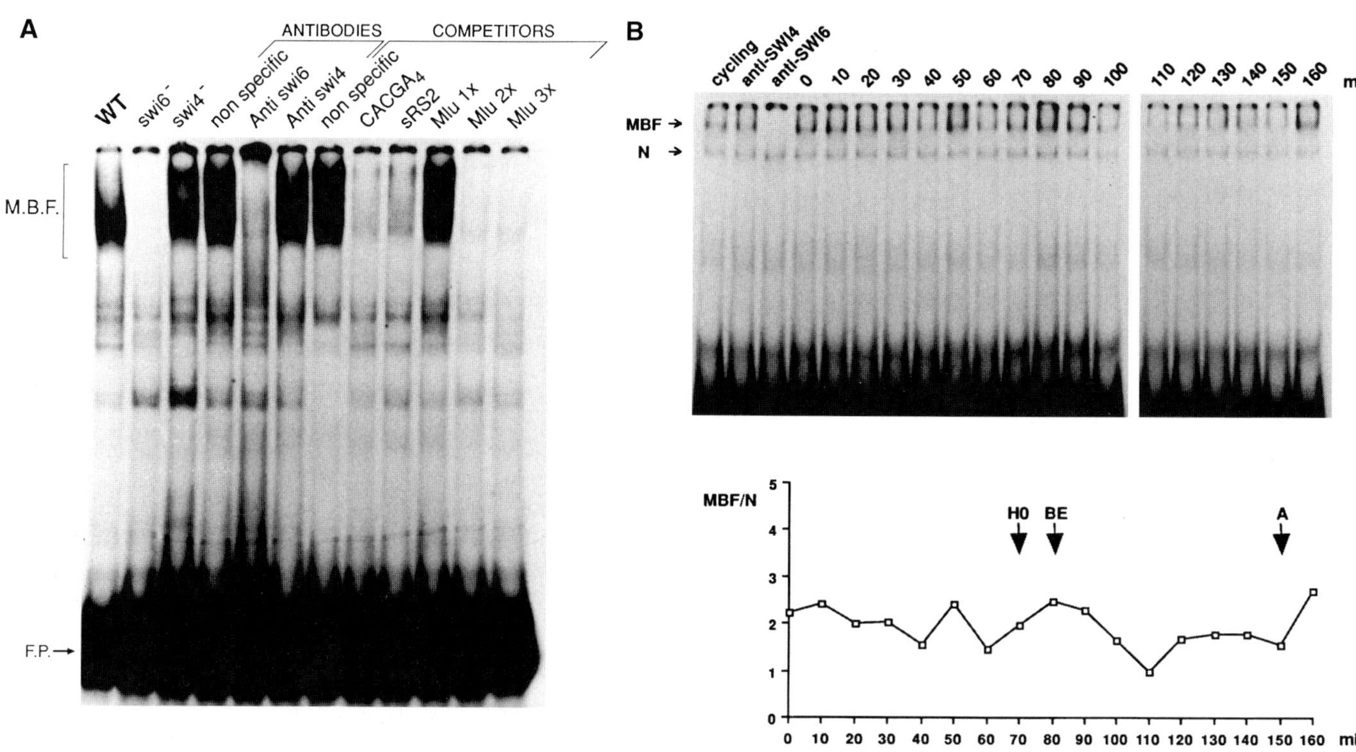

Fig. 1. SWI6 is part of a complex (MBF) that can bind to MCBs at all stages of the cell cycle. (A) Complexes formed on MCB-TMP1 (see Fig. 3) in crude extracts analysed by gel electrophoresis. Whole-cell extracts were prepared from wild-type (K1268), *swi4* mutant (R/H1071) and *swi6* mutant (K1354) strains (Taba et al., 1991). MBF points to the slow migrating complex formed that is present in *swi4* but not in *swi6* mutants (lanes 1-3). Deletion of *SWI4* is lethal for K1268 but such deletions can be obtained if *CLN2* is expressed from the *spADH* promoter. This strain (K2527) also contains wild-type levels of MBF (data not shown). Lanes 4-6 show the effect of adding antibodies (1:20 dilution of sera) directed against DHFR (non specific), SWI6 and SWI4 proteins to extracts from the WT. Lanes 7-12 show competition with cold oligonucleotides (for the sequence of $CACGA_4$ and sRS2, see Taba et al., 1991). Note that oligonucleotides containing several SCBs as well as MCBs can compete. (B) The amount of MBF in α factor-synchronized cells. The graph shows the ratio of MBF to nonspecific (N) complexes in cycling, α factor-arrested, and released cells. In this culture, the START-dependent *HO* gene is fully activated by 70 minutes, buds emerge (BE) by 80 minutes, and the frequency of anaphase cells peaks at 150 minutes (A) (Taba et al., 1991).

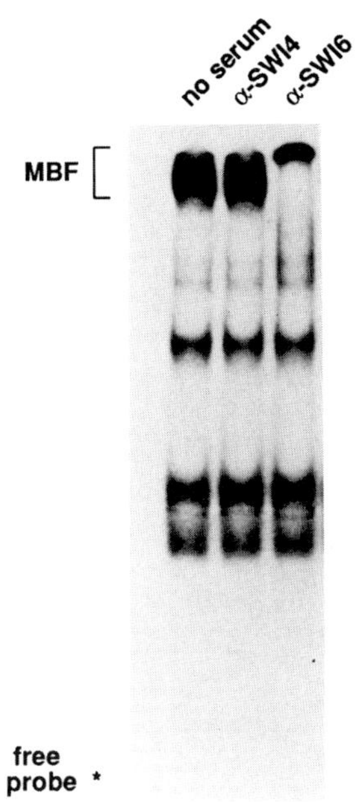

Fig. 2. MBF partially purified by heparin agarose chromatography contains SWI6 but not SWI4. Whole-cell extracts (from a *MAT**a**, prb1, prc1, pep4* strain; K1142) were prepared and fractionated over a heparin-agarose column (Ammerer, 1990). The fractions were then analysed for the formation of gel retardation complexes with the MCB-TMP1 oligonucleotide, as described in Fig. 1. MBF activity elutes from the column at 0.3 M $(NH_4)_2SO_4$ (data not shown). Addition of α-SWI4 (lane 2) and α-SWI6 antibodies (lane 3) to the binding assays confirmed that this complex contains SWI6 but not SWI4.

We also find the same level of MBF in cycling cells compared to a population of unbudded G_1 cells purified by elutriation (data not shown). These data suggest that the level of MBF does not fluctuate noticeably during the cell cycle.

We have partially purified MBF by fractionation on a heparin-agarose column (Ammerer, 1990). An activity that gives rise to a very similar gel retardation complex and contains SWI6 but not SWI4 elutes at 0.3 M ammonium sulphate (Fig. 2). This partially purified form of MBF protects specifically both MCB sites in the *TMP1* promoter from cleavage by copper phrenanthroline (Fig. 3A). We have analysed which G residues when methylated by DMS (Fig. 3B) and which A or G residues when carbethoxylated by diethyl pyrocarbonate (Fig. 3C), interfere with MBF binding. Modification of bases in both MCBs interferes with complex formation, suggesting that MBF binding is cooperative. Consistent with this, oligonucleotides containing only the 5′ MCB cannot compete for binding (Fig. 1A, lane 10) and point mutations in either of the MCBs in MCB-TMP1 abolish complex formation without giving rise to faster migrating complexes in which only a single MCB would be occupied (Fig. 4). The two sites bound are at least two turns of the double helix apart, but this separation is not obligatory since MBF can also bind to MCBs that are separated by only a single turn of the helix (data not shown). The pattern of interference at the two MCBs is similar if we assume that both sites are asymmetric and have opposing orientations (see Fig. 3D). MBF binding requires two A-T base pairs that are outside the *Mlu*I restriction site (positions 3, 4, 8 and 9 in Fig. 3D), as is also found for MCB function in vivo (McIntosh et al., 1991). MBF makes contacts along an extensive stretch of the major groove of each MCB and must therefore bind to both sides of the double helix at each site.

MBF also contains a 120 kDa protein

To determine whether MBF contains proteins other than SWI6, we have analysed which proteins can be cross-linked to BrdU-substituted DNA following UV irradiation. BrdU-substituted ^{32}P-labelled MCB-TMP1 was mixed with partially purified MBF and complexes separated from unbound DNA by gel retardation electrophoresis. The gel was then irradiated, the protein-DNA complexes were digested with DNase I and eluted from the gel, and the ^{32}P-labelled proteins analysed by SDS-PAGE gel electrophoresis. A single protein of approximately 120 kDa (p120) was detected (Fig. 5A1, lane 2), whose recovery is greatly reduced by adding SWI6-specific antibodies to the binding reaction (Fig. 5A1, lane 1). p120 is the wrong size for SWI6, which normally migrates as a 90-95 kDa protein (Taba et al., 1991). To confirm whether or not the cross-linked protein is SWI6, the gel shown in Fig. 5A1 was rehydrated and SWI6 detected by western blotting. SWI6 from a crude extract (Fig. 5A2, lane 3) and from the cross-linked DNase I-treated complexes (Fig. 5A2, lane 2; though barely visible after reproduction) runs alongside the 92.5 kDa marker, well in front of p120. SWI4 is predicted to be 123 kDa but it cannot be p120 because it is not part of MBF (Figs 1A, 2) and because it was shown by western blotting to migrate as a 150 kDa protein in the same gel (data not shown). p120 is also cross-linked to MCBs by UV irradiation when partially purified MBF is mixed with BrdU-substituted MCB-TMP1 in solution in the absence, but not in the presence, of specific competitor (Fig. 5B). In addition, p120 can be co-immunoprecipitated from the reaction with SWI6- but not with SWI4-specific antibodies (Fig. 5C). We therefore suggest that MBF is composed of at least two proteins: SWI6 and a 120 kDa protein that is not SWI4. The detection by silver staining of similar amounts of SWI6 and a 120 kDa protein in highly purified preparations of MBF (data not shown) corroborates the above assertion. We were not able to cross-link SWI6 to the *TMP1* oligonucleotide, suggesting that SWI6 might not be in intimate contact with the DNA.

The effect of swi6 *mutations on transcription from MCBs*

To test whether MBF also binds MCBs in vivo, we have analysed the dependence on *SWI6* of a *LacZ* reporter gene whose transcription is due to the fusion of the MCB-TMP1 oligonucleotide upstream of the *CYC1* TATA box (Guarente and Mason, 1983). We find that the modest activation caused by MCB-TMP1 is noticeably *SWI6*-dependent (Fig. 6). This result tentatively suggests that the MBF:DNA complexes observed by us in vitro also form in vivo and may play a role in transcriptional activation.

It is more relevant, however, to analyse the effect of mutating *SWI6* on the transcription of the *TMP1* gene itself. Using S1 mapping, we find no effect on either the level or the start sites of *TMP1* transcripts (Fig. 7). This is surprising because *TMP1* transcription is completely dependent upon the MCBs within its promoter (McIntosh et al., 1991). In contrast, we do see effects on the *RNR1* gene (Fig. 7), which is transcribed at START and contains MCBs. Mutation of *SWI6* but not of *SWI4* causes an increase in the number of *RNR1* mRNAs that start upstream of the normal cluster of START sites. The latter are somewhat reduced. A similar phenomenon has been observed at the *SWI5* promoter, where mutating either *trans*-acting factors or their binding sites not only reduces correct transcription but also

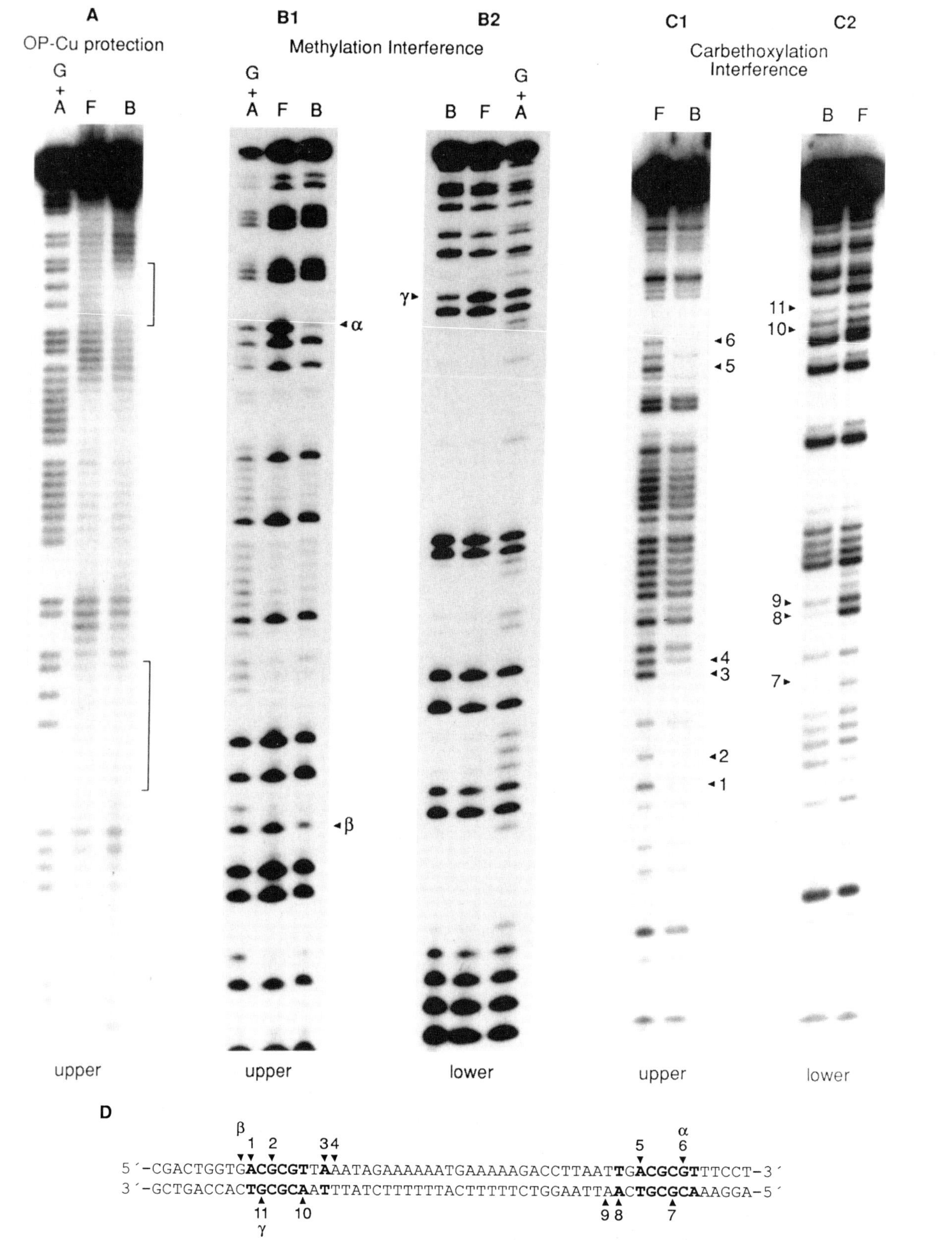

Fig. 3. Footprint and interference analysis of MBF binding to the *TMP1* promoter. (A) Protection from phrenanthroline copper cleavage (Kuwabara and Sigman, 1987); (B) methylation interference (Sturm et al., 1987); (C) carbethoxylation interference (Sturm et al., 1987); (D) summary of interference data and the sequence of MCB-TMP1. Upper and lower refer to the strand analysed and F and B refer to free and bound probe respectively. Base pairs conserved in MCBs are in bold. Note that the two MCBs can be interpreted as having opposite orientations. Greek letters refer to bases whose methylation interferes with binding and numbers refer to bases whose carbethoxylation interferes with binding.

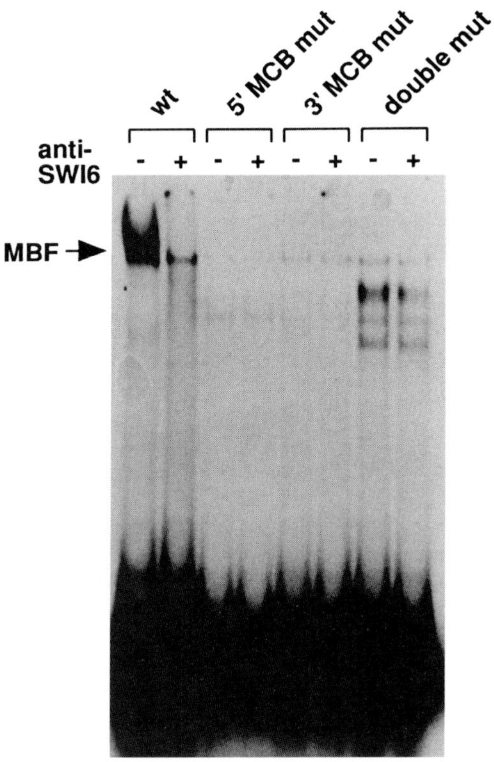

Fig. 4. MBF binding requires both MCBs. Partially purified MBF was mixed with wild-type and mutant oligonucleotide probes in the presence or absence of anti-SWI6 antibodies. The 5′ and 3′ MCBs were mutated from ACGCGTTA to AC<u>T</u>C<u>A</u>TTA and from ACGCGTTT to AC<u>T</u>C<u>A</u>TTT respectively (top strand sequences). Note that both mutations eliminate complexes marked MBF without causing the appearence of faster migrating SWI6-dependent complexes due to the binding of MBF to the remaining single site. Note also that the mutations appear to create a binding site for a new factor.

increases abnormal upstream initiation (Lydall et al., 1991). These data suggest that a SWI6-containing factor (MBF) has a definite role in *RNR1* transcription but leave its role at *TMP1* unresolved.

A regulatory role for SWI6

p120 appears to be the primary DNA-binding moiety of MBF and it may therefore be capable of binding to MCBs without SWI6. Though SWI6 seems necessary for MBF to bind to MCBs in vitro and to activate a *LacZ* reporter gene, it is possible that conditions within nuclei allow p120 to bind without SWI6 in vivo. Thus, *TMP1* may be activated by a SWI6-containing MBF in wild-type cells but by p120 alone in *swi6* mutants. This raises the possibility that, although the overall level in cycling cells is not largely altered, transcription of *TMP1* in the absence of SWI6 might be abnormally regulated.

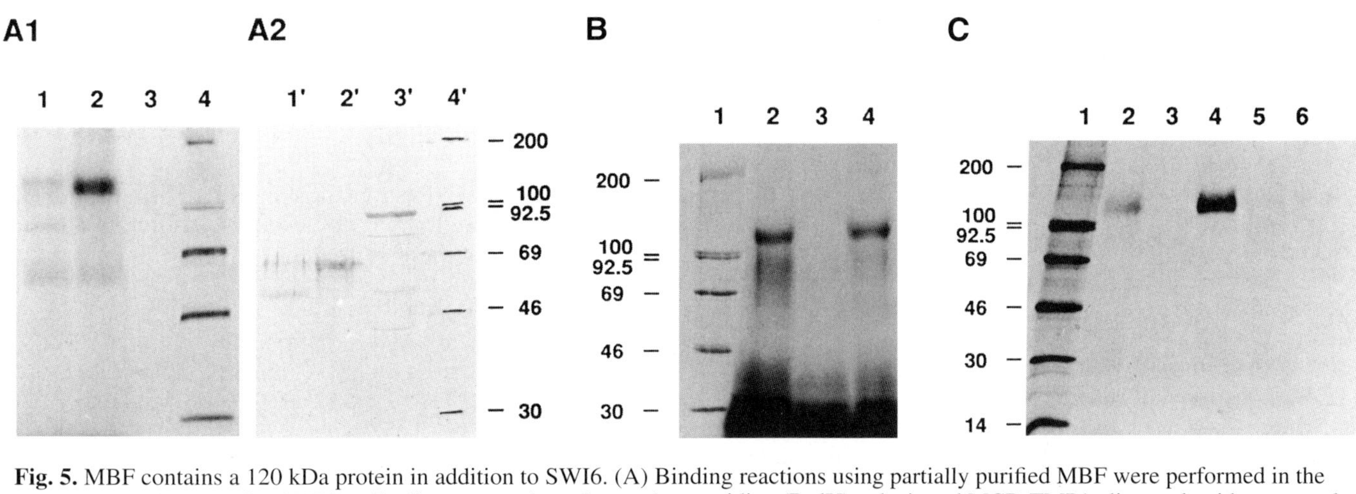

Fig. 5. MBF contains a 120 kDa protein in addition to SWI6. (A) Binding reactions using partially purified MBF were performed in the presence and absence of α-SWI6 antibodies, except that a bromodeoxyuridine (BrdU)-substituted MCB-TMP1 oligonucleotide was used (Barberis et al., 1990). After gel electrophoresis, the protein-DNA complexes were covalently cross-linked by UV irradiation, DNase I-digested, eluted from the gel and covalently bound protein analysed by SDS-PAGE and autoradiography. A single band is detected at approximately 120 kDa (lane 2) that is strongly reduced if α-SWI6 antibodies are present during the binding (lane 1). Yeast whole cell extracts (lane 3) were separated on the same SDS-PAGE gel. An identical UV cross-linking result was obtained using a crude extract (data not shown). The very same gel was rehydrated after autoradiography, proteins were transferred to nitrocellulose and SWI6 was assayed using α-SWI6 antibodies and alkaline phosphatase conjugated to anti-rabbit antibodies. SWI6 is detected not only in the whole cell extract (lane 3′) but also in the cross-linked sample (lane 2′; very faint but visible in original prints). In similar gels, SWI6 from the heparin fraction was shown to co-migrate with that from crude extracts and SWI4 from both the heparin fraction and from crude extracts shown to migrate as a 150 kDa protein (data not shown). The positions of the molecular weight markers (kDa) run in lanes 4 and 4′ are indicated. (B) Binding reaction mixes were directly UV-irradiated, DNase I-digested and analysed by SDS-PAGE and autoradiography. Binding reactions were carried out both in the absence of competitor (lane 2) and in the presence of a 50-fold molar excess of cold specific competitor (MCB-TMP1; lane 3). Addition of a 50-fold molar excess of a cold point-mutated MCB-TMP1 (lane 4; 5′ and 3′ MCBs mutated, see Fig. 4) does not lead to a significant reduction in the amount of p120 detected. The positions of the molecular weight markers (kDa) run in lane 1 are indicated. (C) As in B, except that cross-linked complexes were immunoprecipitated with SWI6-specific (lanes 2-4), SWI4-specific (lane 5) or DHFR-specific (lane 6) antibodies after digestion with DNase I. No competitor (lanes 2, 5 and 6), 50-fold molar excess unlabelled MCB-TMP1 (lane 3) and 50-fold molar excess of unlabelled MCB-TMP1, both of whose MCBs are mutated (lane 4), were added during the binding reactions.

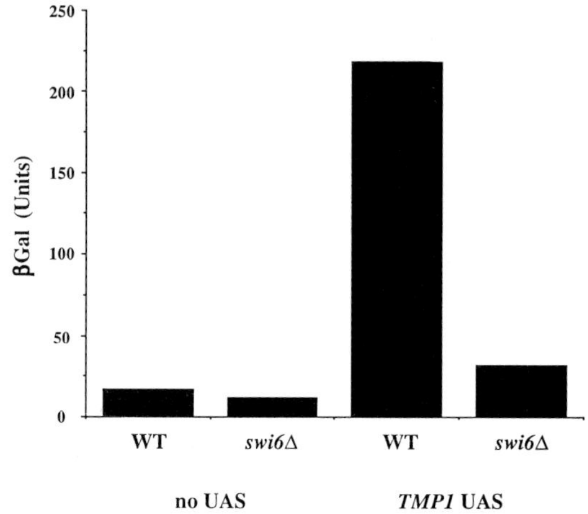

Fig. 6. The levels of β-galactosidase (βGal) activity in wild type (WT) and *swi6*Δ mutant strains containing plasmids expressing *LacZ* either from a UAS-less *CYC1* promoter (pLGΔ178; Guarente and Mason, 1983) or from a derivative in which MCB-TMP1 with *Xho*I sticky ends was cloned in a reverse orientation in the *Xho*I site of pLGΔ178.

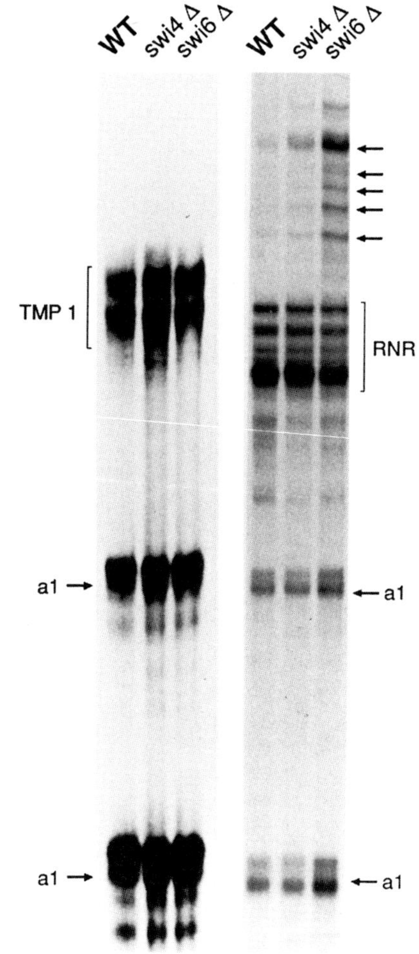

Fig. 7. *TMP1* and *RNR1* RNA levels in *swi4* and *swi6* mutants. *TMP1* and *RNR1* RNAs were measured by an S1 protection assay (Nasmyth, 1983) and quantified using a Molecular Dynamics phospho-imager. The levels of *TMP1* expression (normalized against *MATa1* (a1), which was used as an internal control) were 102% for *swi4*Δ (K1899) and 90% for *swi6*Δ (K1970) relative to wild type (WT) (K1107; 100%). The ratios for "normal" *RNR1* transcripts were 104% for *swi4*Δ and 59% for *swi6*Δ. Strains are as described by Nasmyth and Dirick (1991).

Addressing the regulatory role of SWI6 is complicated by *swi6* mutants being grossly altered in their cell cycle control. They are partially defective in G_1 cyclin function (Nasmyth and Dirick, 1991) and in some aspects of bud formation, cytokinesis or cell separation, with the result that it is difficult to synchronize *swi6* mutants by conventional means. Release from an α factor-induced G_1 arrest is slow and asynchronous and it is difficult to isolate a pure population of unbudded G_1 cells by centrifugal elutriation. We have therefore synchronized cells by directly manipulating G_1 cyclin levels. We have made congenic *SWI6*$^+$ and *swi6*Δ strains whose three G_1 cyclin genes *CLN1, 2* and *3* have been deleted and whose growth is dependent on the conditional expression of *CLN1* from the *GAL1-10* promoter. Removal of galactose from the medium causes such cells to arrest in G_1 and its subsequent re-addition leads, at least in *SWI6*$^+$ cells, to their synchronous passage through START (Cross and Tinkelenberg, 1991). As a measure of synchrony, we have analysed *SWI5* transcription because its S/G_2/M phase-specific transcription is not directly dependent upon *SWI6* (Lydall et al., 1991). *SWI5* transcripts decline upon galactose starvation and transiently increase 40 minutes following its restoration in both *SWI6*$^+$ and *swi6*Δ cells (Fig. 8). Similar results were obtained for *CLB2,* which encodes a B-type cyclin that is expressed at the same stages of the cell cycle as *SWI5* (Surana et al., 1991; Ghiara et al., 1991) (Fig. 8). Histone H2B RNAs which appear somewhat earlier are also cell cycle regulated in both strains (data not shown). By these criteria, then, the synchrony of *swi6* mutants is almost, but not quite, as good as wild type.

In contrast to the above, mutation of *SWI6* has a dramatic effect on the behaviour of genes whose transcription is dependent upon MCBs. The sharp periodicity of *TMP1* RNAs seen in the wild type is abolished in a *swi6* mutant, which has neither a drop following removal of galactose nor any increase following its restoration (Fig. 8). A similar result is seen for the *RNR1* gene whose normal (Fig. 8) and abnormal transcripts (data not shown) are de-regulated. It appears therefore that the cell cycle regulation of several MCB-activated genes is dependent upon *SWI6.*

SWI6 and SCB regulation

SWI6 participates not only in complexes that bind MCBs, but also in complexes with SWI4 that bind SCBs (Andrews and Herskowitz, 1989; Taba et al., 1991). Might the cell cycle periodicity of SCB-driven transcription also depend on SWI6? The *CLN2* promoter contains SCBs and its transcripts are severely reduced in *swi4* mutants and more modestly reduced in *swi6* mutants (Nasmyth and Dirick, 1991). To assess the dependence on *SWI4* of *CLN2* transcription in *swi6* mutants, we have measured the level of *CLN2* transcripts in wild type and *swi4* and *swi6* mutant strains that express *CLN2* from the *S. pombe ADH* promoter (*spADH-CLN2*). This promoter/gene fusion has no effect on the level of endogenous *CLN2* mRNAs (data not shown) and it helps the growth of *swi* mutants, rescuing lethality in the case of *swi4;swi6* double mutants and ameliorating the growth defects of both single mutants. Fig. 9 shows that *swi4;swi6* double mutants have much less endogenous *CLN2* RNA than either single mutants (RNAs from the exogenous and endogenous *CLN2* genes can be distinguished by S1 mapping). Thus, SWI4 is responsible for most of the *CLN2* transcription seen in *swi6* mutants. Fig. 8 shows that the cell cycle periodicity of *CLN2* RNAs is reduced but not entirely

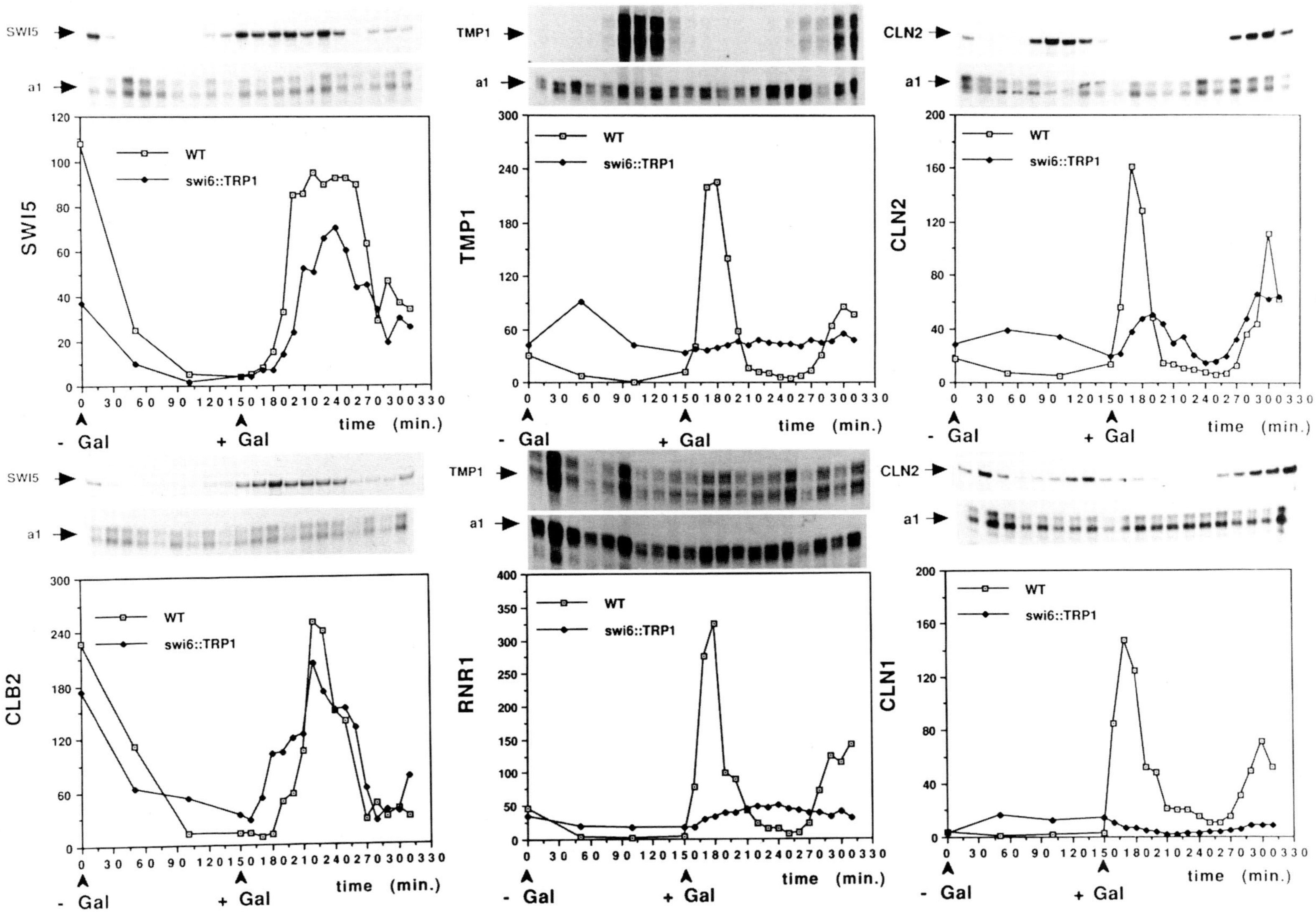

Fig. 8. The effect of mutating *SWI6* on the cell cycle periodicity of genes containing MCBs and SCBs. Congenic *SWI6*$^{+}$ and *swi6*Δ strains whose passage through START is dependent upon expression of *CLN1* from the *GAL1-10* promoter were grown in YEP medium containing 2% raffinose and 2% galactose at 30°C. At time=0, both cultures were harvested by filtration, washed in YEP raff medium and re-inoculated in YEP raff medium (–Gal). After 150 min, galactose was re-added to the medium (+Gal). *SWI5*, *TMP1*, *CLN2*, *RNR1* and *CLN1* RNAs were measured by S1 protection (Nasmyth, 1983), whereas *CLB2* RNAs were measured by northern blotting. All RNAs were quantified using a Molecular Dynamics phospho-imager and normalized using a *MATa1* internal control. The ratio of these RNAs to *MATa1* multiplied by 100 is plotted. Autoradiographs of S1-protected fragments due to *SWI5*, *TMP1* and *CLN2* RNAs are shown above (WT; K2771) and below (*swi6*Δ) the relevant graphs. K2771 (*cln1*Δ, *cln2*Δ, *cln3*Δ, YCpGAL-CLN1; congenic with K1107) was a meiotic segregant from the diploid K2342 (Nasmyth and Dirick, 1991) transformed with YCpGAL-CLN1 (Cross and Tinkelenberg, 1991). The *swi6::TRP1* derivative (K2786) was made by transformation. Both strains grow only in the presence of galactose.

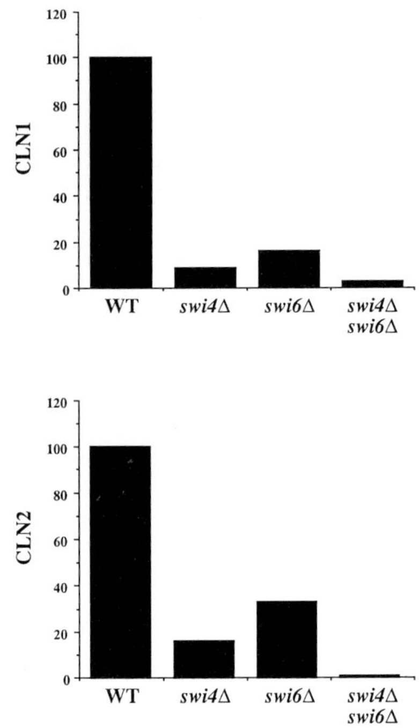

Fig. 9. The levels of *CLN1* and *CLN2* RNAs as measured by S1 mapping in wild-type (WT; K2832), *swi4Δ* (K2833), *swi6Δ* (K2831) and *swi4Δ;swi6Δ* (K2392; Nasmyth and Dirick, 1991) mutants carrying two copies of *spADH-CLN2*. The *spADH-CLN2* fusion has little or no effect on the level of endogenous *CLN2* transcripts. The graph plots the ratio of endogenous *CLN1* and *CLN2* RNAs relative to the unregulated *spADH-CLN2* RNA.

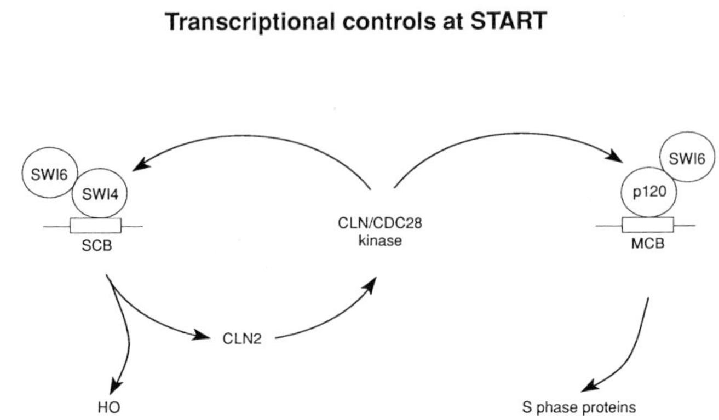

Fig. 10. SWI6 has a central role in modulating START dependent transcription. SWI6 participates in the function of two different START-dependent promoter elements. It forms one type of complex with SWI4 on SCBs found in the *HO* and *CLN2* promoters and another type of complex with a 120 kDa protein on MCBs within the *TMP1* promoter. The activity of both types of complex are dependent upon the CDC28 kinase.

absent in *swi6* cells, suggesting that SWI6 is required for regulating SWI4 activity. Cell cycle-regulation of SWI4-dependent (Fig. 9) *CLN1* transcription is also dependent on *SWI6* (Fig. 8).

Discussion

Most genes involved in DNA replication in yeast are transcribed transiently during late G_1 as cells undergo START. Their promoters all contain one or more versions of an 8-base pair motif (ACGCGTNA) called the *Mlu*I cell cycle box (MCB; Pizzagalli et al., 1988). MCBs have been shown to be both necessary and sufficient for the late G_1-specific transcription of the *TMP1* and *POL1* genes (McIntosh et al., 1991; Gordon and Campbell, 1991). A different late G_1-specific transcription element called the SCB (CACGAAAA; Nasmyth, 1985) is bound by a complex containing the SWI4 and SWI6 proteins (Andrews and Herskowitz, 1989; Taba et al., 1991). We describe here the formation in vitro of complexes on MCBs within a *TMP1* promoter fragment. These complexes contain the SWI6 protein and a 120 kDa protein (p120) that is distinct from SWI4. The 17 kDa protein claimed to bind MCBs (Verma et al., 1991) could be either a proteolytic fragment of p120 or an unrelated protein. Lowndes et al. (1991) have described a factor that binds to an artificial oligonucleotide composed of three tandem *Mlu*I restriction sites. This oligonucleotide did not however contain all the sequences sufficient for the cell cycle regulation characteristic of MCB-containing promoters. The complexes formed on it might nevertheless be the same that we have characterized because they have similar electrophoretic properties.

Three lines of evidence suggest that SWI6/p120 complexes also bind MCBs in vivo and are responsible for their transcriptional regulation. First, bases whose methylation or carbethoxylation interfers with SWI6/p120-binding in vitro all lie within the minimal functional MCB (McIntosh et al., 1991). Second, *SWI6* is required in vivo for the transcriptional activation by *TMP1* MCBs of a *LacZ* reporter gene. Third and most important, transcription of endogenous genes whose promoters contain MCBs is in some cases modestly reduced and in all cases de-regulated in *swi6* mutants. Our demonstration that SWI6/p120 complexes bind cooperatively in vitro to the two *TMP1* MCBs is consistent with the conclusion of McIntosh et al. (1991) that mutation of either MCB interferes with *TMP1* transcription in vivo.

SWI6 may play a central role in late G_1 or START-dependent transcription in yeast because it is the common component of the two different START-dependent transcription factors characterized (Fig. 10). SWI6 forms complexes with SWI4 on SCBs, but with p120 on MCBs. Since SWI4 alone can bind to SCBs in vitro (Primig et al., 1992) and since p120 but not SWI6 can be cross-linked to MCBs, we propose that the slightly different DNA sequence specificities of the two types of SWI6-containing complexes are due to the site-specific DNA-binding domains of SWI4 and p120. These domains may prove to be related because the SCB and MCB are similar sequences and can cross-compete in binding assays (Fig. 1A). SWI6 appears to help SWI4 and p120 in binding SCBs and MCBs in vitro, but this function may be critical in vivo only for some promoters such as *HO* (Breeden and Nasmyth, 1987a), *CLN1* (Fig. 9) and the artificial MCB-TMP1/CYC-LacZ promoter fusion (Fig. 6). To explain why the level of *TMP1* transcripts in cycling cells is unaffected by the deletion of *SWI6* whereas extracts from *swi6* mutants fail to form any specific complexes on *TMP1* MCBs in vitro, we propose that p120 alone may be capable of binding to MCBs, but with a reduced affinity. Thus, it may bind and activate transcription when present at high concentration within intact nuclei but can no longer bind to MCBs in vitro when diluted in extracts. Such a phenomenon would be analogous to the activation of *CLN2* by SWI4 in the absence of SWI6.

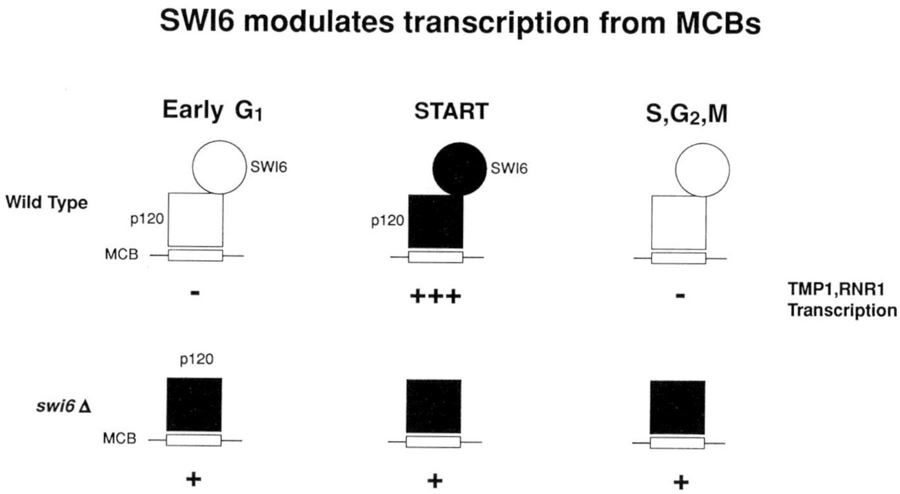

Fig. 11. SWI6 is not essential for transcription due to MCBs but is necessary for its correct regulation. We propose that p120 can bind MCBs without SWI6 in vivo but that it activates transcription at a low level constitutively. SWI6 is required for cell cycle modulation of MCB-driven transcription.

Whereas SWI6 may be dispensable for the binding of p120 and SWI4 to many promoters, it seems to fulfill an essential role in regulating transcription exerted by them. Genes that contain either MCBs (*TMP1*, *RNR1*) or SCBs (*CLN2*) are no longer properly cell cycle regulated in *swi6* mutants. The former become totally aperiodic but the latter merely less periodic. *CLN2* transcription is only modestly reduced in *swi6* mutants but drastically reduced in *swi4;swi6* double mutants, suggesting that SWI6 is required to regulate transcriptional activation by SWI4. We propose by analogy that SWI6 performs the same function for p120 (Fig. 11). This SWI6 regulation must involve not only the repression of transcription at stages of the cell cycle when MCBs and SCBs are normally inactive, but also the activation of transcription during late G_1. SWI6 is somewhat analogous to the GAL80 protein which binds to the GAL4 transcription factor and confers galactose dependence (Lue et al., 1987). Neither GAL80 nor SWI6 alone appear to be capable of binding DNA specifically. An important difference is that GAL80 has a negative role only, whereas SWI6 acts as both repressor and activator. We suggest therefore that it be called a modulator, a name that seems apt for a factor that imposes periodicity on transcription. The tumour-suppressing retinoblastoma gene in humans may perform an analogous function in modulating transcription due to the E2F transcription factor (Devoto et al., 1992).

It is likely that cell cycle changes in the factors that bind SCBs and MCBs are responsible for their cell cycle regulation. The aperiodicity of *swi6* mutants suggests that cell cycle changes in the SWI6 protein could impose cell cycle control. An obvious possibility is that phosphorylation of SWI6 by the CDC28 (or another) kinase transforms it from a repressor to an activator, around the time of START. It is also possible that the changes triggering transcription at START occur on other components of the complexes (eg. SWI4 and p120) and that SWI6 is required to prevent these proteins adopting an active conformation in the absence of such changes. Our detection of SWI6/p120-binding activity throughout the cell cycle suggests that changes in an "activation" function rather than binding per se may trigger MCB transcription at START; a hypothesis that is consistent with our proposal that p120 activates *TMP1* constitutively in the absence of SWI6. Lowndes et al. (1991) have claimed that variations in an *Mlu*I-binding factor are responsible for cell cycle-regulated MCB transcription. Neither our nor their data warrant this conclusion. In vivo footprinting will nevertheless be required to determine whether SWI6/p120 complexes are bound to MCBs throughout the cell cycle.

Further analysis of the phenotype of *swi6* mutants may provide insight into the importance of regulated MCB function. *swi6* mutant cells are large and misshapen even when G_1 cyclins are expressed ectopically (data not shown). This phenotype could stem from de-regulation of MCB transcriptional activation. The proposal (Nasmyth and Dirick, 1991) that the cdc10 protein from *Schizosaccharomyces pombe*, whose C-terminal half is very similar to that of SWI6 (Breeden and Nasmyth, 1987b), may perform similar functions, has recently been confirmed by the observation that cdc10 is a component of an MCB-binding factor (Lowndes et al., 1992). There is, however, at least one crucial difference between the functions of cdc10 and SWI6. cdc10 is an essential gene whereas SWI6 is not. cdc10 could have a more crucial role in DNA-binding than SWI6 and may therefore be indispensable for the transcription of an essential cell cycle gene.

Transcriptional activation by SWI4/SWI6 complexes binding to SCBs appears essential for the accumulation of unstable G_1 cyclins and thereby for further cell cycle progression. In contrast, most of the genes regulated by MCBs in yeast encode stable proteins that do not need to be synthesised immediately prior to S phase (Byers and Sowder, 1980). There may nevertheless be some MCB-regulated genes whose expression in late G_1 is required for subsequent S phase entry. Transcriptional periodicity confered by SWI6 could prove to have an important role in coordinating cell cycle events.

We would like to thank Fred Cross for the YCpGAL-CLN1 plasmid, Gustav Ammerer and Peter Sorger for heparin agarose-purified fractions, Rita Taba for yeast extracts, Hannes Tkadletz for preparing the figures, Robert Kurzbauer and Gotthold Schaffner for oligonucleotide synthesis and DNA sequencing, and Beverley Errede, Christian Koch and Meinrad Busslinger for their comments on the manuscript.

References

Ammerer, G. (1990). Identification, purification and cloning of a polypeptide (PRTF/GRM) that binds to mating-specific promoter elements in yeast. *Genes Dev.* **4**, 299-312.

Andrews, B. J. and Herskowitz, I. (1989). The yeast SWI4 protein contains a motif present in developmental regulators and is part of a complex involved in cell-cycle dependent transcription. *Nature* **342**, 830-833.

Andrews, B. J. and Herskowitz, I. (1990). Regulation of cell-cycle dependent gene expression in yeast. *J. Biol. Chem.* **265**, 14057-14060.

Barberis, A., Widenhorn, K., Vitelli, K. and Busslinger, M. (1990). A novel B-cell lineage-specific transcription factor present at early but not late stages of differentiation. *Genes Dev.* **4**, 849-859.

Breeden, L. and Nasmyth, K. (1987a). Cell cycle control of the yeast *HO* gene: *cis-* and *trans*-acting regulators. *Cell* **48**, 389-397.

Breeden, L. and Nasmyth, K. (1987b). Similarity between cell-cycle genes of budding yeast and fission yeast and the *Notch* gene of *Drosophila*. *Nature* **329**, 651-654.

Byers, B. and Sowder, L. (1980). Gene expression in the yeast cell cycle. *J. Cell Biol.* **87**, 6a.

Cross, F. and Tinkelenberg, A. H. (1991). A potential positive feedback loop controlling CLN1 and CLN2 gene expression at the start of the yeast cell cycle. *Cell* **65**, 875-883.

Devoto, S. M., Mudryj, M., Pines, J., Hunter, T. and Nevins, J. R. (1992). A cyclin A-protein kinase complex possesses sequence specific DNA binding activity: $p33^{cdk2}$ is a component of the E2F-cyclin A complex. *Cell* **66**, 167-176.

Dirick, L. and Nasmyth, K. (1991). Positive feedback in the activation of G_1 cyclins in yeast. *Nature* **351**, 754-757.

Elledge, S. J. and Davis, R. W. (1990). Two genes differentially regulated in the cell cycle and by DNA damaging agents encode alternative regulatory subunits of ribonucleotide reductase. *Genes Dev.* **4**, 740-751.

Ghiara, J. B., Richardson, H. E., Sugimoto, K., Henze, M., Lew, D. J., Wittenberg, C. and Reed, S. I. (1991). A cyclin B homolog in *S. cerevisiae*: chronic activation of the CDC28 protein kinase by cyclin prevents exit from mitosis. *Cell* **65**, 163-174.

Gordon, C. B. and Campbell, J. L. (1991). A cell cycle-responsive transcriptional control element and a negative control element in the gene encoding DNA polymerase apha in *S. cerevisiae*. *Proc. Natl. Acad. Sci. USA* **88**, 6058-6062.

Guarente, L. and Mason, T. (1983). Heme regulates transcription of the *CYC-1* gene of *S. cerevisiae* via an upstream activation site. *Cell* **32**, 1279-1286.

Kuwabara, M. D. and Sigman, D. S. (1987). Foot-printing DNA protein complexes in situ following gel retardation assays using 1, 10-phenanthroline-copper ion: *Escherichia coli* RNA polymerase - *lac* promoter complexes. *Biochemistry* **26**, 7234-7238.

Laemmli, U. K. (1970). Cleavage of structural proteins during the assembly of the head of bacteriophage T4. *Nature* **227**, 680-685.

Lorincz, A. T. and Reed, S. I. (1984). Primary structure homology between the product of yeast division control gene CDC28 and vertebrate oncogenes. *Nature* **307**, 183-185.

Lowndes, N. F., Johnson, A. L. and Johnston, L. H. (1991). Coordination of expression of DNA synthesis genes in budding yeast by a cell-cycle regulated *trans* factor. *Nature* **350**, 247-250.

Lowndes, N. F., McInerny, C. J., Johnson, A. L., Fantes, P. A. and Johnston, L. H. (1992). Control of DNA synthesis genes in fission yeast by the cell-cycle gene *cdc10*+. *Nature* **355**, 449-453

Lue, M. F., Chasman, D. I., Buchman, A. R. and Kornberg, R. D. (1987). Interaction of *GAL4* and *GAL80* gene regulatory proteins in vitro. *Mol. Cell Biol.* **7**, 3446-3451.

Lydall, D., Ammerer, G. and Nasmyth, K. (1991). A new role for MCM1 in yeast: cell cycle regulation of *SWI5* transcription. *Genes Dev.* **5**, 2405-2419.

Maxam, A. M. and Gilbert, W. (1980). Sequencing end-labeled DNA with base-specific chemical cleavages. *Methods Enzymol.* **65**, 499-560.

McIntosh, E. M., Atkinson, T., Storms, R. K. and Smith, M. (1991). Characterization of a short, cis-acting DNA sequence which conveys cell cycle stage-dependent transcription in *S. cerevisiae*. *Mol. Cell Biol.* **11**, 329-337.

Nash, R., Tokowa, G., Anand, S., Erickson, K. and Futcher, A. B. (1988). The WHI+ gene of *Saccharomyces cerevisiae* tethers cell division to cell size and is a cyclin homolog. *EMBO J.* **7**, 4335-4346.

Nasmyth, K. (1983). Molecular analysis of a cell lineage. *Nature* **302**, 670-676.

Nasmyth, K. (1985). A repetitive DNA sequence that confers cell-cycle *START (CDC28)*-dependent transcription of the *HO* gene in yeast. *Cell* **42**, 225-235.

Nasmyth, K. and Dirick, L. (1991). The role of *SWI4* and *SWI6* in the activity of G_1 cyclins in yeast. *Cell* **66**, 995-1013.

Ogas, J., Andrews, B. J. and Herskowitz, I. (1991). Transcriptional activation of CLN1, CLN2 and a putative new G_1 cyclin (HCS26) by SWI4, a positive regulator of G_1-specific transcription. *Cell* **66**, 1015-1026.

Pizzagalli, A., Valsasnini, P., Plevani, P. and Lucchini, G. (1988). DNA polymerase I gene of *S. cerevisiae*: nucleotide sequence, mapping of a temperature-sensitive mutation and protein homology with other DNA polymerases. *Proc. Natl. Acad. Sci. USA* **85**, 3772-3776.

Price, C., Nasmyth K. and Schuster T. (1991). A general approach to the isolation of cell cyle-regulated genes in the budding yeast, *Saccharomyces cerevisiae*. *J. Mol. Biol.* **218**, 543-556.

Primig, M., Sockanathan, S., Auer, H. and Nasmyth, K. (1992). Anatomy of a transcription factor important for the START of the cell cycle in *Saccharomyces cerevisiae*. *Nature* **358**, 593-596.

Pringle, J. R. and Hartwell, L. H. (1981). The *Saccharomyces cerevisiae* cell cycle. In *The Molecular Biology of the Yeast Saccharomyces: Life Cycle and Inheritance*, (ed. J. N. Strathern, E. W. Jones and J. R. Broach) pp. 97-142. Cold Spring Harbor.

Reid, B. and Hartwell, L. H. (1977). Regulation of mating in the cell cycle of *S. cerevisiae*. *J. Cell Biol.* **75**, 355-365.

Richardson, H. E., Wittenberg, C., Cross, F. and Reed, S. I. (1989). An essential G_1 function for cyclin-like proteins in yeast. *Cell* **59**, 1127-1133.

Sturm, R., Baumruker, T., Franua, B. R. and Herr, W. (1987) A 100-kD HeLa cell octamer binding protein (OBP100) interacts differently with two separate octamer-related sequences within the SV40 enhancer. *Genes Dev.* **1**, 1147-1160.

Surana, U., Robitsch, H., Price, C., Schuster, T., Fitch, I., Futcher, A. B. and Nasmyth, K. (1991). The role of CDC28 and cyclins during mitosis in the budding yeast *S. cerevisiae*. *Cell* **65**, 145-161.

Taba, M. R., Muroff, I., Lydall, D., Tebb, G. and Nasmyth, K. (1991). Changes in a SWI4, 6-DNA-binding complex occur at the time of *HO* gene activation in yeast. *Genes Dev.* **5**, 2000-2013.

Verma, R., Patapoutian, A., Gordon, C. B. and Campbell, J. L. (1991). Identification and purification of a factor that binds to the *Mlu*I cell cycle box of yeast DNA replication genes. *Proc. Natl. Acad. Sci. USA* **88**, 7155-7159.

Wittenberg, C., Sugimoto, K. and Reed, S. I. (1990). G_1-specific cyclins of *S. cerevisiae*: cell cycle periodicity, regulation by mating pheromone and association with the $p34^{CDC28}$ protein kinase. *Cell* **62**, 225-237.

Journal of Cell Science, Supplement 16, 97-109 (1992)
Printed in Great Britain © The Company of Biologists Limited 1992

Identification of Fos target genes by the use of selective induction systems

SYLVIA BRASELMANN, GABRIELE BERGERS, CHRISTOPHER WRIGHTON[1], PAULA GRANINGER, GIULIO SUPERTI-FURGA[2] and MEINRAD BUSSLINGER*

Institute of Molecular Pathology, Dr Bohr-Gasse 7, 1030 Vienna, Austria

Present addresses: [1]Laboratory of Transplantation Immunology, Vienna International Research Cooperation Center, Brunner Strasse 59, A-1235 Vienna, Austria and [2]European Molecular Biology Laboratory, Meyerhofstrasse 1, D-6900 Heidelberg, Germany
*Author for correspondence

Summary

c-Fos is a major component of the transcription factor AP-1 which has been implicated in the control of cell proliferation and differentiation as well as in transformation. In order to identify Fos target genes involved in these processes, we have taken advantage of the regulatory properties of the hormone-binding domain of the human estrogen receptor to develop transcriptional and post-translational induction systems, both of which allow selective elevation of Fos activity within a cell. Using this approach we have searched for Fos-responsive genes in rat fibroblasts and PC12 cells. Here we describe the identification and regulation of five Fos-responsive genes encoding a transcription factor (Fra-1), a secreted protein (Fit-1), a biosynthetic enzyme (ODC) and two membrane-associated proteins (annexin II and V), respectively. The post-translational induction system was also used to study the Fos-mediated block of neuronal differentiation of PC12 cells. These experiments demonstrate that Fos activity is dominant over NGF function and interferes with the expression of late NGF-inducible genes.

Key words: selective Fos induction systems, AP1-dependent transcription, block of neuronal differentiation, Fos target genes, Fit-1, ODC, annexin, fra-1.

Introduction

The c-*fos* gene codes for an important component of the transcription factor AP-1, which consists of three members of the Jun family (c-Jun, JunB and JunD) and four members of the Fos family (c-Fos, FosB, Fra-1 and Fra-2). These proteins share a stretch of uninterrupted homology containing the "leucine zipper", which mediates dimerization, and an adjacent basic region, which constitutes the DNA-binding domain. Jun proteins are able to homodimerize, while Fos proteins are only capable of forming heterodimeric complexes with Jun proteins. In the AP-1 complex, the basic regions of both partner proteins are brought into close proximity to form a bipartite DNA-binding domain which recognizes the consensus sequence TGA G/C TCA. The Fos-Jun heterodimers are thermodynamically more stable than Jun homodimers. The presence of Fos protein therefore forces Jun into heterodimeric complexes which bind to DNA more effectively. Jun and Fos both contribute to the transactivation function of the AP-1 complex, which stimulates transcription by binding to promoter sequences (reviewed by Angel and Karin, 1991; Kerppola and Curran, 1991).

c-Fos, as part of the transcription factor AP-1, is thought to play a central role in signal transduction by coupling short-term stimulation of the cell to long-term alterations in gene expression. The c-*fos* mRNA level is low or undetectable in most unstimulated cells, in contrast to the relatively high basal level of *jun* gene transcripts, indicating that c-Fos is a limiting component of AP-1 activity under these conditions. Transcription of the c-*fos* gene is, however, rapidly and transiently induced in many different cell types by a large variety of signals involving protein kinase A- and C-dependent pathways. The overall regulation of c-*fos* expression is complex, being tightly controlled at the level of transcription initiation and elongation as well as mRNA and protein stability. Moreover, the c-Fos protein is part of a negative autoregulatory loop, as it represses transcription from its own promoter by inactivating the serum response element. Intact C-terminal sequences of c-Fos are required for this trans-repression activity, while the DNA-binding and dimerization domains are dispensable (reviewed by Rivera and Greenberg, 1990).

Fos/AP-1 has been implicated in the control of cell proliferation by the following arguments. Firstly, the *fos* gene was originally identified as the oncogenic principle of the

FBJ and FBR murine sarcoma viruses which cause osteosarcoma tumours in mice (Finkel et al., 1966; Curran et al., 1984). It was then later shown that deregulated expression of the c-*fos* gene also leads to transformation of rat fibroblasts in vitro (Miller et al., 1984) and to bone and cartilage tumours in transgenic mice in vivo (Rüther et al., 1987; Wang et al., 1991). Secondly, the expression of the c-*fos* gene is rapidly increased in response to mitogens in many cell types. Thirdly, AP-1 activity correlates with cell growth, as indicated by higher levels of AP-1 activity in cycling compared with serum-starved cells (Angel et al., 1987; Piette et al., 1988). Fourthly, microinjection of anti-Fos antibodies and expression of *fos* antisense RNA prevent DNA synthesis in tissue culture cells (Holt et al., 1986; Nishikura et al., 1987; Riabowol et al., 1988, 1992). Fifthly, ligand-activated steroid receptors which are known to antagonize cell proliferation also interfere with AP-1 activity (Jonat et al., 1990). Other observations, however, argue against an exclusive role of Fos/AP-1 in cellular proliferation. For instance, TPA-induced differentiation of the human myeloid U937 cells to macrophages is accompanied both by cessation of cell division and by a large increase in c-*fos* and c-*jun* expression (Williams et al., 1990). Fos/AP-1 is in addition involved in the autogenous induction of TGFβ-1, which is a potent growth inhibitor for several cell types (Kim et al., 1990). Furthermore, a single point mutation in the intracellular domain of the CSF-1 receptor prevents mitogenesis without abrogating c-*fos* induction in mouse fibroblasts (Roussel et al., 1990). As a possible explanation for these conflicting observations it has been suggested that the transcription factor Fos/AP-1 plays a general role in signal transduction and that the cell-type specific regulatory environment specifies the response to AP-1 activation (Angel and Karin, 1991).

The targets of Fos/AP-1 which control proliferation, transformation or differentiation are still largely unkown. In order to facilitate the identification of such target genes, we have established two different, selective induction systems for Fos. Here we describe the characterization of five Fos-responsive genes (*Fit-1*, *annexin II* and *V*, *ODC*, *fra-1*), and furthermore demonstrate the utility of one of these induction systems to study the effects of Fos on neuronal differentiation of PC12 cells.

A selective transcriptional induction system based on Gal4-estrogen receptor fusion proteins

The most commonly used induction systems for mammalian cells consist of promoters that are regulated by endogenous transcription factors in response to glucocorticoid hormone (Hynes et al., 1981), heat shock (Topol et al., 1985), heavy metal ions (Mayo et al., 1982) or interferon (Goodbourn et al., 1985), respectively. Although these promoters can give rise to high levels of gene expression in the induced state, they suffer from two disadvantages. Their basal activity is often relatively high, and the physiological inducers used trigger an endogenous programme of gene expression which may interfere with or mask the specific effects elicited by the gene product of interest. This is best illustrated by a c-*fos* gene placed under the control of the MMTV LTR promoter. Transcription from this promoter is stimulated by the activated glucocorticoid receptor which then, however, antagonizes the function of the induced Fos protein (Jonat et al., 1990).

We have recently established a novel and selective transcriptional induction system for mammalian cells (Fig. 1A). Central to this system is the use of the hybrid transcription factor Gal-ER (74/282) which is composed of the DNA-binding region of the yeast Gal4 protein (amino acids 1-74) and of the hormone-binding domain of the human estrogen receptor (amino acids 282-595; Webster et al., 1988b). Both the DNA-binding and transactivation functions of this transcription factor are under strict hormonal control (Webster et al., 1988b). A characteristic feature of Gal-ER is its exceptional target gene specificity, which is determined by the inducer estrogen as well as the Gal4 DNA-binding domain. Most mammalian cell types do not express endogenous estrogen receptor. Estrogen is therefore an inert signal for these cells and is only able to stimulate the exogenous Gal-ER transcription factor. Furthermore, mammalian cells lack any Gal4-like activity (Webster et al., 1988a; Kakidani and Patshne, 1988; Ornitz et al., 1991). In addition, the yeast Gal4 protein recognizes a complex palindromic sequence of 17 base pairs (Gininger et al., 1985; Bram et al., 1986) which is unlikely to occur by chance in a mammalian gene promoter. Based on these considerations, Gal-ER should exclusively regulate a transfected Gal4-responsive gene in mammalian cells.

Synthetic Gal4-responsive promoters were optimized for transactivation by Gal-ER in NIH 3T3 cells. The best promoter, consisting of four Gal4-binding sites, an inverted CCAAT element, a TATA-box and the adenovirus major late transcription initiation region, was linked to the mouse c-*fos* gene. This Gal4-responsive *fos* gene was stimulated ~50-fold by Gal-ER in an estrogen-dependent manner (Fig. 2A; Braselmann et al., 1993). The utility of the Gal-ER induction system was demonstrated in rat fibroblast cell lines which were stably transfected with exogenous *Gal-ER* and Gal4-responsive *fos* genes. These cell lines were untransformed in the absence of estrogen, but clearly transformed in the presence of hormone as judged by their morphology (Fig. 4) and growth in soft agar (Braselmann et al., 1993). Transcription of the exogenous c-*fos* gene was rapidly induced by estrogen to reach maximal levels between one and two hours (Braselmann et al., 1993). These rapid kinetics of induction are similar to those of many immediate early genes in response to signal transduction (Almendral et al., 1988) and are in marked contrast to the slow induction of the *E.coli* lac repressor-operator system which has so far been used to regulate heterologous genes in mammalian cells (Labow et al., 1990; Baim et al., 1991).

The post-translational Fos-ER induction system

The hormone-binding region of the human estrogen receptor contains, in addition to the ligand-inducible transactivation (Webster et al., 1988b) and dimerization domains (Kumar and Chambon, 1988), a general "protein inactivation" function which, in the absence of hormone, can repress other activities on the same polypeptide chain

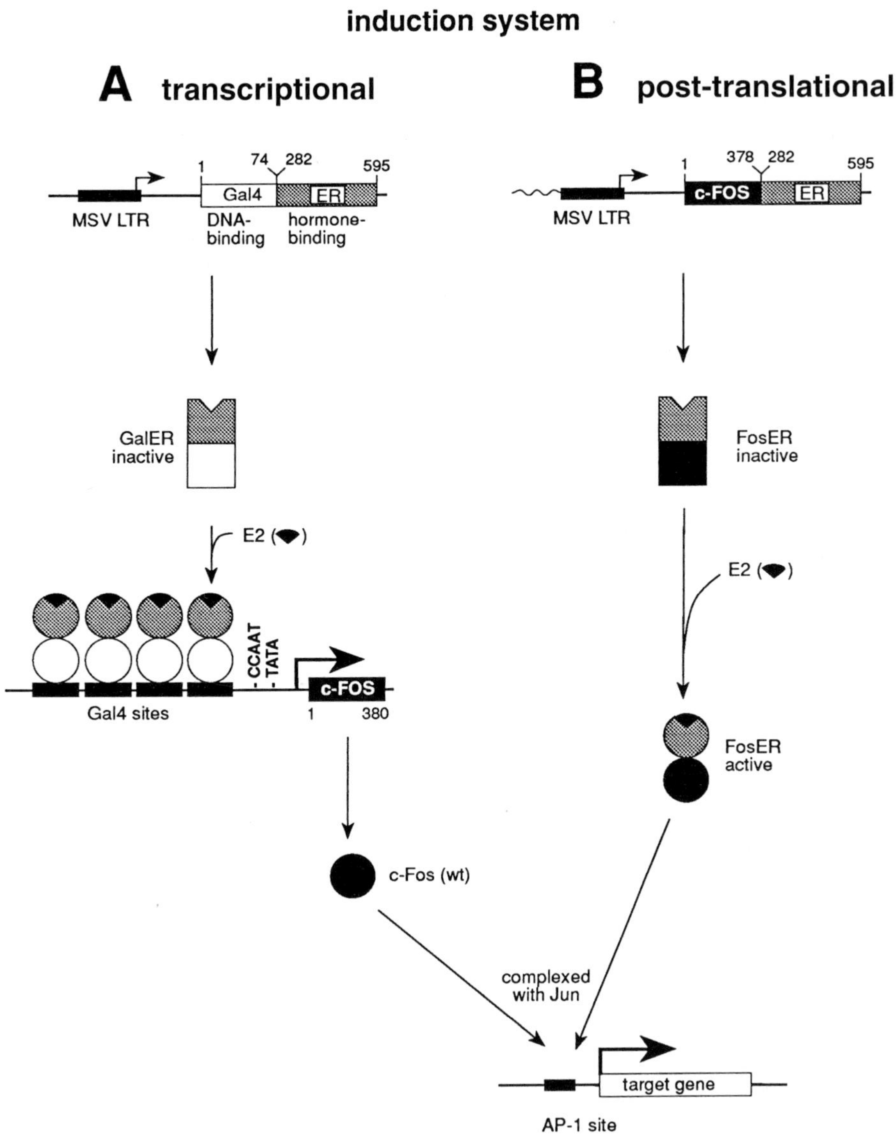

Fig. 1. Schematic diagram of the Gal-ER and Fos-ER induction systems. The chimaeric genes coding for Gal-ER and Fos-ER have been described by Webster et al. (1988b) and Superti-Furga et al. (1991), respectively. ER, estrogen receptor; MSV LTR, Moloney murine sarcoma virus long terminal repeat; E2, β-estradiol.

(Picard et al., 1988). It has been postulated that this repression may be caused by the interaction of the abundant heat shock protein hsp90 with the unliganded hormone-binding domain (Picard et al., 1988). This "protein inactivation" function has been successfully used to subject several transcription factors to hormonal regulation by fusing them to the ligand-binding domain of the human estrogen receptor (Picard et al., 1988; Eilers et al., 1989; Superti-Furga et al., 1991; Umek et al., 1991; Burk and Klempnauer, 1991). As shown in Fig. 1B, we have linked the c-Fos protein at its C terminus to the hormone-binding domain of the estrogen receptor and expressed this fusion protein from the constitutive MSV LTR promoter. In this way three activities of c-Fos, i.e. AP-1-dependent transactivation, repression of the endogenous c-*fos* promoter and cellular transformation, were brought under hormonal control (Superti-Furga et al., 1991). Hormonal regulation of AP-1-dependent transactivation and morphological transformation by Fos-ER is documented by the experiments shown in Figs 2B and 4, respectively. The ligand-binding region of the estrogen receptor also contains a hormone-dependent dimerization domain. It is therefore conceivable that dimerization of the Fos-ER protein via this domain may allow Fos-ER homodimers to form and bind to AP-1 sites, thus by-passing the requirement for Jun proteins. To test this possibility we synthesized c-Jun, c-Fos and Fos-ER proteins by in vitro translation and analyzed their DNA-binding properties by electrophoretic mobility shift assay. As shown in Fig. 3, the Fos-ER protein is unable to bind to the AP-1 sequence on its own and, like the native c-Fos protein, requires complex formation with c-Jun for DNA-binding.

Both the transcriptional Gal-ER/Fos and the post-translational Fos-ER induction systems are equally well suited for regulating Fos activity in rat fibroblasts (Fig. 4). From a practical point of view, the Fos-ER system appears to be simpler, as it requires transfer of a single gene into the cell. However, the hormone-binding domains of steroid receptors harbour many different functions including a potent transactivation domain (Beato, 1989) which may modify the activity of Fos in the fusion protein. So far we have not

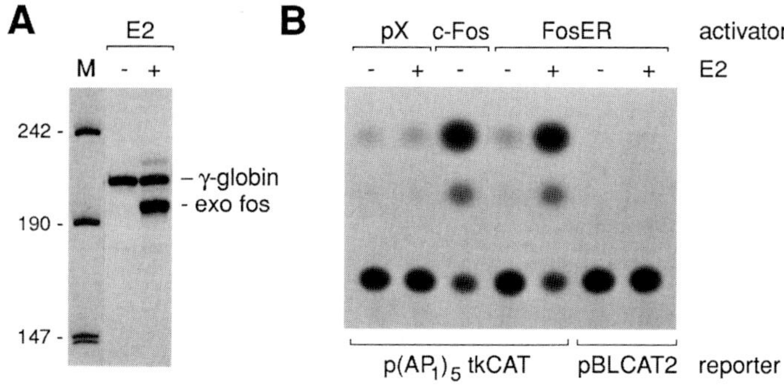

Fig. 2. Hormone-dependent stimulation of transcription by Gal-ER and Fos-ER. (A) Estrogen-dependent transactivation of a Gal4-responsive *fos* gene by Gal-ER. The transactivator plasmid pCMV-GalER (1 μg), the reporter construct pfos-5 (5 μg) and the control plasmid pAγ-3′E (3 μg) were transiently transfected into NIH 3T3 tk⁻ cells and cytoplasmic RNA was analyzed by S1 nuclease mapping with γ-globin- and fos-specific DNA probes. pUC18 digested with *Msp*I was used as end-labelled DNA size marker (lane M). For details see Braselmann et al. (1993). (B) Estrogen-dependent transactivation by Fos-ER. The effect of c-Fos and Fos-ER on transcription of a thymidine kinase promoter lacking (pBLCAT2) or containing five upstream AP-1-binding sites (p[AP1]$_5$tkCAT) was studied by transient transfection and CAT assay in HeLa cells. pX, expression vector without insert; E2, β-estradiol (1 μM). For details see Superti-Furga et al. (1991).

observed any significant qualitative and quantitative differences between c-Fos and the Fos-ER fusion protein in transactivation, repression and transformation assays (Superti-Furga et al., 1991). The Gal-ER induction system facilitates, however, expression of native c-Fos and in this respect is complementary to the Fos-ER system. In addition, the Gal-ER system is more versatile as it allows regulated expression of native protein from any gene that is cloned downstream of the Gal4-responsive promoter. Having established these two selective Fos induction systems we used them to study the effects of Fos on neuronal differentiation of PC12 cells and to identify transformation-relevant target genes of Fos in rodent fibroblasts.

Fos blocks neuronal differentiation of PC12 cells

The rat adrenal pheochromocytoma cell line PC12 provides an attractive model system for studying the action of nerve growth factor (NGF) and the mechanism of neuronal differentiation. In response to NGF, these cells undergo growth arrest and acquire properties of sympathetic neuron-like cells including neurite outgrowth, increased electrical excitability and changes in neurotransmitter synthesis (Greene and Tischler, 1982). Interaction of NGF with its surface receptor is followed by transcriptional activation of specific sets of early and late genes whose protein products are responsible for the differentiated neuronal phenotype. Oncoproteins can be divided into two classes according to their effects on this process. Constitutive expression of viral *src* and activated *ras* genes, which code for membrane-bound and cytoplasmic components of the signal transduction pathway, induce neuronal differentiation and thus mimic the effect of NGF (Alema et al., 1985; Bar-Sagi and Feramisco, 1985; Noda et al., 1985). In contrast, all nuclear oncogenes tested so far (c-*myc*, *E1A*, c-*fos*, *fra-1*) block PC12 cell differentiation (Maruyama et al., 1987; Ito et al., 1989, 1990). In the latter case it has been postulated that the differentiation block results from a perturbation of NGF-induced immediate early gene expression (Ito et al., 1990; Heasley et al., 1991).

We have established Fos-ER-expressing PC12 cell lines in order to gain insight into the molecular mechanism by which Fos interferes with neuronal differentiation. As shown in Fig. 5, the unliganded Fos-ER protein did not affect PC12 cell morphology under normal growth conditions and in response to NGF treatment. Estrogen activation of the Fos-ER protein, however, blocked NGF-induced differentiation as evidenced by the lack of neurite outgrowth. Moreover, activated Fos-ER completely prevented cell cycle arrest of NGF-treated cells (H. Ully, unpublished data) in agreement with the proposed role of Fos in the control of cell proliferation. Normal differentiation was observed with a cell line expressing high levels of a mutant Fos-ER protein that lost its ability to bind DNA. This clearly indicates that Fos-ER must interact with DNA to exert its inhibitory effect. Order of addition experiments revealed that Fos activity is dominant over NGF function, as differentiated Fos-ER-expressing PC12 cells rapidly dedifferentiate upon estrogen addition. Immediate early genes were not induced under these conditions. This result demonstrates therefore that the transient immediate early gene response is not the critical target of Fos for blocking neuronal differentiation as has been proposed by Ito et al. (1990) and Heasley et al. (1991). Activation of Fos-ER in the absence of NGF induced cycling PC12 cells to assume an epithelial-like morphology that resembled the chromaffin phenotype of dexamethasone-treated PC12 cells (C. W., unpublished data).

Prolonged exposure to NGF stimulates expression of the so-called late genes in PC12 cells. In agreement with the observed block of morphological differentiation, Fos activity prevents the induction of all genes of this class analyzed, including those coding for GAP-43, peripherin, neurofilament-M, Thy-1, N-CAM and the chromogranins A and B (C. W., unpublished data). We regard it as unlikely that Fos represses each of these genes individually. Instead we hypothesize that Fos activity interferes with one or a few master regulators of neuronal differentiation that are required for activation of late gene expression. The recently cloned *MASH* genes were of particular interest in this respect as they encode helix-loop-helix proteins homologous to the neuronal regulators of the *Drosophila* achaete-scute locus (Johnson et al., 1990). We have shown that expression of the *MASH-1* gene is also down-regulated by Fos activity in PC12 cells. However, forced expression of the *MASH-1* gene from a retroviral promoter did not rescue neuronal differentiation of PC12 cells that were blocked by Fos activity (C. W., unpublished data). This is in contrast to the situation described for MyoD in myoblasts (Lassar et al., 1989), but is consistent with recent data suggesting that the MASH-1 protein is an important regulator at early, but not late stages of neuronal differentiation (Lo et al., 1991).

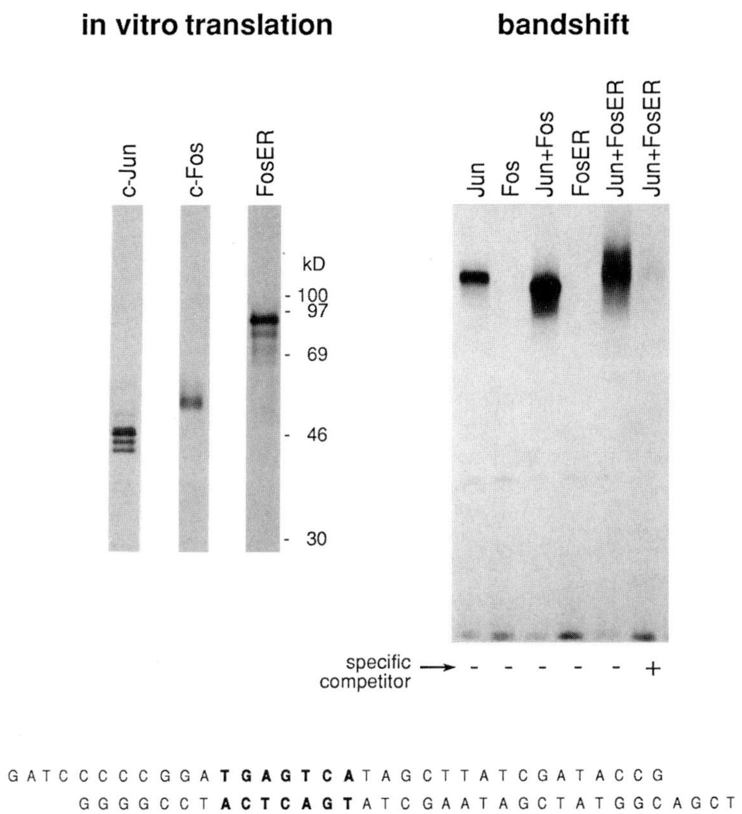

Fig. 3. DNA-binding specificity of Fos-ER. c-Jun, c-Fos and Fos-ER proteins were synthesized by in vitro translation in a rabbit reticulocyte lysate in the presence of [^{35}S]methionine followed by 10% SDS-PAGE and autoradiography. The positions and molecular sizes of the marker proteins used are indicate to the left. c-Fos and Fos-ER were complexed with c-Jun by incubation at 30°C for 15 min. The end-labelled oligonucleotide shown in the bottom part was used for bandshift analysis of the individual proteins and complexes as previously described (Barberis et al., 1987). A 500-fold excess of the same unlabelled oligonucleotide was used for specific competition.

The Fos-responsive gene *Fit-1* codes for a secreted protein

We have used Fos-ER-expressing rat fibroblasts to identify endogenous Fos-responsive genes by differential cDNA cloning (Superti-Furga et al., 1991). We referred to the first gene cloned in this way as Fos-induced transcript 1 or *Fit-1*. *Fit-1* was the only cDNA that was independently isolated many times, suggesting that it represents the most abundant Fos-responsive transcript of rat fibroblasts. *Fit-1* mRNA increases ~10-fold in response to estrogen in Fos-ER-expressing FR 3T3 cells, as shown by S1 nuclease analysis (Fig. 6). In agreement with this result, *Fit-1* expression is low in parental FR 3T3 cells, but high in c-Fos-expressing fibroblasts (Superti-Furga et al., 1991). The relatively rapid accumulation of *Fit-1* transcripts in response to estrogen suggests direct transcriptional regulation of the *Fit-1* gene by Fos-ER. Indeed, nuclear run-on analysis revealed a dramatic increase in the rate of *Fit-1* gene transcription within one hour after estrogen addition. This increase even occurred in the presence of protein synthesis inhibitors demonstrating a direct involvement of Fos-ER in the regulation of *Fit-1* gene transcription (G.B., unpublished data).

Analysis of full-length *Fit-1* cDNA indicated that *Fit-1* is the rat homologue of the mouse *T1* gene. The predicted amino acid sequences of both genes are 80% identical. The extensive sequence homology between these two genes even extends into their 5′ flanking sequences (G.B., unpublished data; Tominaga et al., 1991). Moreover, both genes are equivalent with regard to Fos inducibility, as the endogenous *T1* gene of mouse BALB/c 3T3 cells is also stimulated by ligand-actived Fos-ER (Fig. 6). The *T1* gene was originally identified as a gene that is stimulated by conditional expression of an activated *Ha-ras* or viral *mos* oncogene (Klemenz et al., 1989). Ras and Mos were previously shown to mediate most of their effects on gene expression by activating the transcription factor AP-1 (Schönthal et al., 1988; Schönthal and Feramisco, 1990; Binetruy et al., 1991). Taken together these findings suggest therefore that Fos/AP-1 is involved in Ras- and Mos-mediated regulation of *T1* gene expression. Both the rat *Fit-1* and mouse *T1* genes are strongly induced by serum. In fact, a cDNA clone (ST2), which was isolated as a serum-inducible transcript by Tominaga (1989), proved to be identical with *T1*. Serum stimulation of the *Fit-1* gene is prevented in the presence of cycloheximide (G.B., unpublished data). The requirement for on-going protein synthesis is consistent with our hypothesis that immediate early gene products such as c-*Fos* are involved in the regulation of this gene. Little is yet known about the *T1* and *Fit-1* expression pattern during embryogenesis and in the adult animal. So far abundant *T1* gene expression has only been detected in cultured fibroblasts, but not in any of the adult mouse tissues tested (Tominaga, 1989).

The Fit-1 and T1 proteins are members of the immunoglobulin superfamily. They are most closely related to the extracellular domain of the mouse interleukin-1 receptor and to the immunoglobulin-like repeats of the human carcinoembryonic antigen (Klemenz et al., 1989; Tominaga, 1989). Both the Fit-1 and T1 proteins are heavily glycosylated, contain a signal peptide at their N termi-

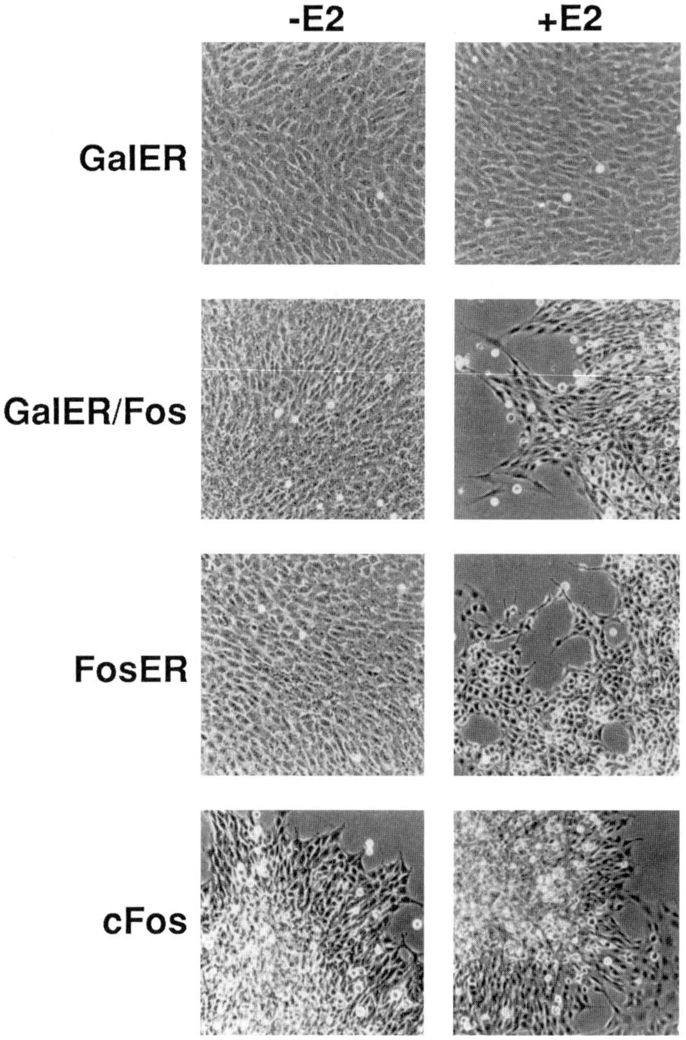

Fig. 4. Estrogen-dependent transformation of Rat-1A fibroblasts. Stable Rat-1A cell lines were established that contained either an integrated *Gal-ER* gene (GalER), the *Gal-ER* and Gal4-responsive *fos* genes (GalER/Fos), the *fos-ER* gene (FosER) or a constitutively expressed c-*fos* gene (cFos). Cells of these four Rat-1A lines were plated at the same low density with or without 1 μM β-estradiol (E2). Photographs were taken after 5 days. For details see Braselmann et al. (1993) and Superti-Furga et al. (1991).

nus, but lack a membrane anchor (Klemenz et al., 1989; G.B., unpublished data). Immunoprecipitation experiments with specific antibodies demonstrated that the Fit-1 protein is indeed secreted by fibroblasts (G.B., unpublished data). It is therefore an attractive hypothesis that Fit-1/T1 may be a peptide hormone which is involved in cell-to-cell communication, either by inducing an autocrine loop in fibroblasts or by paracrine stimulation of other cell types. So far we have been unable to overexpress the *Fit-1* gene ectopically in rat fibroblasts. It therefore remains an open question whether Fos-induced expression of the *Fit-1* gene is, at least in part, responsible for the Fos transformation phenotype of fibroblasts.

Induction of *annexin II* and *V* gene expression by Fos-ER

By differential cDNA screening we have identified two additional cDNA clones which correspond to Fos-inducible transcripts in rat fibroblasts, as shown by the northern blot analysis of Fig. 7A. The 210-bp cDNA insert of the first clone codes for part of the rat *annexin II* gene, as its cDNA sequence is 90% homologous to the corresponding region of the mouse *annexin II* gene (also known as *lipocortin II*, *calpactin I*, *p36*: Saris et al., 1986). The 1460-bp long cDNA insert of the second clone is completely identical with and comprises the entire protein-coding sequence of the rat *annexin V* gene (also known as *lipocortin V*, *endonexin II*: Pepinsky et al., 1988). The annexins are a family of eight related proteins that bind to cellular membranes and phosopholipids in a Ca^{2+}-dependent manner. These proteins are expressed in many different tissues and contain four or eight conserved repeat units of ~75 amino acids which constitute the Ca^{2+} and lipid-binding domains. Although no clear biological function has been defined for any of the annexins, they have been implicated in the regulation of membrane traffic and exocytosis, mediation of cytoskeleton-membrane interactions and mitogenic signal transduction (reviewed by Crompton et al., 1988; Klee, 1988; Haigler et al., 1989).

Annexin II (p36), which was first identified as a major cellular substrate for the $pp60^{src}$ kinase (Huang et al., 1986), was found to be very abundant in fibroblastic cell lines, accounting for 0.1% to 0.4% of total protein (Isacke et al., 1986). In agreement with this finding, we have observed that the mRNA of the rat *annexin II* gene is also abundantly expressed in our Fos-ER fibroblasts (see legend to Fig. 7A). Activation of Fos-ER resulted in a consistent two-fold increase of annexin II mRNA within 17 hours of estrogen addition which, in view of its high mRNA level, must be brought about by a dramatic stimulation of the transcription rate (Fig. 7A). Interestingly, phorbol esters (TPA) also induce a two-fold increase of annexin II mRNA levels in the human myeloid cell line U-937 (Isacke et al., 1989). Moreover, a putative AP-1-binding site is present in the promoter of the human *annexin II* gene at position −125 upstream of the transcription start site (Spano et al., 1990; Fig. 7D). The electrophoretic mobility shift experiment of Fig. 7C demonstrates that this promoter sequence is efficiently and specifically recognized by in vitro-translated c-Fos/c-Jun complexes. It remains to be seen whether this high affinity AP-1 site has also been conserved in the rat annexin II promoter. If so, it is likely that this AP-1 recognition sequence mediates stimulation of the rat annexin II promoter by Fos/AP-1.

Annexin V (endonexin II) was initially identified as an in vitro inhibitor of blood coagulation and phospholipase A_2, although in retrospect both inhibitory activities may not reflect physiological in vivo functions (reviewed by Haigler et al., 1989). Recent biophysical and crystal structure analyses have revealed that annexin V forms voltage-dependent calcium channels upon association with lipid bilayers (Rojas et al., 1990; Huber et al., 1990). Annexin V mRNA is about ten times less abundant than annexin II mRNA in our Fos-ER-expressing rat fibroblasts and is induced ~5-

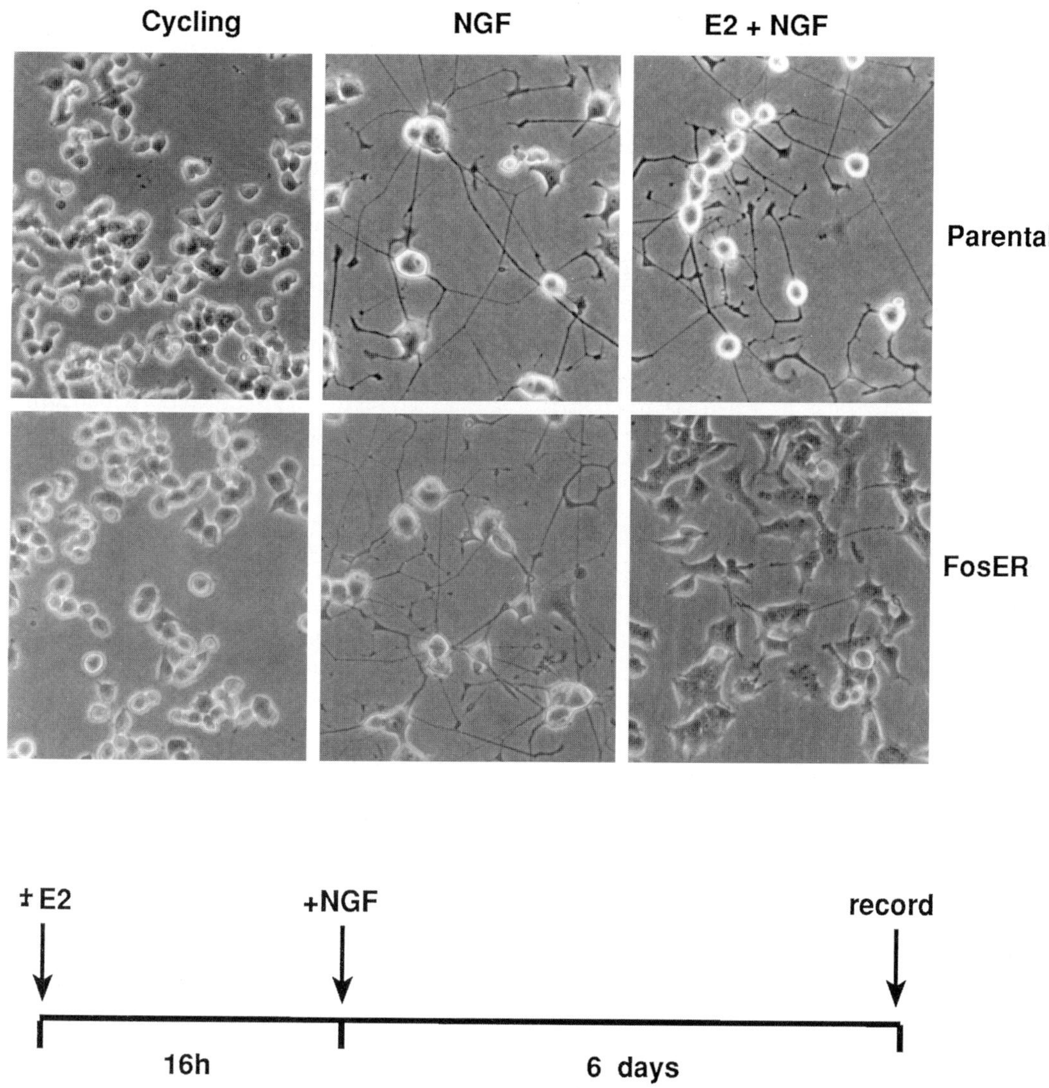

Fig. 5. Hormone-dependent block of neuronal differentiation of Fos-ER-expressing PC12 cells. Parental and Fos-ER-expressing PC12 cells were compared with regard to morphological differentiation in response to NGF (50 ng/ml) under standard conditions or after estrogen treatment (1 μM, E2).

fold in response to estrogen activation of Fos-ER (Fig. 7A). A similar degree of annexin V mRNA induction is also seen in Fos-ER-expressing PC12 cells (Fig. 7B). Annexin V expression has previously been reported to increase five-fold in PC12 cells in response to nerve growth factor (Schlaepfer and Haigler, 1990). It is therefore conceivable that stimulation of AP-1 activity by nerve growth factor could be responsible for increased expression of annexin V in these cells.

Fos activity is sufficient for transcriptional stimulation of the ornithine decarboxylase gene in PC12 cells, but not in fibroblasts

Ornithine decarboxylase (ODC) is the rate-limiting enzyme in the biosynthetic pathway of polyamines which are known to be indispensable for optimal growth of mammalian cells. A large number of signals including hormones, growth factors, tumour promoters and regenerative stimuli cause a rapid increase in ODC activity both in vivo and in tissue culture cells. Increased ODC gene expression is also characteristic of many tumours and transformed cell lines (for review, see Tabor and Tabor, 1984). In particular, activation of the c-*Ha-ras* gene was shown to result in rapid induction of the *ODC* gene in fibroblasts (Sistonen et al., 1987, 1989; Hölttä et al., 1988). Moreover, the *ODC* gene is induced transiently and with delayed kinetics relative to the c-*fos* gene in NGF-treated PC12 cells (Greenberg et al., 1985) and in neuronal cells of the brain after electrical stimulation (Zawia and Bondy, 1990; Vendrell et al., 1991). Based on this evidence, it has been postulated that Fos/AP-1 may be involved in the regulation of *ODC* gene expression in response to signal transduction (Hölttä and Sistonen, 1989; Vendrell et al., 1991).

We have tested this hypothesis using our Fos-ER-

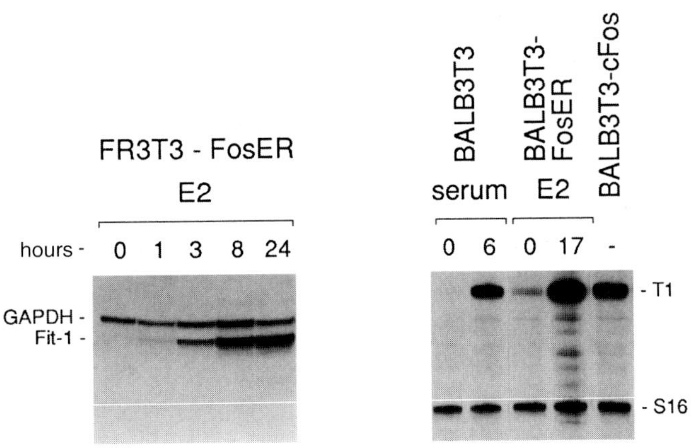

Fig. 6. Induction of *Fit-1* and *T1* gene expression by Fos-ER. Fos-ER-expressing rat (FR 3T3) and mouse (BALB/c 3T3) fibroblasts were treated with estrogen (1 µM, E2) for the indicated time. Serum-stimulated and constitutively c-Fos-expressing BALB/c 3T3 cells were used as controls. RNA isolated from the FR 3T3 cells was analyzed by S1 mapping for rat glyceraldehyde dehydrogenase (GAPDH) and Fit-1 mRNA, while the BALB/c 3T3 RNA was analyzed for mouse T1 and ribosomal protein S16 gene transcripts.

expressing PC12 and FR 3T3 cell lines. Estrogen activation of Fos-ER resulted in a rapid increase of ODC mRNA in PC12 cells, as shown by S1 nuclease analysis (Fig. 8). Both the rate of accumulation and the maximal level of ODC mRNA were comparable to that obtained by NGF treatment of these cells. Induction of the *ODC* gene by Fos-ER was also seen in the presence of protein synthesis inhibitors, showing this to be a direct effect of Fos activation. Consistent with this, nuclear run-on analysis revealed that the transcription rate of the *ODC* gene was rapidly increased within one hour after estrogen addition. Moreover, the *ODC* gene could not be induced in a PC12 cell line expressing a mutant Fos-ER protein with a non-functional DNA-binding domain. In summary, we conclude therefore that the transcription factor Fos, as part of the AP-1 complex, is directly involved in the regulation of the *ODC* gene in PC12 cells possibly by binding to enhancer or promoter sequences (Wrighton and Busslinger, 1993). These regulatory sequences, which still remain to be identified, may include one or both of the putative AP-1-binding sites present in the 5′ region of the rat *ODC* gene (Wen et al., 1989; van Steeg et al., 1990).

No increase in ODC mRNA synthesis was observed in Fos-ER-expressing fibroblasts (FR 3T3) following estrogen treatment (Fig. 8). ODC expression was, however, strongly up-regulated in these fibroblasts by serum stimulation indicating that the endogenous *ODC* gene is transcriptionally competent and fully inducible. Furthermore, the Fos-ER protein is functional in these fibroblasts as shown by the induction of the *Fit-1* gene in response to estrogen (Fig. 6). These data indicate therefore that Fos activity alone is not sufficient to stimulate *ODC* transcription in fibroblasts. Transcriptional stimulation by AP-1 is thought to be brought about by cooperative interaction with other upstream transcription factors. The best documented cases

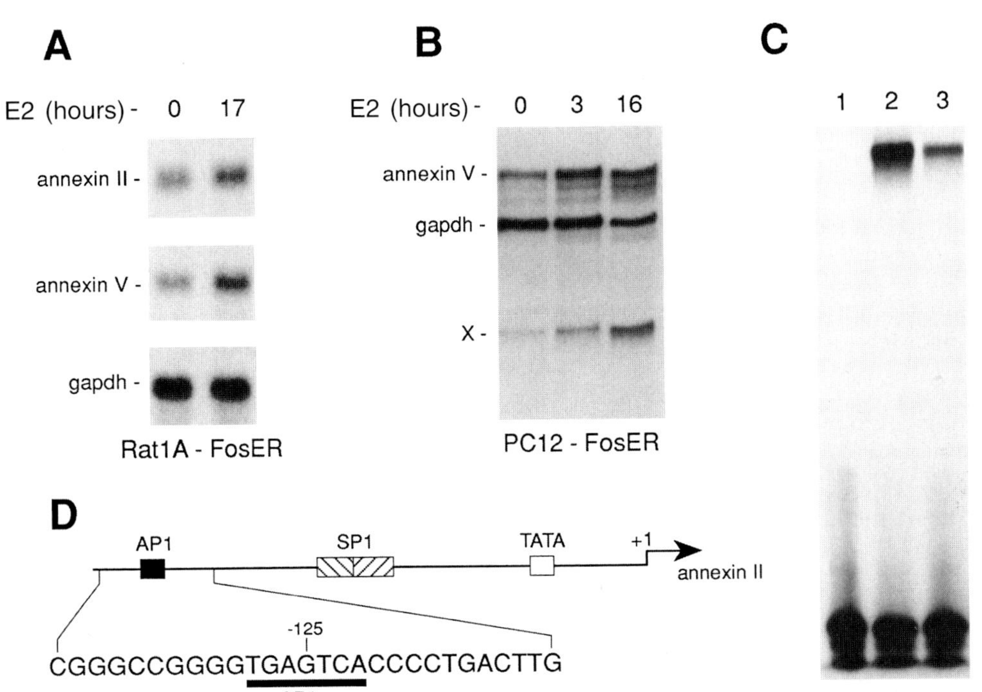

Fig. 7. Fos induction of *annexin II* and *V* gene expression. (A) Northern blot analysis. Poly(A)$^+$ RNA (3 µg) isolated from untreated and estrogen-stimulated Fos-ER-expressing Rat-1A cells was analysed by northern blot hybridization with annexin II and V cDNAs which were cloned from the same cell line by differential screening. The autoradiographs shown correspond to an exposure time of 90 min (annexin II) and 15 hours (annexin V), respectively. As a control for RNA loading, the same filters were rehybridized with a rat glyceraldehyde dehydrogenase probe (gapdh). (B) Fos induction of the *annexin V* gene in PC12 cells. Fos-ER-expressing PC12 cells were treated with estrogen for the indicated time followed by S1 nuclease analysis with annexin V- and gapdh-specific DNA probes. X denotes an annexin V-related transcript that gave rise to partial protection of the annexin V DNA probe. (C) Bandshift analysis. The oligonucleotide shown in panel D was end-labelled and used for bandshift analysis with complexes of in vitro translated c-Fos and c-Jun proteins (lane 2 and 3). A 100-fold excess of the unlabelled oligonucleotide containing the high affinity AP-1-binding site shown in Fig. 3 was used for competition (lane 3). No protein was added in lane 1. (D) Schematic diagram of the human *annexin II* gene promoter (see Spano et al., 1990).

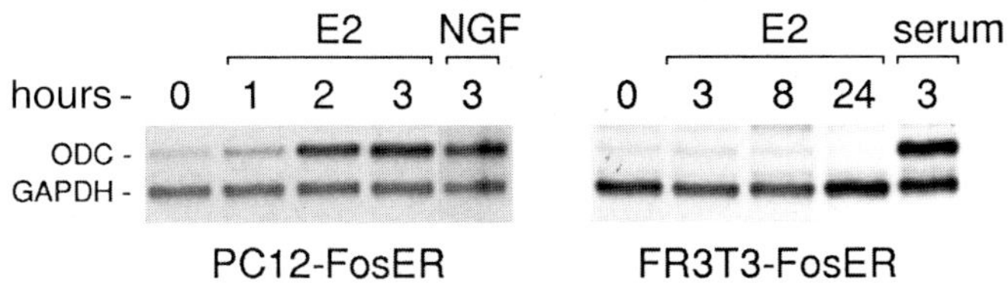

Fig. 8. Induction of the ornithine decarboxylase gene by Fos-ER in PC12 cells. Fos-ER-expressing PC12 cells were treated with estrogen (1 µM, E2) or NGF (50 ng/ml) for up to 3 hours. Fos-ER-expressing FR 3T3 cells were serum stimulated or treated with estrogen (1 µM in growth medium) for the time indicated. Total RNA was isolated and transcripts of the *ornithine decarboxylase* (ODC) and *glyceraldehyde dehydrogenase* (GAPDH) genes were analyzed by S1 nuclease mapping.

include the polyoma enhancer and the promoters of the *stromelysin*, *collagenase* and *interleukin-2* genes, whose transcriptional activities entirely rely on synergism between AP-1 and transcription factors of the Ets oncoprotein family (Wasylyk et al., 1990, 1991; Gutman and Wasylyk, 1990; Jain et al., 1992). Moreover, transcription of the myelomonocytic growth factor gene has recently been shown to depend on the cooperative action of AP-1 and a myeloid-specific factor, NF-M (Sterneck et al., 1992). In light of these observations, our results could be best explained by assuming that transcription of the *ODC* gene also depends on synergism between AP-1 and a hitherto unidentified transcription factor(s). This cooperating factor(s) may be constitutively expressed in PC12 cells and hence an increase in AP-1 activity would be sufficient to activate the *ODC* gene. In fibroblasts however, AP-1 is not the only limiting component of *ODC* transcription, suggesting that the activity of the hypothetical cooperating factor(s), like that of c-Fos, is regulated in response to signal transduction in these cells.

The *fra-1* gene is positively autoregulated by AP-1 activity

Both the c-*jun* and c-*fos* genes are known to be under the control of AP-1 activity itself. The c-*jun* gene is positively autoregulated by its own gene product in embryonal carcinoma cells (Angel et al., 1988), while the c-Fos protein represses the activity of its own promoter in all cell types analyzed so far (Sassone-Corsi et al., 1988; König et al., 1989; Gius et al., 1990). We therefore asked whether induction of Fos activity in our Fos-ER-expressing fibroblasts also leads to up- or down-regulation of members of the AP-1 gene family and hence to an overall change in the composition of AP-1. Northern blot analysis revealed that the levels of all three *jun* gene transcripts (c-*jun*, *junB*, *junD*) and of *fra-2* mRNA were unaffected by estrogen activation of Fos-ER. *FosB* gene expression was undetectable in these cells and the endogenous c-*fos* gene was repressed by exogenous Fos-ER activity as previously described by Superti-Furga et al. (1991). The expression of the *fra-1* gene was, however, significantly activated in response to estrogen as shown by the northern blot analysis of Fig. 9. The

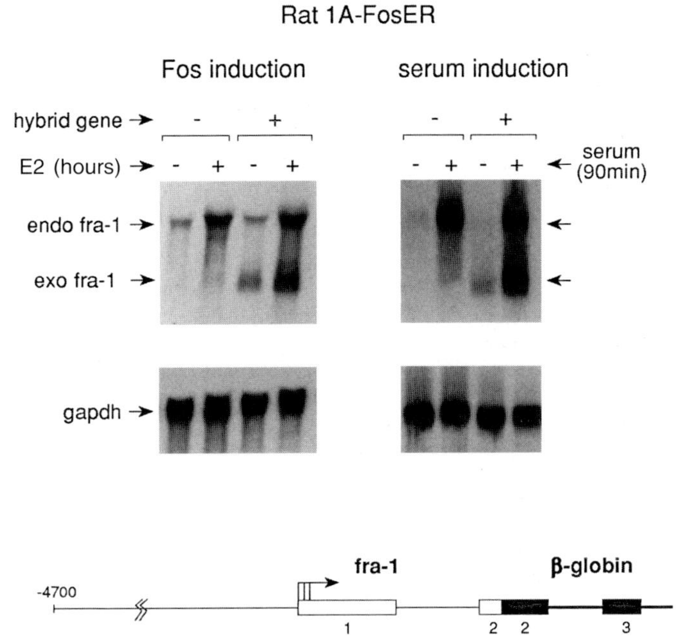

Fig. 9. Fos and serum induction of the fra-1 gene in rat fibroblasts. The rat *fra-1* gene containing 4700 bp of 5′ flanking sequences was fused in the second exon to the rabbit β-*globin* gene. This chimaeric construct was transfected together with a hygromycin resistance gene into Fos-ER-expressing Rat-1A cells. Parental cells and a cell pool derived from ~100 resistant colonies were treated with estrogen (in growth medium) for 15 hours or were serum-starved for two days followed by serum stimulation (90 minutes). Endogenous and exogenous *fra-1* gene transcripts were detected with labelled fra-1 cDNA by northern blot analysis of total RNA. The same filters were rehybridized with a gapdh probe.

same increase in *fra-1* mRNA synthesis was also observed in rat fibroblast lines that expressed the native c-Fos protein under the control of the transcriptional Gal-ER induction system (Braselmann et al., 1993). In addition, the *fra-1* gene was strongly induced by estrogen in Fos-ER-expressing PC12 cells (C.W., unpublished data). This evidence indicates therefore that up-regulation of *fra-1* gene expression by Fos/AP-1 is a more general phenomenon (Bergers et al., 1993b).

Fra-1 gene induction by Fos/AP-1 occurs at the level of transcription initiation, as shown by nuclear run-on experiments with Fos-ER-expressing fibroblast and PC12 cells. Furthermore, a Fos-ER protein lacking a functional DNA-binding domain was unable to stimulate *fra-1* gene transcription (G.B. and C.W., unpublished data). These data suggest that Fos/AP-1 activity directly regulates the *fra-1* gene by binding to *cis*-regulatory regions. To test this hypothesis we have isolated the *fra-1* gene from the rat genome. Transcription of the *fra-1* gene was shown to start heterogeneously within a 10-bp element. Consistent with this finding there is no TATA sequence in the immediate 5′ flanking region. Moreover, the promoter and first exon of the *fra-1* gene are both located within a CpG island which is a characteristic feature of housekeeping genes (see Bird et al., this volume). For promoter mapping we have linked the *fra-1* gene in the second exon to the rabbit β-globin gene (see Fig. 9). A construct containing 4700 bp of 5′ flanking sequences was stably transfected into Fos-ER-expressing Rat1A cells and a cell pool derived from ~100 hygromycin-resistant colonies was analyzed for Fos and serum induction of the transfected *fra-1* gene by northern blot hybridization (Fig. 9). This experiment demonstrated that sequences in the 5′ region of the *fra-1* gene from position –4700 to the second exon are sufficient for induction by Fos/AP-1 activity and to a lesser extent also by serum.

Elevated levels of endogenous *fra-1* mRNA were also observed in Rat-1A cells constitutively expressing c-Jun or Fra-1 (G.B., unpublished data). This evidence raises the possibility that the *fra-1* gene may be regulated by other members of the AP-1 family. In particular, it suggests that the Fra-1 protein may positively autoregulate transcription from its own promoter. Autoregulation has previously been described for the c-*jun* gene and was proposed to be responsible for prolonged c-*jun* expression in response to stimulation of cell surface receptors (Angel et al., 1988). The kinetics of *fra-1* induction following serum or phorbol ester stimulation are delayed compared with those of the c-*fos* gene (Cohen and Curran, 1988; Matsui et al., 1990). This protracted expression pattern may reflect transcriptional upregulation of the *fra-1* gene by increased Fos/AP-1 activity in stimulated cells.

What may be the significance of *fra-1* induction by c-Fos/AP-1 activity? The dimerization and DNA-binding potential of Fra-1 is similar, if not identical, to that of other members of the Fos family (c-Fos, FosB, Fra-2; Cohen et al., 1989; Ryseck and Bravo, 1991). However, the transcriptional properties of Fra-1 differ from those of c-Fos, as was recently shown by transient transfection experiments in F9 cells (Suzuki et al., 1991). In contrast to c-Fos, the Fra-1 protein was able to stimulate AP-1 dependent transcription only in conjunction with JunD, but failed to do so in the presence of c-Jun and JunB. More importantly, Fra-1 efficiently suppressed transactivation by c-Fos in a dose-dependent manner. The induction of *fra-1* expression by c-Fos may therefore be a mechanism to down-modulate the transcripitional activity of c-Fos, quite by analogy to the naturally occurring truncated form of FosB which also exerts an inhibitory effect on AP-1 activity (Nakabeppu and Nathans, 1991; Wisdom et al., 1992). Alternatively, Fra-1 may control only a subset of Fos-responsive genes or regulate an entirely distinct class of target genes. In support of this idea, we have shown Fra-1 to be a much more weakly transforming oncoprotein than c-Fos, as constitutive expression of the *fra-1* gene in rat fibroblasts promoted growth in soft agar, but did not result in overt morphological transformation (G.B., unpublished data). It is therefore conceivable that Fos-induced expression of the *fra-1* gene may be reponsible for part of the Fos transformation phenotype, i.e. for anchorage-independent growth of fibroblasts.

Conclusions

During the last decade we have experienced the discovery of a wealth of transcriptional regulators that are involved in the control of biological processes such as early development, differentiation and signal transduction. Paradoxically it has proven more difficult to identify the important downstream target genes which are responsible for the biological effects elicited by these transcription factors. To date three different approaches have been successfully used for this purpose. The so-called enhancer detection method was developed in the *Drosophila* system to search for genes with an expression pattern that is similar to that of the developmental regulator studied (Wagner-Bernholz et al., 1991). An alternative procedure depends on immunopurification of those chromatin fragments which are bound by the transcription factor under study (Gould et al., 1990; White et al., this volume). Here we have demonstrated the feasibility of selective induction systems to identify Fos target genes. A similar approach has recently led to the isolation of the first Myc- and Myb-regulated genes (Eilers et al., 1991; Burk and Klempnauer, 1991).

Functional AP-1-binding sites have been identified as regulatory sequences in many cellular genes by in vitro mutagenesis and transient co-transfection experiments using c-*fos* expression vectors. Potential Fos target genes identified in this way include the genes coding for stromelysin (Kerr et al., 1988), collagenase (Angel et al., 1987), metallothionein (Angel et al., 1987), proliferin (Mordacq and Linzer, 1989) and nerve growth factor (Hengerer et al., 1990), all of which are expressed in fibroblasts. None of the endogenous genes of this family could, however, be induced by Fos-ER activity in Rat-1A fibroblasts suggesting that they are not critically involved in the Fos transformation phenotype of these cells (G. B., unpublished data). The putative Fos-responsive genes expressed in PC12 cells code for tyrosine hydroxylase (Gizang-Ginsberg and Ziff, 1990), proenkephalin (Sonnenberg et al., 1989), stromelysin (Machida et al., 1989), neurotensin/neuromedin N (Kislauski and Dobner, 1990) and prodynorphin (Naranjo et al., 1991). Of all these genes tested only the *tyrosine hydroxylase* gene was induced to physiological levels in Fos-ER-expressing PC12 cells by estrogen treatment (Wrighton and Busslinger, 1993).

The Fos-ER induction system was also used to study the effects of Fos on differentiation of PC12 cells, myoblasts and mammary epithelial cells (Reichmann et al., 1992). In all three cases, Fos activity interfered with the normal differentiated phenotype. The results obtained with epithelial cells were particularly interesting as they uncovered a novel function of Fos/AP-1. Short-term stimulation of Fos activity induced a reversible loss of epithelial cell polarity. In contrast, long-term stimulation caused these cells to irreversibly lose their polarized phenotype and to undergo conversion to fibroblastoid cells, suggesting that members of the Fos family are important for both transient and permanent changes of the epithelial cell phenotype (Reichmann et al., 1992).

We are grateful to G. Schaffner and his team for oligonucleotide synthesis and DNA sequencing, to H. Tkadletz for graphical work and to M. Nicklin for critical reading of the manuscript. This work was supported by the I. M. P. and, in part, by a grant from the Austrian Industrial Research Promotion Fund.

References

Alema, S., Casalbore, P., Agostini, E. and Tato, F. (1985). Differentiation of PC12 phaeochromocytoma cells induced by v-src oncogene. *Nature* **316**, 557-560.

Almendral, J. M., Sommer, D., MacDonald-Bravo, H., Burckhardt, J., Perera, J. and Bravo, R. (1988). Complexity of the early genetic response to growth factors in mouse fibroblasts. *Mol. Cell. Biol.* **8**, 2140-2148.

Angel, P., Imagawa, M., Chiu. R., Stein, B., Imbra, R. J., Rahmsdorf, H. J., Jonat, C., Herrlich, P. and Karin, M. (1987). Phorbol ester-inducible genes contain a common cis element recognized by a TPA-modulated trans-acting factor. *Cell* **49**, 729-739.

Angel, P., Hattori, K., Smeal, T. and Karin, M. (1988). The jun proto-oncogene is positively autoregulated by its product, Jun/AP-1. *Cell* **55**, 875-885.

Angel, P. and Karin, M. (1991). The role of Jun and Fos and the AP-1 complex in cell proliferation and transformation. *Biochim. Biophys. Acta* **1072**, 129-157.

Baim, S. B., Labow, M. A., Levine, A. J. and Shenk, T. (1991). A chimeric mammalian transactivator based on the lac repressor that is regulated by temperature and isopropyl β-D-thiogalactopyranoside. *Proc. Nat. Acad. Sci. USA* **88**, 5072-5076.

Barberis, A., Superti-Furga, G. and Busslinger, M. (1987). Mutually exclusive interaction of the CCAAT-binding factor and of a displacement protein with overlapping sequences of a histone gene promoter. *Cell* **50**, 347-359.

Bar-Sagi, D. and Feramisco, J. R. (1985). Microinjection of the ras oncogene protein into PC12 cells induces morphological differentiation. *Cell* **42**, 841-848.

Beato, M. (1989). Gene regulation by steroid hormones. *Cell* **56**, 335-344.

Binetruy, B., Smeal, T. and Karin, M. (1991). Ha-Ras augments c-Jun activity and stimulates phosphorylation of its activation domain. *Nature* **351**, 122-127.

Bram, R. J., Lue, N. F. and Kornberg, R. D. (1986). A GAL family of upstream activating sequences in yeast: roles in both induction and repression of transcription. *EMBO J.* **5**, 603-608.

Braselmann, S., Graninger, P. and Busslinger, M. (1993). A selective transcriptional induction system for mammalian cells based on Gal4-estrogen receptor fusion proteins. *Proc. Nat. Acad. Sci. USA*, in press.

Burk, O. and Klempnauer, K. -H. (1991). Estrogen-dependent alterations in differentiation state of myeloid cells caused by a v-myb/estrogen receptor fusion protein. *EMBO J.* **10,** 3713-3719.

Cohen, D. R. and Curran, T. (1988). Fra-1: a serum-inducible, cellular immediate-early gene that encodes a Fos-related antigen. *Mol. Cell. Biol.* **8**, 2063-2069.

Cohen, D. R., Ferreira, P. C. P., Gentz, R., Franza, R. B. Jr. and Curran, T. (1989). The product of a fos-related gene, fra-1, binds cooperatively to the AP-1 site with Jun: transcription factor AP-1 is comprised of multiple protein complexes. *Genes Dev.* **3**, 173-184.

Crompton, M. R., Moss, S. E. and Crumpton, M. J. (1988). Diversity in the lipocortin/calpactin family. *Cell* **55**, 1-3.

Curran, T., Miller, A. D., Zokas, L. and Verma, I. M. (1984). Viral and cellular fos proteins: A comparative analysis. *Cell* **36**, 259-268.

Eilers, M., Picard, D., Yamamoto, K. R. and Bishop, M. J. (1989). Chimaeras of myc oncoprotein and steroid receptors cause hormone-dependent transformation of cells. *Nature* **340**, 66-68.

Eilers, M., Schirm, S. and Bishop, M. J. (1991). The Myc protein activates transcription of the α-prothymosin gene. *EMBO J.* **10**, 133-141.

Finkel, M. P., Biskis, B. O. and Tinkins, P. B. (1966). Virus induction of osteosarcomas in mice. *Science* **151**, 698-701.

Gininger, E., Varnum, S. M. and Ptashne, M. (1985). Specific DNA binding of GAL4, a positive regulatory protein of yeast. *Cell* **40**, 767-774.

Gius, D., Cao, X., Rauscher III, F. J., Cohen, D. R., Curran, T. and Sukhatme, V. P. (1990). Transcriptional activation and repression by Fos are independent functions: the C terminus represses immediate-early gene expression via CArG elements. *Mol. Cell. Biol.* **10**, 4243-4255.

Gizang-Ginsberg, E. and Ziff, E. B. (1990). Nerve growth factor regulates tyrosine hydroxylase gene transcription through a nucleoprotein complex that contains c-Fos. *Genes Dev.* **4**, 477-491.

Goodbourn, S., Zinn, K. and Maniatis, T. (1985). Human β-interferon gene expression is regulated by an inducible enhancer element. *Cell* **41**, 509-520.

Gould, A. P., Brookman, J. J., Strutt, D. I. and White, R. A. H. (1990). Targets of homeotic gene control in Drosophila. *Nature* **348**, 308-312.

Greenberg, M. E., Greene, L. A. and Ziff, E. B. (1985). Nerve growth factor and epidermal growth factor induce rapid transient changes in proto-oncogene transcription in PC12 cells. *J. Biol. Chem.* **260**, 14101-14110.

Greene, L. A. and Tischler, A. S. (1982). PC12 pheochromocytoma cultures in neurobiologicalresearch. *Adv. Cell. Neurobiol.* **3**, 373-414.

Gutman, A. and Wasylyk, B. (1990). The collagenase gene promoter contains a TPA and oncogene-responsive unit encompassing the PEA3 and AP-1 binding sites. *EMBO J.* **9**, 2241-2246.

Haigler, H. T., Fitch, J. M., Jones, J. M. and Schlaepfer, D. D. (1989). Two lipocortin-like proteins, endonexin II and anchorin CII, may be alternate splices of the same gene. *Trends Biochem. Sci.* **14**, 48-50.

Heasley, L. E., Benedict, S., Gleavy, J. and Johnson, G. L. (1991). Requirement of the adenovirus E1A transformation domain 1 for inhibition of PC12 cell neuronal differentiation. *Cell Regul.* **2**, 479-489.

Hengerer, B., Lindholm, D., Heumann, R., Rüther, U., Wagner, E. F. and Thoenen, H. (1990). Lesion-induced increase in nerve growth factor mRNA is mediated by c-fos. *Proc. Nat. Acad. Sci. USA* **87**, 3899-3903.

Holt, J. T., Gopal, T. V., Moulton, A. D. and Nienhuis, A. W. (1986). Inducible production of c-fos antisense RNA inhibits 3T3 cell proliferation. *Proc. Nat. Acad. Sci. USA* **83**, 4794-4798.

Hölttä, E. and Sistonen, L. (1989). Ornithine decarboxylase gene expression in normal and transformed cells. In *Biological and Chemical Polyamines* (ed. S. H. Goldemberg and I. D. Algranati), pp. 193-202. Oxford: IRL, UK.

Hölttä, E., Sistonen, L. and Alitalo, K. (1988). The mechanism of ornithine decarboxylase deregulation in c-Ha-ras oncogene-transformed NIH 3T3 cells. *J. Biol. Chem.* **263**, 4500-4507.

Huang, K.-S., Wallner, B. P., Mattaliano, R. J., Tizard, R., Burne, C., Frey, A., Hession, C., McGray, P., Sinclair, L. K., Chow, P. E., Browning, J. L., Ramachandran, K. L., Tang, J., Smart, J. E. and Pepinsky, R. B. (1986). Two human 35 kd inhibitors of phospolipase A_2 are related to substrates of $pp60^{v\text{-}src}$ and of the epidermal growth factor receptor/kinase. *Cell* **46**, 191-199.

Huber, R., Römisch, J. and Paques, E.-P. (1990). The crystal and molecular structure of human annexin V, an anticoagulant protein that binds to calcium and membranes. *EMBO J.* **9**, 3867-3874.

Hynes, N. E., Kennedy, N., Rahmsdorf, U. and Groner, B. (1981). Hormone-responsive expression of an endogenous proviral gene of mouse mammary tumor virus after molecular cloning and gene transfer into cultured cells. *Proc. Nat. Acad. Sci. USA* **78**, 2038-2042.

Isacke, C. M., Lindberg, R. A. and Hunter, T. (1989). Synthesis of p36 and p35 is increased when U-937 cells differentiate in culture but expression is not inducible by glucocorticoids. *Mol. Cell. Biol.* **9**, 232-240.

Isacke, C. M., Trowbridge, I. S. and Hunter, T. (1986). Modulation of p36 phosphorylation in human cells: studies using anti-p36 monoclonal antibodies. *Mol. Cell. Biol.* **6**, 2745-2751.

Ito, E., Sonnenberg, J. L. and Narayanan, R. (1989). Nerve growth factor-induced differentiation of PC12 cells is blocked by fos oncogene. *Oncogene* **4**, 1193-1199.

Ito, E., Sweterlitsch, L. A., Tran, P. B.-V., Rauscher, F. J. III and Narayanan, R. (1990). Inhibition of PC12 cell differentiation by the immediate early gene fra-1. *Oncogene* **5**, 1755-1760.

Jain, J., McCaffrey, P. G., Valge-Archer, V. E. and Rao, A. (1992). Nuclear factor of activated T cells contains Fos and Jun. *Nature* **356**, 801-804.

Johnson, J. E., Birren, S. J. and Anderson, D. J. (1990). Two rat homologues of Drosophila achaete-scute specifically expressed in neuronal precursors. *Nature* **346**, 858-861.

Jonat, C., Rahmsdorf, H. J., Park, K.-K., Cato, A. C. B., Gebel, S., Ponta, H. and Herrlich, P. (1990). Antitumor promotion and antiinflammation: down-modulation of AP-1 (Fos/Jun) activity by glucocorticoid hormone. *Cell* **62**, 1189-1204.

Kakidani, H. and Patshne, M. (1988). GAL4 activates gene expression in mammalian cells. *Cell* **52**, 161-167.

Kerppola, T. K. and Curran, T. (1991). Transcription factor interactions: basics on zippers. *Curr. Opin. Struct. Biol.* **1**, 71-79.

Kerr, L. D., Holt, J. T. and Matrisian, L. M. (1988). Growth factors regulate transin gene expression by c-fos-dependent and c-fos-independent pathways. *Science* **242**, 1424-1427.

Kim, S.-J., Angel, P., Lafyatis, R., Hattori, K., Kim, K. Y., Sporn, M. B., Karin, M. and Roberts, A. B. (1990). Autoinduction of transforming growth factor β1 is mediated by the AP-1 complex. *Mol. Cell. Biol.* **10**, 1492-1497.

Kislauski, E. and Dobner, P. R. (1990). Mutually dependent response elements in the cis-regulatory region of the neurotensin/neuromedin N gene integrate environmental stimuli in PC12 cells. *Neuron* **4,** 783-795.

Klee, C. D. (1988). Ca^{2+}-dependent phospholipid- (and membrane-) binding proteins. *Biochemistry* **27**, 6645-6653.

Klemenz, R., Hoffmann, S. and Werenskiold, A.-K. (1989). Serum- and oncoprotein-mediated induction of a gene with sequence similarity to the gene encoding carcinoembryonic antigen. *Proc. Nat. Acad. Sci. USA* **86**, 5708-5712.

König, H., Ponta, H., Rahmsdorf, U., Büscher, M., Schönthal, A., Rahmsdorf, H. J. and Herrlich, P. (1989). Autoregulation of fos: the dyad symmetry element as the major target of repression. *EMBO J.* **8**, 2559-2566.

Kumar, V. and Chambon, P. (1988). The estrogen receptor binds tightly to its responsive element as as ligand-induced homodimer. *Cell* **55**, 145-156.

Labow, M. A., Baim, S. B., Shenk, T. and Levine, A. J. (1990). Conversion of the lac repressor into an allosterically regulated transcriptional activator for mammalian cells. *Mol. Cell. Biol.* **10**, 3343-3356.

Lassar, A. B., Thayer, M. J., Overell, R. W. and Weintraub, H. (1989). Transformation of activated Ras and Fos prevents myogenesis by inhibiting expression of MyoD. *Cell* **58**, 659-667.

Lo, L.-C., Johnson, J. E., Wuenschell, C. W., Saito, T. and Anderson, D. J. (1991). Mammalian achaete-scute homolog 1 is transiently expressed by spatially restricted subsets of early neuroepithelial and neural crest cells. *Genes Dev.* **5**, 1524-1537.

Machida, C. M., Rodland, K. D., Matrisian, L., Magun, B. E. and Ciment, G. (1989). NGF induction of the gene encoding the protease transin accompanies neuronal differentiation in PC12 cells. *Neuron* **2**, 1587-1596.

Maruyama, K., Schiavi, S. C., Huse, W., Johnson, G. L. and Ruley, H. E. (1987). Myc and E1A oncogenes alter the response of PC12 cells to nerve growth factor and block differentiation. *Oncogene* **1**, 361-367.

Matsui, M., Tokuhara, M., Konuma, Y., Nomura, N. and Ishzaki, R. (1990). Isolation of human fos-related genes and their expression during monocyte-macrophage differentiation. *Oncogene* **5**, 249-255.

Mayo, K. E., Warren, R. and Palmiter, R. D. (1982). The mouse metallothionein-I gene is transcriptionally regulated by cadmium following transfection into human or mouse cells. *Cell* **29**, 99-108.

Meehan, R., Lewis, J., Cross, S., Nan, X., Jeppeson, P. and Bird, A. (1992). Transcriptional repression by methylation of CPG. *J. Cell Sci.* (suppl.) **16**, 9-14.

Miller, D. A., Curran, T. and Verma, I. M. (1984). c-Fos protein can induce cellular transformation: a novel mechanism of activation of a cellular oncogene. *Cell* **36**, 51-60.

Mordacq, J. C. and Linzer, D. I. H. (1989). Co-localization of elements required for phorbol ester stimulation and glucocorticoid repression of proliferin gene expression. *Genes Dev.* **3**, 760-769.

Nakabeppu, Y. and Nathans, D. (1991). A naturally occurring truncated form of FosB that inhibits Fos/Jun transcriptional activity. *Cell* **64**, 751-759.

Naranjo, J. R., Mellström, B., Achaval, M. and Sassone-Corsi, P. (1991). Molecular pathways of pain: Fos/Jun-mediated activation of a noncanonical AP-1 site in the prodynorphin gene. *Neuron* **6**, 607-617.

Nishikura, K. and Murray, J. M. (1987). Antisense RNA of proto-oncogene c-fos blocks renewed growth of quiescent 3T3 cells. *Mol. Cell. Biol.* **7**, 639-649.

Noda, M., Ko, M., Ogura, A., Liu, D. G., Amano, T., Takano, T. and Ikawa, Y. (1985). Sarcoma viruses carrying ras oncogenes induce differentiation-associated properties in a neuronal cell line. *Nature* **318**, 73-75.

Ornitz, D. M., Moreadith, R. W. and Leder, P. (1991). Binary system for regulating transgene expression in mice: Targeting int-2 gene expression with yeast GAL4/UAS control elements. *Proc. Nat. Acad. Sci. USA* **88**, 698-702.

Pepinsky, B. R., Tizard, R., Mattaliano, R. J., Sinclair, L. K., Miller, G. T., Browning, J. L., Chow, P. E., Burne, C., Huang, K.-S., Pratt, D., Wachter, L., Hession, C., Frey, A. Z. and Wallner, B. P. (1988). Five distinct calcium and phospholipid binding proteins share homology with lipocortin I. *J. Biol. Chem.* **263**, 10799-10811.

Picard, D., Salser, S. J. and Yamamoto, K. R. (1988). A movable and regulable inactivation function within the steroid binding domain of the glucocorticoid receptor *Cell* **54**, 1073-1080.

Piette, J., Hirai, S.-I. and Yaniv, M. (1988). Constitutive synthesis of activator protein 1 transcription factor after viral transformation of mouse fibroblasts. *Proc. Nat. Acad. Sci. USA.* **85**, 3401-3405.

Reichmann, E., Schwarz, H., Deiner, E. M., Leitner, I., Eilers, M., Busslinger, M. and Beug, H. (1992). Activation of an inducible c-FosER fusion protein causes loss of epithelial polarity and triggers epithelial-fibroblastoid cell conversion. *Cell*, in press.

Riabowol, K. T., Vosatka, R. J., Ziff, E. B., Lamb, N. J. and Feramisco, J. R. (1988). Microinjection of fos-specific antibodies blocks DNA synthesis in fibroblasts. *Mol. Cell. Biol.* **8**, 1670-1676.

Riabowol, K., Schiff, J. and Gilman, M. Z. (1992). Transcription factor AP-1 activity is required for initiation of DNA synthesis and is lost during cellular aging. *Proc. Nat. Acad. Sci. USA* **89**, 157-161.

Rivera, V. M. and Greenberg, M. E. (1990). Growth factor-induced gene expression: the ups and downs of c-fos regulation. *New Biologist* **2**, 751-758.

Rojas, E., Pollard, H. B., Haigler, H. T., Parra, C. and Burns, A. L. (1990). Calcium-activated endonexin II forms calcium channels across acidic phospolipid bilayer membranes. *J. Biol. Chem.* **265**, 21207-21215.

Roussel, M. F., Shurtleff, S. A., Downing, J. R. and Sherr, C. J. (1990). A point mutation at tyrosine-809 in the human colony-stimulating factor 1 receptor impairs mitogenesis without abrogating tyrosine kinase activity, association with phosphatidylinositol 3-kinase, or induction of c-fos and junB genes. *Proc. Nat. Acad. Sci. USA* **87**, 6738-6742.

Rüther, U., Garber, C., Komitowski, D., Müller, R. and Wagner, E. F. (1987). Deregulated c-fos expression interferes with normal bone development in transgenic mice. *Nature* **325**, 412-416.

Ryseck, R.-P. and Bravo, R. (1991). c-Jun, JunB and JunD differ in their binding affinities to AP-1 and CRE consensus sequences: effect of Fos proteins. *Oncogene* **6**, 533-542.

Saris, C. J. M., Tack, B. F., Kristensen, T., Glenney, J. R. Jr. and Hunter, T. (1986). The cDNA sequence for the protein-tyrosine kinase substrate p36 (calpactin I heavy chain) reveals a multidomain protein with internal repeats. *Cell* **46**, 201-212.

Sassone-Corsi, P., Sisson, J. C. and Verma, I. M. (1988). Transcriptional autoregulation of the proto-oncogene fos. *Nature* **334**, 314-319.

Schlaepfer, D. D. and Haigler, H. T. (1990). Expression of annexins as a function of cellular growth state. *J. Cell Biol.* **111**, 229-238.

Schönthal, A., Herrlich, P., Rahmsdorf, H. J. and Ponta, H. (1988). Requirement for fos gene expression in the transcriptional activation of collagenase by other oncogenes and phorbol esters. *Cell* **54**, 325-334.

Schönthal, A. and Feramisco, J. R. (1990). Different promoter elements are required for the induced expression of c-fos and c-jun proto-oncognes by the v-mos oncogene product. *New Biologist* **2**, 143-150.

Sistonen, L., Hölttä, E., Mäkelä, T. P., Keski-Oja, J. and Alitalo, K. (1989). The cellular response to induction of the $p21^{c\text{-}Ha\text{-}ras}$ oncoprotein includes stimulation of jun gene expression. *EMBO J.* **8**, 815-822.

Sistonen, L., Keski-Oja, J., Ulmanen, I., Hölttä, E., Wikgren, B.-J. and Alitalo, K. (1987). Dose effects of transfected $c\text{-}Ha\text{-}ras^{Val\ 12}$ oncogene in transformed cell clones. *Exp. Cell Res.* **168**, 518-530.

Spano, F., Raugei, G., Palla, E., Colella, C. and Melli, M. (1990). Characterization of the human lipocortin-2-encoding multigene family: its structure suggests the existence of a short amino acid unit undergoing duplication. *Gene* **95**, 243-251.

Sonnenberg, J. L., Rauscher III, F. J., Morgan, J. I. and Curran, T. (1989). Regulation of proenkephalin by Fos and Jun. *Science* **246**, 1622-1625.

Sterneck, E., Müller, C., Katz, S. and Leutz, A. (1992). Autocrine growth induced by kinase type oncogens in myeloid cells requires AP-1 and NF-M, a myeloid specific C/EBP-like factor. *EMBO J.* **11**, 115-126.

Superti-Furga, G., Bergers, G. , Picard, D. and Busslinger, M. (1991). Hormone-dependent transcriptional regulation and cellular transformation of Fos-steroid receptor fusion proteins. *Proc. Nat. Acad. Sci. USA* **88**, 5114-5118.

Suzuki, T., Okuno, H., Yoshida, T., Endo, T., Nishina, H. and Iba, H.

(1991). Difference in transcriptional regulatory function between c-Fos and Fra-2. *Nucl. Acids Res.* **19**, 5537-5542.

Tabor, C. W. and Tabor, H. (1984). Polyamines. *Ann. Rev. Biochem.* **53**, 749-790.

Tominaga, S.-I. (1989). A putative protein of a growth-specific cDNA from BALB/c-3T3 cells is highly similar to the extracellular protion of mouse interleukin 1 receptor. *FEBS Lett.* **258**, 301-304.

Tominaga, S.-I., Jenkins, N. A., Gilbert, D. J., Copeland, N. G. and Tetsuka, T. (1991). Molecular cloning of the murine ST2 gene: characterization and chromosomal mapping. *Biochem. Biophys. Acta* **1090**, 1-8.

Topol, J., Ruden, D. M. and Parker, C. S. (1985). Sequences required for in vitro transcriptional activation of a Drosophila hsp 70 gene. *Cell* **42**, 527-537.

Umek, R. M., Friedman, A. D. and McKnight, S. L. (1991). CCAAT-enhancer binding protein: a component of a differentiation switch. *Science* **251**, 288-292.

van Steeg, H., van Oostrom, C. T. M., Hodemaekers, H. M. and van Kreyl, C. F. (1990). Cloning and functional analysis of the rat ornithine decarboxylase-encoding gene. *Gene* **93**, 249-256.

Vendrell, M., Zawia, N. H., Serratosa, J. and Bondy, S. C. (1991). c-fos and ornithine decarboxylase gene expression in brain as early markers of neurotoxicity. *Brain Res.* **544**, 291-296.

Wagner-Bernholz, J. T., Wilson, C., Gibson, G., Schuh, R. and Gehring, W. J. (1991). Identification of target genes of the homeotic gene Antennapedia by enhancer detection. *Genes Dev.* **5**, 2467-2480.

Wang, Z. -Q., Grigoriadis, A. E., Möhle-Steinlein, U. and Wagner, E. F. (1991). A novel target cell for c-fos-induced oncogenesis: development of chondrogenic tumours in embryonic stem cell chimeras. *EMBO J.* **10**, 2437-2450.

Wasylyk, B., Wasylyk, C., Flores, P., Begue, A., Leprince D. and Stehelin, D. (1990). The c-ets proto-oncogenes encode transcription factors that cooperate with c-Fos and c-Jun for transcriptional activation. *Nature* **346**, 191-193.

Wasylyk, C., Gutman, A., Nicholson, R. and Wasylyk, B. (1991). The c-Ets oncoprotein activates the stromelysin promoter through the same elements as several non-nuclear oncoproteins. *EMBO J.* **10**, 1127-1134.

Webster, N., Jin, J. R., Green, S., Hollis, M. and Chambon, P. (1988a). The yeast UAS_G is a transcriptional enhancer in human HeLa cells in the presence of the GAL4 transactivator. *Cell* **52**, 169-178.

Webster, N. J. G., Green, S., Jin, J. R. and Chambon, P. (1988b). The hormone-binding domains of the estrogen and glucocorticoid receptors contain an inducible transcription activation function. *Cell* **54**, 199-207.

Wen, L., Huang, J.-K. and Blackshear, P. J. (1989). Rat ornithine decarboxylase gene. *J. Biol. Chem.* **264**, 9016-9021.

White, R. J., Rigby, P. W. J. and Jackson, S. P. (1992). The TATA-binding protein is a general transcription factor for RNA polymerase III. *J. Cell Sci.* (suppl.) **16**, 1-7.

Williams, F., Wagner, F., Karin, M. and Kraft, A. S. (1990). Multiple doses of diacylglycerol and calcium ionophore are necessary to activate AP-1 enhancer activity and induce markers of macrophage differentiation. *J. Biol. Chem.* **265**, 18166-18171.

Wisdom, R., Yen, J., Rashid, D. and Verma, I. M. (1992). Transformation by FosB requires a trans-activation domain missing in FosB2 that can be substituted by heterologous activation domains. *Genes Dev.* **6**, 667-675.

Wrighton, C. and Busslinger, M. (1993). Direct transcriptional stimulation of the ornithine decarboxylase gene by Fos in PC12 cells, but not in fibroblasts. *Mol. Cell. Biol.*, in press.

Zawia, N. H. and Bondy, S. C. (1990). Electrically stimulated rapid gene expression in the brain: ornithine decarboxylase and c-fos. *Mol. Brain Res.* **7**, 243-247.

Journal of Cell Science, Supplement 16, 111-115 (1992)
Printed in Great Britain © The Company of Biologists Limited 1992

Functional conservation of nematode and vertebrate myogenic regulatory factors

MICHAEL KRAUSE[1], ANDREW FIRE[2], SUSAN WHITE-HARRISON[2], HAROLD WEINTRAUB[1] and STEPHEN TAPSCOTT[1]

[1]*Fred Hutchinson Cancer Research Center, Seattle, Washington 98104, USA*
[2]*Carnegie Institute of Washington, Baltimore, Maryland 21210, USA*

Summary

The *Caenorhabditis elegans* protein, CeMyoD, is related to the vertebrate myogenic regulatory factors MyoD, myogenin, MRF-4 and Myf-5. Like its vertebrate counterparts, CeMyoD accumulates in the nucleus of striated muscle cells prior to the onset of terminal differentiation. CeMyoD also shares functional similarities with the vertebrate myogenic regulatory factors. Viral LTR driven expression of CeMyoD in mouse 10T1/2 cells can convert this cell line into myoblasts as well as efficiently *trans*-activate mouse muscle-specific promoters. Furthermore, mouse MyoD expression can activate a CeMyoD-β-galactosidase reporter construct in a 10T1/2 co-transfection assay.

Key words: *C. elegans*, mouse, myogenic regulatory factors, MyoD.

Introduction

The myogenic regulatory genes encode a subgroup of the helix-loop-helix (HLH) family of transcription factors (Murre et al., 1989b). Originally described in vertebrates, the myogenic subgroup consists of the gene products MyoD (Davis et al., 1987), myogenin (Wright et al., 1989), MRF-4/Herculin/Myf-6 (Rhodes and Konieczny, 1989; Braun et al., 1990a; Miner and Wold, 1990), and Myf-5 (Braun et al., 1989). All share a high degree of sequence conservation across a 60 amino acid stretch that includes the HLH region and an adjacent basic (B) domain (Benezra et al., 1990). The basic region has been shown to constitute the DNA-binding motif of these proteins whereas the HLH region is a dimerization motif that allows multimerization with other HLH proteins (Murre et al., 1989a,b; Davis et al., 1990).

The expression of each of these myogenic factors by itself is sufficient to convert many different cultured cells into myoblasts that will, in turn, differentiate into striated muscle when placed in the appropriate growth conditions (Weintraub et al., 1989; Choi et al., 1990). The myogenic activity of these factors is due to their direct action as transcription factors; they can positively auto-activate themselves, cross-activate each other, and bind to and activate enhancer elements regulating muscle-specific promoters (Braun et al., 1989; Thayer et al., 1989; Edmondson and Olson, 1989; Lassar et al., 1989; Murre et al., 1989a; Braun et al., 1990b; Davis et al., 1990). This ability to convert non-myogenic cells to myoblasts, coupled with an early onset of expression during development, has led to the notion that these myogenic factors play a key role in regulating myogenesis in vertebrates (Weintraub et al., 1991).

CeMyoD and developmental expression

The nematode *Caenorhabditis elegans* has a gene, *hlh-1* (*h*elix-*l*oop-*h*elix), that encodes a protein, CeMyoD, that is related to the myogenic regulatory factors of vertebrates (Krause et al., 1990). To date, no other related genes have been identified in the nematode. With in the B-HLH region, 80% of the amino acid residues of CeMyoD are identical to the vertebrate myogenic factors (Table 1), and CeMyoD preferably binds to the E-box (CANNTG) motif (K. Blackwell, M. K. and H. W., unpublished data). Outside of the B-HLH region, there is no obvious similarity between CeMyoD and any other myogenic regulatory factor yet described.

As in the vertebrates, CeMyoD is expressed in striated muscle cells and their precursors and begins to accumulate prior to any overt signs of muscle differentiation. CeMyoD expression occurs in two waves during *C. elegans* development. The first wave is transient, beginning at the 28-cell stage of development and lasting only for two cell divisions. During this stage, low levels of CeMyoD are detected in only one founder cell lineage, the MS (Krause et al., 1990; Chen et al., 1992). Although some descendants of these CeMyoD positive blastomeres will become striated muscle cells, others will become pharyngeal cells, neurons and non-muscle mesodermal cell types. Thus, low level

Table 1. *Comparison of the basic-helix-loop-helix region of the myogenic factors*

Factor	Amino acid residues: Basic — Helix 1 — Loop — Helix 2
MyoD subfamily	DRRKAATMRERRRLSKVNEAFETLKR-TS-NPNQRLPKVEILRNAI-YIE-LQ
Myf-5 subfamily	::::::::::::::K:::Q:::::::C:TT::::::::::::::::-:::S::
Myogenin subfamily	:::R:::L::K:::K:::::::A:::S:LL:::::::::::::-::Q:::R::
MRF-4 subfamily	:::::::L::::::K:I:::::A:::R:VA:::::::::::::S::-:::R::
C. elegans MyoD	::::::::::::::R:::::::VV:QR:CP:::::::::::::S::D::NN:E

A comparison of consensus B-HLH region sequences of the vertebrate myogenic regulatory genes and CeMyoD. All sequences are arbitrarily compared to the vertebrate MyoD subfamily. Colons (:), amino acid identity; dashes (–), no consensus residue exists for the position; amino acid differences are as indicated below the corresponding residue in the MyoD subfamily sequence.

expression of CeMyoD is not sufficient, at this stage of development, to ensure that these cells become committed to the striated muscle cell fate.

A second wave of CeMyoD expression begins at about the 90-cell stage of embryogenesis; expression at this stage is stable, that is, once a blastomere becomes CeMyoD positive, it and its descendants remain CeMyoD positive throughout development (Krause et al., 1990). With one exception, all stable CeMyoD positive blastomeres will become striated muscle cells and only striated muscle cells. Moreover, stable CeMyoD expression precedes terminal differentiation (expression of muscle structural proteins) by up to four cell divisions in some lineages. This early onset of expression is analogous to that seen for the myogenic factors during mammalian development (Bober et al., 1991).

There is one exception to the rule that all stable CeMyoD positive cells become striated muscle. Late in embryogenesis a set of six cells (glial-like cells) near the nerve ring become CeMyoD positive; these cells are not muscle cells and do not stain with antibodies raised against muscle structural proteins. Interestingly, these six cells form gap junctions with some of the striated muscle cells in the head of the animal. It is possible that intercellular signaling through these gap junctions results in CeMyoD accumulation in these six non-muscle cells. Accumulation of CeMyoD in these six cells appears to occur only after they have been born and begun differentiating as glial-like cells. CeMyoD expression, therefore, appears to be unable to alter the fate of a previously differentiated cell type.

CeMyoD can convert murine 10T1/2 cells to muscle

The similarities between the nematode and vertebrate myogenic factors suggested they might also be functionally related. A transient transfection assay of the murine fibroblast cell line C3H10T1/2 (10T1/2) (Taylor and Jones, 1979) has been used to address the question of functional conservation.

A viral LTR expression vector, EMSVscribe (Davis et al., 1987), was engineered to transcribe a full length cDNA encoding CeMyoD in either the sense or anti-sense orientation. These constructs, along with the vector alone or an EMSV mouse MyoD sense construct, were transfected independently into the murine fibroblast cell line 10T1/2. After two days, cells were placed in differentiation media (DMEM supplemented with insulin and transferin) for an additional two days and assayed for myogenesis as previously described (Davis et al., 1987; Davis et al., 1990).

Transfection of the EMSV mouse MyoD construct resulted in a high frequency of myogenic conversion as expected (2%-5% of the cell population). In contrast, both the EMSV vector alone and the EMSV CeMyoD anti-sense construct resulted in no myogenesis as assayed by visual inspection and staining for myosin heavy chain gene expression. However, transfection of the EMSV CeMyoD sense construct resulted in low level myogenesis with approximately 0.1% of the cells forming bipolar cells expressing myosin heavy chain gene. Although weak, this myogenic activity of the EMSV CeMyoD sense construct is significant as both negative controls give no myogenic conversion.

The level of myogenesis observed following transfection of 10T1/2 cells with EMSV CeMyoD sense could be attributable to activation of the mouse myogenic regulatory genes. Steady state mRNA levels of endogenous mouse myogenic gene products were determined by reverse transcriptase-polymerase chain reaction (RT-PCR) analysis of total cellular RNA isolated from cells maintained in either proliferative or differentiation media. Specifically, levels of mRNA for transfected CeMyoD and the endogenous genes encoding mouse MyoD, myogenin and muscle creatine kinase (MCK) were compared (Fig. 1).

Although EMSV mouse MyoD transfection led to high level expression from the three endogenous genes monitored, EMSV CeMyoD sense resulted in only low levels of expression (Fig. 1A,B). For both transfected expression constructs, the level of endogenous myogenic regulatory gene expression correlates with their respective strengths as myogenic activators; high for EMSV mouse MyoD and low for EMSV CeMyoD sense. Immunological staining of EMSV CeMyoD transfected cells shows that mouse MyoD is detected within the nuclei of the myotubes that arise. Taken together, these results suggest that the CeMyoD induced myogenic conversion events are mediated by activation of endogenous myogenic regulatory genes.

It is important to note that regardless of whether CeMyoD directly, or indirectly, activates the myogenic program in 10T1/2 cells, it clearly can function as a myogenic factor within these mouse cells. This myogenic ability is not a general feature of B-HLH proteins but rather is unique to the myogenic subgroup of B-HLH proteins. In fact, replacement of the mouse MyoD basic residues with the corresponding basic region from the HLH protein E12, that

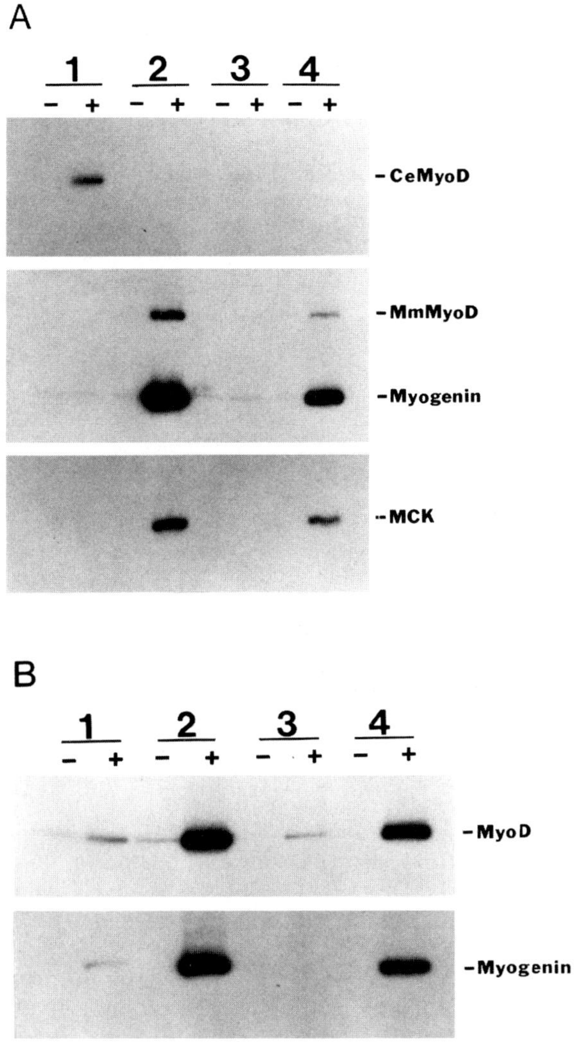

Fig. 1. Reverse transcriptase-polymerase chain reaction (RT-PCR) analysis of gene expression in transfected 10T1/2 cells. (A) 10T 1/2 cell cultures were transfected with either EMSV-CeMyoD (lanes 1 and 3) or EMSV-mouse MyoD (lanes 2 and 4) and grown in differentiation (lanes 1 and 2) or proliferation (lanes 3 and 4) conditions; note similar results are obtained regardless of growth conditions. Total RNA isolated from each culture was used to prime a cDNA reaction with (+) or without (−) the addition of reverse transcriptase. One fifth of the cDNA reaction was subsequently used in a standard PCR reaction to detect the presence or absence of cDNA products corresponding to the following mRNAs: CeMyoD, the transfected nematode MyoD gene; MmMyoD, the endogenous mouse MyoD gene; Myogenin, the endogenous mouse myogenin gene; MCK, the endogenous mouse muscle creatine kinase gene. Transfected mouse MyoD gene products were not assayed and are not detected by the primer pair used to assay the endogenous mouse MyoD gene products. (B) Same as in A except only the endogenous mouse MyoD and myogenin gene products were assayed. CeMyoD expression elicits a weak, but detectable, level of endogenous MyoD and myogenin expression (lanes 1 and 3). For comparison, notice the high levels of endogenous mouse gene expression activated by transfection of the EMSV-mouse MyoD construct (lanes 2 and 4).

also binds the E-box (CANNTG) motif, completely eliminates myogenic activity (Davis et al., 1990).

CeMyoD activates mouse muscle-specific promoters

The EMSV CeMyoD constructs have also been used in a co-transfection assay of 10T1/2 cells with two mouse muscle-specific reporter gene constructs linked to bacterial chloramphenicol acteyltransferase (CAT): MCK-CAT (Jaynes et al., 1988) and desmin-CAT (Pieper et al., 1987). CAT activity was quantitated biochemically using a ^{14}C-labelled acetyl CoA incorporation assay (Sleigh, 1986).

Table 2 shows the results of CAT activity from this type of transient co-transfection experiment. Both MCK-CAT and desmin-CAT are activated to comparable levels (200- to 300-fold above background) by either EMSV mouse MyoD or EMSV CeMyoD in the sense orientation. In contrast, EMSV CeMyoD anti-sense gives only background levels of activation of both CAT reporters.

As discussed above, EMSV CeMyoD activates very low levels of endogenous mouse MyoD and myogenin. To eliminate the possibility that these endogenous mouse myogenic gene products were responsible for the high levels of MCK-CAT and desmin-CAT activity in these experiments, a CAT reporter construct driven by a tandemly repeated mouse MyoD binding site derived from the MCK enhancer was tested. This reporter construct, called 4R-CAT, is efficiently *trans*-activated by mouse MyoD and myogenin and serves as an indicator of either exogenous or endogenous mouse myogenic B-HLH protein activity (Weintraub et al., 1990). If, for example, CeMyoD-induced low level expression of the endogenous mouse MyoD was activating high levels of MCK-CAT and desmin-CAT, then it should similarly activate high levels of 4R-CAT. However, when EMSV CeMyoD constructs are co-transfected into 10T1/2 cells with 4R-CAT in a transient assay, the 4R-CAT reporter is not activated. Although it is formally possible that EMSV CeMyoD is specifically inhibiting 4R-CAT expression in these co-transfections, a more likely explanation is that CeMyoD positively activates the high level expression of MCK-CAT and desmin-CAT independently of the endogenous B-HLH myogenic regulatory genes. It is not known if CeMyoD activates the MCK and desmin promoters directly or works by interacting with, or activating, other endogenous mouse transcription factors.

Table 2. Trans-*activation of CAT reporter gene constructs*

Expression construct	Reporter construct	
	MCK-CAT	Desmin-CAT
EMSVscribe	0.4	0.4
EMSV mouse MyoD	107	150
EMSV CeMyoD anti-sense	0.7	0.3
EMSV CeMyoD sense	122	134

10T1/2 cells were co-transfected with an EMSV expression construct and either MCK-CAT or desmin-CAT reporter constructs. CAT activity was quantitated by a ^{14}C-labelled acetyl CoA incorporation assay. Numbers represent raw counts in thousands.

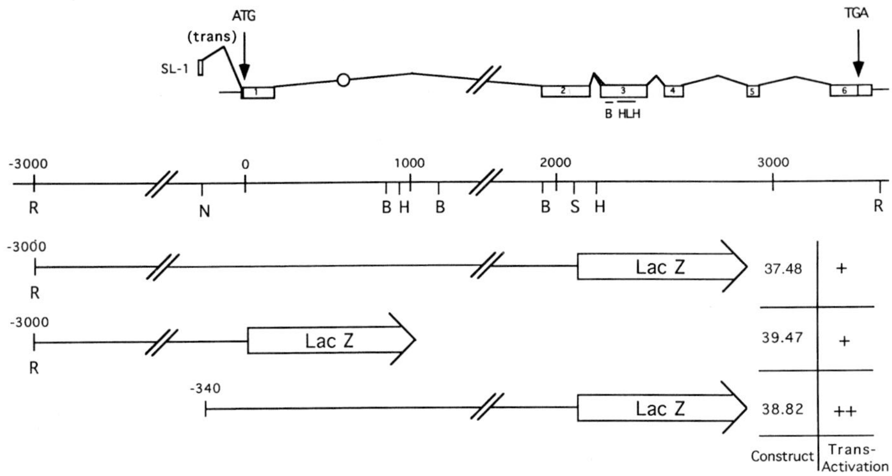

Fig. 2. *C. elegans hlh-1* gene structure and β-galactosidase reporter gene fusions. The intron-exon structure of the *C. elegans* gene *hlh-1*, that encodes CeMyoD, is shown at top with the exons numbered 1 to 6. The 5′ end of the *hlh-1* transcript is generated by a *trans*-splicing reaction indicated by the offset, and unnumbered, exon box labeled SL-1. A potential autoregulatory enhancer site is indicated by a circle within the large intron between exons 1 and 2. Distance (in base pairs) and representative restriction enzyme sites are depicted along a line beneath the gene structure. Three *hlh-1*-β-galactosidase reporter gene constructs are also diagrammed and labeled 37.48, 39.47 and 38.82. The relative efficiency of *trans*-activation of these reporter genes in murine 10T1/2 cells following co-transfection of EMSV-mouse MyoD is indicated by plus (+) signs.

Mouse myoD *trans*-activates CeMyoD

One characteristic of the vertebrate myogenic regulatory factors is that they auto-activate their own expression and can *trans*-activate expression of at least some of the other myogenic genes. For example, transfection of EMSV mouse MyoD into 10T1/2 cells will activate the endogenous MyoD and myogenin genes (Braun et al., 1989a; Thayer et al., 1989). To determine if mouse MyoD could positively act on sequences of the nematode gene *hlh-1* that encodes CeMyoD, EMSV mouse MyoD was co-transfected with three *hlh-1* β-galactosidase reporter constructs into 10T1/2 cells. All three *hlh-1* reporter constructs were activated by EMSV mouse MyoD but not by the EMSVscribe vector alone (Fig. 2). A β-galactosidase reporter gene construct lacking *hlh-1* sequences was not *trans*-activated suggesting that one or more sites within the *hlh-1* region is responsible for the positive effect by mouse MyoD.

Studies of the nematode *hlh-1* promoter elements has shown that the 3 kb upstream of the CeMyoD coding region is sufficient, in transgenic nematode strains, to give the correct temporal and spatial pattern of CeMyoD expression during development (Krause et al., 1990). It was surprising then that the *hlh-1* reporter construct that lacks much of these upstream promoter sequences (38.82) responded as well, if not better, than those constructs with an intact promoter region (37.48 and 39.47) (Fig. 2). An inspection of the *hlh-1* intron sequences that are retained in construct 38.82 revealed a cluster of seven E-box (CANNTG) elements spanning a 100 bp region; this region could serve as a site for mouse MyoD binding and *trans*-activation. Preliminary evidence (gel mobility shift and inter-species *hlh-1* gene sequence comparisons) supports this notion and suggests that this 100 bp region is a site for CeMyoD auto-activation (T. Gerber, M. K. and H. W., unpublished data).

Conclusions

A comparison of the vertebrate myogenic regulatory factors with the related nematode protein CeMyoD reveals a remarkable degree of evolutionary conservation. In addition to the sequence similarity across the B-HLH domain of these proteins, they share similar temporal and spatial patterns of expression during development. Vertebrate and nematode myogenic proteins also share functional similarities; CeMyoD is able to convert the murine cell line 10T1/2 to myoblasts and efficiently *trans*-activate mouse muscle-specific promoters in transient transfection assays. Furthermore, mouse MyoD appears to recognize, and activate, a CeMyoD autoregulatory site within the nematode *hlh-1* gene.

This is not the first reported instance of a non-mammalian myogenic factor acting in the 10T1/2 cell line. Venuti et al. (1991) isolated a sea urchin analogue of the vertebrate myogenic factors called SUM-1. Transfection of EMSV SUM-1 can convert 10T1/2 cells, and *trans*-activate muscle-specific reporter gene expression, as efficiently as EMSV mouse MyoD. More recently, the *Xenopus laevis MyoDb* gene has been used to convert this same cell line to muscle (R. Rupp and H. W., unpublished data).

The ability of these non-mammalian myogenic factors to convert 10T1/2 cells to myoblasts should, however, be kept in perspective; 10T1/2 cells are a cultured line that grow outside the developmental context in which these factors normally operate. One should be cautious, therefore, in ascribing developmental functions to these myogenic factors based solely on this assay.

Related to this point, recent results in *C. elegans* demonstrate that myogenesis can occur in the absence of *zygotic* CeMyoD (Chen et al., 1992). This was shown by generating chromosomal deficiencies spanning the *hlh-1* (and flanking) genes. Mutant embryos homozygous for these deficiencies arrest at the two-fold stage of embryogenesis but still make striated muscle. Although it is still possible that maternal *hlh-1* products or a second myogenic gene (as yet undiscovered) is functioning in these homozygous deficiency animals, these unanticipated results suggest that we do not yet understand the role, if any, of CeMyoD in determining muscle cell fate in the nematode.

Given the similarities between the nematode and vertebrate myogenic factors, one might also reconsider the role of the myogenic regulatory factors in vertebrate myogenesis. Is the developmental process of myogenesis absolutely conserved through evolution or have these distantly related organisms diverged to the point that their respective myogenic factors function in fundamentally different programs? Presently one can make circumstantial arguments in support of either side of this issue. For example, based on the sequence, DNA binding, 10T1/2 cell conversion assay, and temporal and spatial pattern of expression one would conclude that the vertebrate and nematode myogenic proteins are very similar. However, no other B-HLH protein has been uncovered in *C. elegans*, leading one to wonder if the CeMyoD dimerization partners (E2a and Id gene products) exist in the nematode. Furthermore, no muscle structural gene promoter studied to date in *C. elegans* has been shown to require an E-box (CANNTG) site for expression.

Circumstantial arguments do not answer the question of whether developmentally relevant functional similarities exist between the vertebrate and nematode myogenic regulatory factors. Fortunately, these are questions that can be answered experimentally through a combination of nematode genetics and vertebrate gene disruptions.

References

Benezra, R., Davis, R. L., Lockshon, D., Turner, D. L. and Weintraub, H. (1990). The protein Id: a negative regulator of helix-loop-helix DNA binding proteins. *Cell* **61**, 49-59.

Bober, E., Lyons, G. E., Braun, T., Cossu, G., Buckingham, M. and Arnold, H. H. (1991). The muscle regulatory gene, Myf-6, has a biphasic pattern of expression during early mouse development. *J. Cell Biol.* **113**, 1255-1265.

Braun, T., Bober, E., Buschhausen-Denker, G., Kotz, S., Grzeschik, K-H. and Arnold, H. H. (1989). Differential expression of myogenic determination genes in muscle cells: possible autoactivation by Myf gene products. *EMBO J.* **8**, 3617-3625.

Braun, T., Bober, E., Winter, B., Rosenthal, N. and Arnold, H. H. (1990a). Myf-6, a new member of the human gene family of myogenic determination factors: evidence for a gene cluster on chromosome 12. *EMBO J.* **9**, 821-831.

Braun, T., Winter, B., Bober, E. and Arnold, H. H. (1990b). Transcriptional activation domain of the muscle-specific gene-regulatory protein myf5. *Nature* **346**, 663-665.

Chen, L., Krause, M., Draper, B., Weintraub, H. and Fire, A. (1992). Body-wall muscle formation in *Caenorhabditis elegans* embryos that lack the MyoD homolog *hlh-1*. *Science* **256**, 240-243.

Choi, J., Costa, M. L., Mermelstein, C. S., Chagas, C., Holtzer, S. and Holtzer, H. (1990). MyoD converts primary dermal fibroblasts, chondroblasts, smooth muscle and retinal pigment epithelial cells into striated, mononucleated myoblasts and multinuceated myotubes. *Proc. Nat. Acad. Sci. USA* **87**, 7988-7992.

Davis, R. L., Weintraub, H. and Lassar, A. B. (1987). Expression of a single transfected cDNA converts fibroblasts to myoblasts. *Cell* **51**, 987-1000.

Davis, R. L., Chen, P. F., Lassar, A. B. and Weintraub, H. (1990). The MyoD DNA binding domain contains a recognition code for muscle-specific gene activation. *Cell* **60**, 733-746.

Edmondson, D. G. and Olson, E. N. (1989). A gene with homology to the *myc* similarity region of MyoD1 is expressed during myogenesis and is sufficient to activate the muscle differentiation program. *Genes Dev.* **3**, 628-640.

Jaynes, J. B., Johnson, J. E., Buskin, J. N., Gartside, C. L. and Hauschka, S. (1988). The muscle creatine kinase gene is regulated by multiple upstream elements, including a muscle-specific enhancer. *Mol. Cell Biol.* **8**, 62-70.

Krause, M., Fire, A., White-Harrison, S., Priess, J. and Weintraub, H. (1990). CeMyoD accumulation defines the body wall muscle cell fate during *C. elegans* embryogenesis. *Cell* **63**, 907-919.

Lassar, A. B., Buskin, J. N., Lockshon, D., Davis, R. L., Apone, S., Hauschka, S. D. and Weintraub, H. (1989). MyoD is a sequence-specific DNA binding protein requiring a region of *myc* homology to bind to the muscle creatine kinase enhancer. *Cell* **58**, 823-831.

Miner, J. H. and Wold, B. (1990). Herculin, a fourth member of the MyoD family of myogenic regulatory genes. *Proc. Nat. Acad. Sci. USA* **87**, 1089-1093.

Murre, C., McCaw, P. S. and Baltimore, D. (1989a). The new DNA binding and dimerization motif in immunoglobulin enhancer binding *daughterless*, MyoD, and *myc* proteins. *Cell* **56**, 777-783.

Murre, C., McCaw, P. S., Vassin, H., Caudy, M., Jan, L. Y., Jan, Y. N., Cabrera, C. V., Buskin, J. N., Hauschka, S. D., Lassar, A. B., Weintraub, H. and Baltimore, D. (1989b). Interactions between heterologous helix-loop-helix proteins generate complexes that bind specifically to a common DNA sequence. *Cell* **58**, 537-544.

Pieper, F. R., Slobbe, R. L., Ramaekers, F. C. S., Cuypers, H. T. and Bloemendal, H. (1987). Upstream sequences of the hamster demin and vimentin genes regulate expression during *in vitro* myogenesis. *EMBO J.* **6**, 3611-3618.

Rhodes, S. J. and Konieczny, S. F. (1989). Identification of MRF4: a new member of the muscle regulatory factor gene family. *Genes Dev* **3**, 2050-2061.

Sleigh, M. J. (1986). A non-chromatographic assay for expression of the chloramphenicol acetyltransferase gene in eucaryotic cells. *Ann. Biochem.* **156**, 254-256.

Taylor, S. M. and Jones, P. A. (1979). Multiple new phenotypes induced in 10T1/2 cells treated with 5-azacytidine. *Cell* **17**, 771-779.

Thayer, M. J., Tapscott, S. J., Davis, R. L., Wright, W. E., Lassar, A. B. and Weintraub, H. (1989). Positive autoregulation of the myogenic determination gene MyoD1. *Cell* **58**, 241-248.

Venuti, J. M., Goldberg, L., Chakraborty, T., Olson, E. N. and Klein, W. H. (1991). A myogenic factor from sea urchin embryos cabable of programming muscle differentiation in mammalian cells. *Proc. Nat. Acad. Sci. USA* **88**, 6219-6223.

Weintraub, H., Tapscott, S. J., Davis, R. L., Thayer, M. J., Adam, M. A., Lassar, A. B. and Miller, A. D. (1989). Activation of muscle-specific genes in pigment, nerve, fat, liver and fibroblast cell lines by forced expression of MyoD. *Proc. Nat. Acad. Sci. USA* **86**, 5434-5438.

Weintraub, H., Davis, R., Lockshon, D. and Lassar, A. (1990). MyoD binds cooperatively to two sites in a target enhancer sequence: Occupancy of two sites is required for activation. *Proc. Nat. Acad. Sci. USA* **87**, 5623-5627.

Weintraub, H., Davis, R., Tapscott, S. J., Thayer, M., Krause, M., Benezra, R., Blackwell, T. K., Turner, D., Rupp, R., Hollenberg, S., Shuang, Y. and Lassar, A. (1991). The MyoD gene family: Nodal point during specification of the muscle cell lineage. *Science* **251**, 761-766.

Wright, W. E., Sassoon, D. A. and Lin, V. K. (1989). Myogenin, a factor regulating myogenesis, has a domain homologous to MyoD1. *Cell* **56**, 607-617.

Journal of Cell Science, Supplement 16, 117-122 (1992)
Printed in Great Britain © The Company of Biologists Limited 1992

Multiple effects on liver-specific gene expression in albino lethal mice caused by deficiency of an enzyme in tyrosine metabolism

GAVIN KELSEY*, SIEGFRIED RUPPERT†, ANDREAS SCHEDL, ERIKA SCHMID, EDDA THIES and GÜNTHER SCHÜTZ

Institute of Cell and Tumor Biology, German Cancer Research Center, Im Neuenheimer Feld 280, 6900 Heidelberg, Germany

*Author for correspondence

†Present address: Howard Hughes Medical Institute, Department of Molecular and Cell Biology, University of California at Berkeley, 401 Barker Hall, Berkeley, CA 94720, USA

Summary

***alf/hsdr-1* is a locus in the mouse defined by albino deletions to be essential for neonatal viability. Homozygous deletion of *alf/hsdr-1* leads to a pleiotropic phenotype in liver and kidney, including impaired perinatal activation of hormone-dependent genes, and the induction of detoxifying enzymes and early-response genes. To elucidate the molecular basis of this complex phenotype, we have identified the gene mapping at *alf/hsdr-1* by positional cloning, using overlapping albino locus deletions to define the location of *alf/hsdr-1*. The gene encodes fumarylacetoacetate hydrolase, FAH, an enzyme of tyrosine metabolism. Genetically determined FAH deficiency in man leads to a severe liver failure in infants. In mice, we find that the normal sites of expression of FAH correlate tightly with cell-types which display abnormalities in albino lethal mice. The identification of the *Fah* gene as a candidate for *alf/hsdr-1* offers a novel explanation for the complex phenotype, one into which all aspects can be accommodated. The phenotype can now be understood as a sequence of responses to toxic electrophilic metabolites.**

Key words: albino-deletion complex, liver-specific gene expression, tyrosine metabolism.

Multiple effects on gene expression in mice carrying neonatally lethal albino deletions

The albino-deletion complex on chromosome 7 in the mouse comprises at least 37 radiation-induced, recessive lethal deletions which define several loci essential for normal development before or after birth (Russell et al., 1982; Rinchik and Russell, 1990). One locus is necessary for survival beyond birth and has been proposed to play an important, regulatory role in the perinatal differentiation of hepatocytes (Gluecksohn-Waelsch, 1979). The locus is designated *alf* (*f*actor indicated by the *a*lbino *l*ethal mutation, Ruppert et al., 1990) or *hsdr-1* (*h*epatocyte-*s*pecific *d*evelopmental *r*egulation *1*, McKnight et al., 1989) and is referred to as *alf/hsdr-1* here.

Mice homozygous for deletions that include *alf/hsdr-1* die a few hours after birth. Lethality is associated with severe hypoglycaemia and pleiotropic effects on gene expression in the liver and kidney. The most extensively investigated defect is the failure of perinatal activation of several enzymes involved in gluconeogenesis, notably glucose-6-phosphatase (G6Pase), phosphoenolpyruvate carboxykinase (PEPCK) and tyrosine aminotransferase (TAT), (Erickson et al., 1968; Gluecksohn-Waelsch, 1979; Schmid et al., 1985; Loose et al., 1986). Concomitant with the enzyme deficiencies are reductions in the abundance of a limited number of mRNAs in liver and kidney and decreased transcription of the respective genes (Loose et al., 1986; Morris et al., 1988; Ruppert et al., 1990). The severity of the deficit of TAT mRNA in late fetal and newborn albino lethal liver is illustrated by the northern analysis in Fig. 1. Analysis of a set of affected mRNAs identified by subtractive cDNA hybridization revealed that expression of most of the affected genes was normally inducible in liver by glucocorticoids and/or cAMP around birth (Ruppert et al., 1990). Since expression of G6Pase and TAT cannot be induced by exogenous hormones in liver of albino lethal mice (Thorndike et al., 1973; Schmid et al., 1985), it seemed likely that lack of hormone inducibility was the major component of the reduced expression. This conclusion led to models proposing that *alf/hsdr-1* encoded a factor participating in hormonal activation of gene transcription (Gluecksohn-Waelsch, 1987). A specific deficit of factors in the signal transduction pathways for glucocorticoids and cAMP has not been found (Ruppert et al., 1990; DeFranco et al., 1991). Recently, however, transcripts for some transcription factors involved in liver-specific gene expression have been shown to be present at reduced levels. These include the CCAAT/enhancer binding protein C/EBP, and hepatic nuclear factors HNF-1 and HNF-4

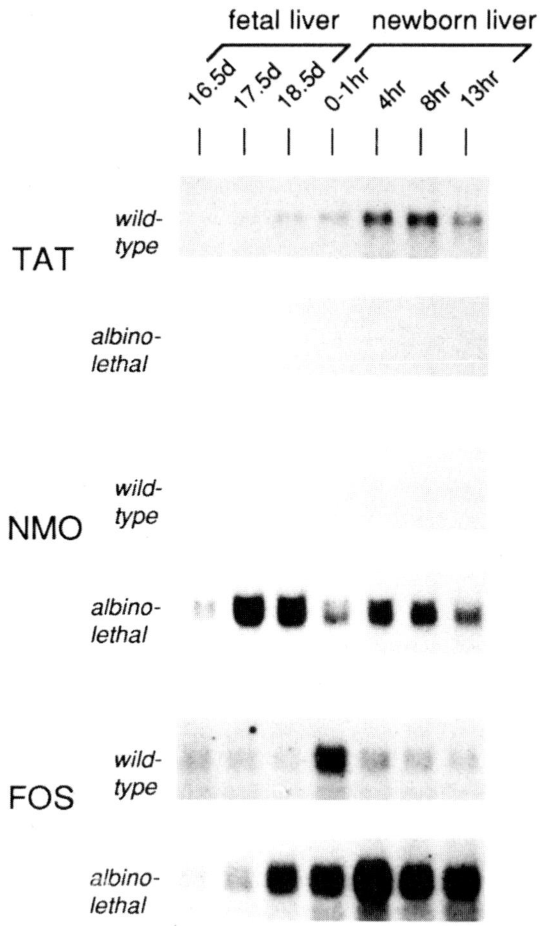

Fig. 1. Multiple effects on gene expression in mice carrying neonatally lethal albino deletions. Northern blot analysis of the expression of TAT, NMO and FOS mRNAs in fetal and newborn liver from albino lethal and wild-type mice. The fetal stages are indicated as days (d) post-coitum, and newborn as hours (hr) after birth. Each lane contains 10 μg total RNA.

(McKnight et al., 1989; Ruppert et al., 1990; Gonzalez et al., 1990; Tönjes et al., 1992).

In addition to the reduced expression of specific genes in liver and kidney, some genes have been found to be over-expressed. These appear to fall into two categories. The first set encodes enzymes involved in oxidative and non-oxidative detoxification reactions: UDP-glucoronyltransferase (Ugt-1), glutathione *S*-transferase B (Gt-1) and NAD(P)H:menadione oxido-reductase (NMO), (Thaler et al., 1976; Gatmaitan et al., 1977; Petersen et al., 1989). NMO mRNA is highly induced by fetal day 16.5 in albino lethal liver, whereas the transcript is not detected in wild-type liver (Fig. 1). These enzymes belong to the aryl hydrocarbon [*Ah*] battery (Nebert et al., 1990), and it has been proposed that a locus encoding a negative regulator of the *Ah* battery, designated *Nmo-1n*, is removed by lethal albino deletions (Nebert et al., 1990). The second category appears to comprise genes with the common property of induction in response to DNA damage. This includes three *gadd* genes (*g*rowth *a*rrest and *D*NA *d*amage inducible, Fornace et al., 1989), and c-*fos* (Ruppert *et al.*, 1992). Whereas FOS mRNA is readily detected only 0-1 h after birth in wild-type liver, it is present at high levels from fetal day 18.5 and maintained up to 13 h after birth in albino lethal liver (Fig. 1). A sustained induction of FOS mRNA accompanies DNA damage by alkylating agents (Hollander and Fornace, 1989). Of the *gadd* transcripts induced in albino lethal liver, one, *gadd153*, has recently been identified to encode a member of the C/EBP family, CHOP-10, a putative inhibitor of C/EBP-like transcription factors (Ron and Habener, 1992).

As well as effects on gene expression, ultrastructural abnormalities visible by electron microscopy have been described in liver and kidney (Trigg and Gluecksohn-Waelsch, 1973). They develop late in gestation and affect the membranes of the rough endoplasmic reticulum, Golgi apparatus and nuclear membrane. These abnormalities may account for a reduced synthesis of serum proteins by the albino lethal liver (Garland et al., 1976).

Identification of a candidate gene for *alf/hsdr-1* by positional cloning

The diverse nature of the phenotype made it difficult to predict a simple function for the *alf/hsdr-1* gene product upon which a biochemical approach to its identification could be based. Instead, we attempted to isolate the gene mapping at *alf/hsdr-1* by positional cloning, taking advantage of the fact that the locus is defined by a large number of chromosomal deletions around the albino locus, *c*. Complementation analysis of *c*-locus deletions indicates that *alf/hsdr-1* is the first essential locus proximal to *c* (Russell et al., 1982). A panel of overlapping deletions that do or do not involve *alf/hsdr-1* was used to order molecular markers. In parallel, pulsed-field gel electrophoresis was employed to link the markers physically and create a long-range restriction map of the albino-deletion complex. This revealed that lethal albino deletions encompassing *alf/hsdr-1* were large; c^{14CoS}, for example, a deletion which has been the subject of much analysis at the phenotypical level, proved to be approx. 3800 kb (Fig. 2A; Kelsey et al., unpublished data). However, by comparing with deletions such as c^{11DSD} that complement c^{14CoS} for neonatal lethality and do not, therefore, extend to *alf/hsdr-1*, it was possible to exclude most of the c^{14CoS} deletion and define an approx. 310 kb minimal interval for the candidate gene (Fig. 2A; Schedl et al., unpublished data). To obtain new probes within the interval, chromosome jumping was undertaken from two flanking markers, RN226.1 proximally (Klebig et al., 1992) and palb18 distally (Niswander et al., 1991). Eight jumps in libraries based on rare-cutting enzymes *Xma*I, *Sal*I and *Bss*HII were made, and more than 100 kb of the approx. 310 kb interval cloned in phage contigs (Fig. 2B; Schedl et al., unpublished data). Transcription units mapping within the contigs were identified by zoo blot and northern analysis. The contig designated RN which overlaps the proximal end of the c^{14CoS} deletion was found to contain the gene for fumarylacetoacetate hydrolase (FAH; Fig. 2B; Klebig et al., 1992; Ruppert et al., 1992). FAH catalyses the final step in the degradation of tyrosine, the hydrolysis of fumarylacetoacetate, FAA. FAH deficiency is

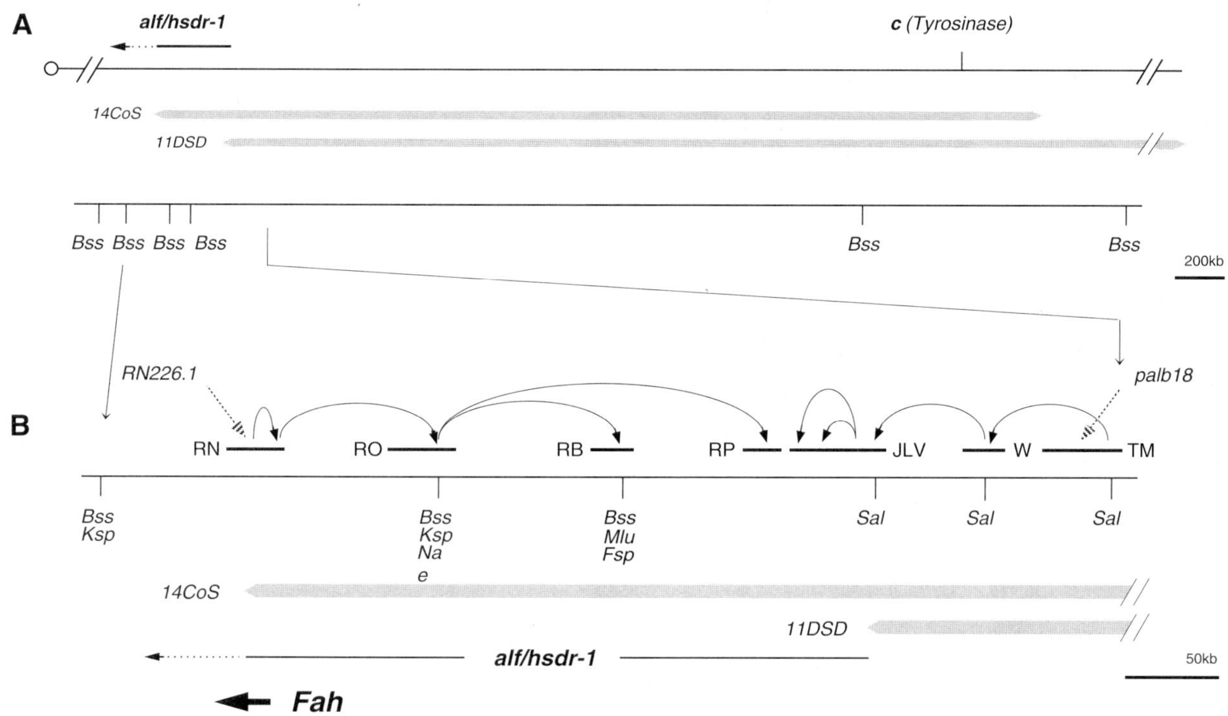

Fig. 2. Identification of a candidate gene for *alf/hsdr-1* by positional cloning. In (A), the location of *alf/hsdr-1* is shown on the chromosome proximal to *c*, and defined as lying between the ends of albino deletions c^{14CoS} and c^{11DSD}, which are represented by the stippled bars. The distribution of *Bss*HII sites (*Bss*) over the region spanned by the c^{14CoS} deletion is shown on the lower line. (B) shows a high resolution physical analysis of the minimal region for *alf/hsdr-1*. The series of chromosome jumps made from flanking markers RN226.1 and palb18 is shown by the arcs. The solid lines indicate contigs cloned across the region, and the location of a selection of rare-cutter sites is shown. The stippled bars represent the c^{14CoS} and c^{11DSD}deletions. The minimal region for the location of *alf/hsdr-1* is shown; the broken arrow indicates the possibility that the gene for *alf/hsdr-1* could extend beyond the c^{14CoS} deletion. The thick arrow demotes the location of the *Fah* gene.

known to be the underlying defect in an inborn error of metabolism in man, hereditary tyrosinaemia type I (HT; Kvittingen, 1986; Goldsmith and Laberge, 1989). HT presents as a severe liver failure which is fatal in the first months of life if liver transplantation cannot be performed. Liver damage is thought to come about from the accumulation of the highly electrophilic tyrosine metabolite FAA, which is considered to be toxic (Laberge et al., 1986). The existence of this human disease suggests that absence of FAH might account for or contribute to the liver phenotype of albino lethal mice.

The expression pattern of FAH implicates FAH deficiency in the *alf/hsdr-1* phenotype

The positive and negative effects on gene expression that characterize the *alf/hsdr-1* phenotype occur in liver and kidney. In situ hybridization analysis with probes for TAT and PEPCK identified the affected cell-types as parenchymal cells in the liver and proximal convoluted tubule cells in the kidney (Ruppert et al., 1990). The same two cell-types manifest the ultrastructural abnormalities (Trigg and Gluecksohn-Waelsch, 1973). To obtain an indication that absence of the *Fah* gene could contribute to this pleiotropic phenotype, the sites of expression of FAH mRNA were examined. Northern blot analysis of wild-type mouse tissues detected high levels of FAH mRNA specifically in liver and kidney. Furthermore, expression was found to commence by fetal day 16.5 in both organs (Ruppert et al., 1992). In situ hybridization was carried out to refine the correlation between sites of expression and the phenotype. Fig. 3A shows the detection of FAH transcripts in wild-type kidney (fetal day 18): hybridization is localized to the proximal convoluted tubules and is absent from structures such as glomeruli and collecting ducts. PEPCK mRNA, whose expression is strongly reduced in lethal albino liver and kidney (Ruppert et al., 1990), is also detected specifically in proximal convoluted tubules in wild-type kidney (Fig. 3B). The elevated expression of NMO mRNA in the lethal albino kidney occurs in the same cell-type (Fig. 3C). In liver, expression of FAH and PEPCK mRNAs in wild-type, and NMO mRNA in lethal albino mice, is restricted to hepatocytes (data not shown). The precise coincidence between the sites of FAH expression and the *alf/hsdr-1* phenotype is strong correlative evidence that absence of FAH underlies the phenotype. Functional evidence has come from our ability to reproduce changes in gene expression similar to those seen in the albino lethal liver in primary hepatocyte cultures treated with tyrosine metabolites (Ruppert et al., 1992). In conclusive experiments in transgenic mice, we have been able to show that an FAH

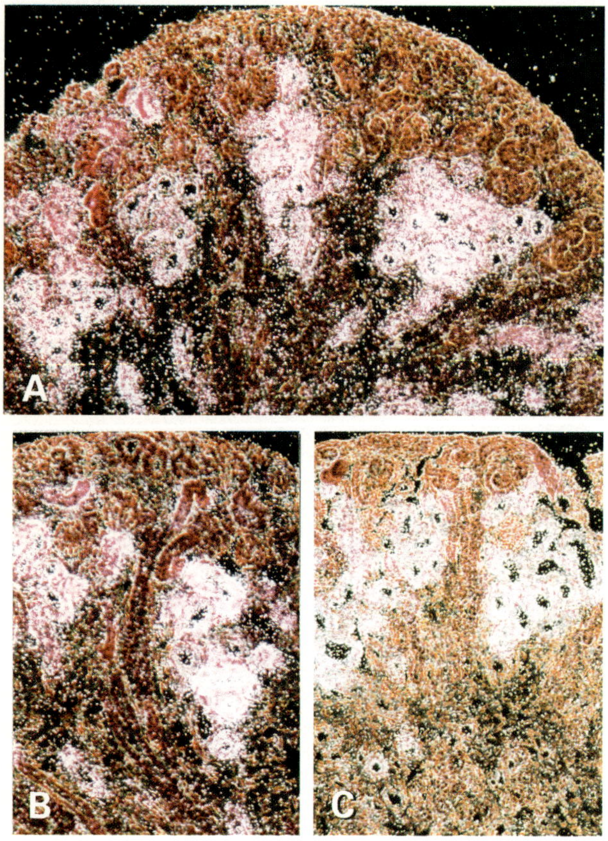

Fig. 3. The expression pattern of FAH correlates with the *alf/hsdr-1* phenotype. In situ hybridization of sections of fetal day-18 kidney with probes for FAH (A), PEPCK (B) and NMO-1 (C). (A) and (B) are sections of wild-type kidney, (C) is a section of albino lethal kidney. Dark field illumination is shown.

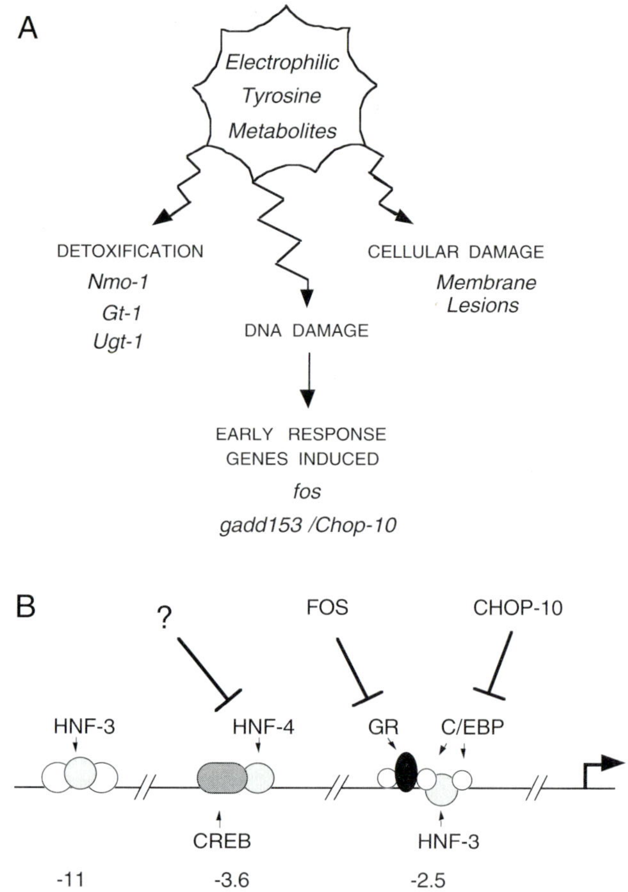

Fig. 4. Toxic tyrosine metabolites and specific effects on gene expression. Our current working model for the routes by which electrophilic tyrosine metabolites accumulating as a result of FAH deficiency could induce the diverse phenomena that characterize the *alf/hsdr-1* phenotype is shown schematically in (A). (B) illustrates the mechanisms by which the perinatal, hormone induction of the TAT gene could be blocked in the albino lethal liver, as discussed in the text.

cDNA transgene restores viability to lethal albino c^{14CoS} mice (G. K., S. R. and F. Beermann, unpublished observations).

Toxic tyrosine metabolites and specific effects on gene expression

The discovery that the *Fah* gene is disrupted and that tyrosine metabolism, in consequence, is blocked in albino lethal mice, offers a means for rationalizing the multiple components of the neonatal lethal phenotype (Fig. 4A). Contrary to the prediction that *alf/hsdr-1* encodes a regulatory factor, the characteristic changes in gene expression must be viewed as phenomena secondary to the absence of FAH. It is possible to speculate how these secondary effects come about, and these hypotheses can be tested. FAA, the tyrosine metabolite likely to accumulate, is highly electrophilic and a potential alkylating agent (Laberge et al., 1986). Induction of NMO, Gt-1 and Ugt-1 (Thaler et al., 1976; Gatmaitan et al., 1977; Petersen et al., 1989), can be considered to be attempts to detoxify the liver and kidney. NMO and Gt-1 expression is inducible by a variety of electrophilic agents (Talalay et al., 1988) and, consistent with a role in detoxification, over-expression of NMO mRNA is an early component of the phenotype and occurs in the same cell-types that should normally express FAH. Therefore, we would reject the hypothesis that a negative regulatory locus for NMO expression, *Nmo-1n*, maps within the albino deletions (Nebert et al., 1990). The ultrastructural abnormalities (Trigg and Gluecksohn-Waelsch, 1973) are likely to be a manifestation of the presence of highly reactive compounds. Induction of the *gadd* and c-*fos* genes might be a response to DNA or other intracellular damage precipitated by FAA (Fornace et al., 1989; Holbrook and Fornace, 1991; Ruppert et al., 1992). These genes are induced before birth, which could be considered to be early responses to the electrophilic metabolite(s).

The failure to activate hormone-dependent genes is the component of the phenotype that becomes pronounced around birth. Taking the TAT gene as prototype for this set of genes, several mechanisms can be envisaged by which activation is impaired, considering that the signal transduction pathways for glucocorticoids and cAMP are apparently intact (Ruppert et al., 1990; DeFranco et al., 1991).

They are outlined in Fig. 4B. Activation of TAT gene expression around birth is dependent upon permissive effects of glucocorticoids and cAMP, and the elements mediating this regulation have been defined for the rat gene (Nitsch et al., 1991). Three far-upstream enhancers each contribute a unique aspect of tissue-specificity and/or hormonal regulation (Nitsch et al., 1990). The enhancer at −2.5kb is responsible for regulation by glucocorticoids (Jantzen et al., 1987), and binding sites for the glucocorticoid receptor and transcription factors C/EBP and HNF-3 have been identified (Becker et al., 1986; Jantzen et al., 1987; Grange et al., 1991; D. Nitsch, personal communication). The *gadd153* gene, highly induced in albino lethal liver, encodes the C/EBP-like protein CHOP-10 (Ron and Habener, 1992). CHOP-10 is a putative dominant inhibitor of C/EBP proteins; it contains a leucine zipper domain capable of dimerizing with C/EBP and preventing recognition of C/EBP sites. CHOP-10 over-expression, therefore, could contribute to down-regulation of the glucocorticoid enhancer of the TAT gene. C/EBP may also play a role in PEPCK gene expression (Park et al., 1990). The sustained expression of FOS in lethal albino liver may also affect glucocorticoid induction. Many studies have indicated that FOS can antagonize glucocorticoid induction (Jonat et al., 1990; Lucibello et al., 1990; Yang-Yen et al., 1990; Shemshedini et al., 1991), although the mechanism remains to be established. Finally, of the transcription factors whose expression is down-regulated in albino lethal liver (Tönjes et al., 1992), HNF-4 is an important factor for TAT expression. The cAMP-inducible enhancer at −3.6 kb contains an HNF-4 site as one of the crucial elements (Boshart et al., 1990; D. Nitsch, personal communication). It remains to be determined how expression of the HNF-4 gene is reduced in the albino lethal liver, but it is conceivable that over-expression of CHOP-10, FOS or other early response genes might be involved.

In conclusion, with the identification that the *Fah* gene is disrupted, the components of the neonatal lethal albino phenotype become explicable as a sequence of events stemming from the accumulation of a reactive metabolite(s). These include induction of detoxification mechanisms and early-response genes, the latter possibly in response to DNA damage, and altered activity of key transcription factors. Future experiments will be designed to deepen the understanding of these events.

We should like to thank Doris Nitsch for sharing with us her views on the regulation of the rat TAT gene. We are grateful to G. Withers for photographic work and C. Schneider for excellent secretarial assistance. This work was supported by the Deutsche Forschungsgemeinschaft through SFB 229 and the Leibniz Programm, and the Fonds der Chemischen Industrie.

References

Becker, P. B., Gloss, B., Schmid, W., Strähle, U. and Schütz, G. (1986). *In vivo* protein-DNA interactions within a glucocorticoid response element require the presence of the hormone. *Nature* **324**, 686-688.

Boshart, M., Weih F., Schmidt, A., Fournier, R. E. K. and Schütz, G. (1990). A cyclic AMP response element mediates repression of tyrosine aminotransferase gene transcription by tissue-specific extinguisher locus *Tse-1*. *Cell* **61**, 905-916.

DeFranco, D., Bali, D., Torres, R., DePinho, R. A., Erickson, R. P. and Gluecksohn-Waelsch, S. (1991). The glucocorticoid hormone signal transduction pathway in mice homozygous for chromosomal deletions causing failure of cell type-specific inducible gene expression. *Proc. Nat. Acad. Sci. USA* **88**, 5607-5610.

Erickson, R. P., Gluecksohn-Waelsch, S. and Cori, C. (1968). Glucose-6-phosphatase deficiency caused by radiation-induced alleles at the albino locus in the mouse. *Proc. Nat. Acad. Sci. USA* **59**, 437-444.

Fornace Jr., A. J., Nebert, D. W., Hollander, M. C., Luethy, J. D., Papathanasiou, M., Fargnoli, J. and Holbrook, N. J. (1989). Mammalian genes coordinately regulated by growth arrest signals and DNA-damaging agents. *Mol. Cell. Biol.* **9**, 4196-4203.

Garland, R. C., Satrustegui, J., Gluecksohn-Waelsch, S. and Cori, C. F. (1976). Deficiency in plasma protein synthesis caused by x-ray-induced lethal albino alleles in mouse. *Proc. Nat. Acad. Sci. USA* **73**, 3376-3380.

Gatmaitan, Z., Lewis, S., Turchin, H. and Arias, I. M. (1977). Premature development of ligandin (GSH transferase B) in mice with an inherited defect in endoplasmic reticulum-golgi structure and function. *Biochem. Biophys. Res. Comm.* **75**, 337-341.

Gluecksohn-Waelsch, S. (1979). Genetic control of morphogenetic and biochemical differentiation: lethal albino deletions in the mouse. *Cell* **16**, 225-237.

Gluecksohn-Waelsch, S. (1987). Regulatory genes in development. *Trends Genet.* **3**, 123-127.

Goldsmith, L. A. and Laberge, C. (1989). Tyrosinemia and related disorders. In *The Metabolic Basis of Inherited Disease*, 6th ed. (ed. C. R. Scriver, A. L. Beaudet, W. S. Sly and D. Valle), pp. 547-562. New York: McGraw-Hill.

Gonzalez, F. J., Liu, S-Y., Kozak, C. A. and Nebert, D. W. (1990). Decreased *Hnf-1* gene expression in mice homozygous for a 1.2-centiMorgan deletion on chromosome 7. *DNA and Cell Biol.* **9**, 771-776.

Grange, T., Roux, J., Rigaud, G. and Pictet, R. (1991). Cell-type specific activity of two glucocorticoid responsive units of rat tyrosine aminotransferase gene is associated with multiple binding sites for C/EBP and a novel liver-specific nuclear factor. *Nucl. Acids Res.* **19**, 131-139.

Holbrook, N. J. and Fornace Jr., A. J. (1991). Response to adversity: molecular control of gene activation following genotoxic stress. *New Biologist* **3**, 825-833.

Hollander, M. C. and Fornace Jr., A. J. (1989). Induction of *fos* RNA by DNA-damaging agents. *Cancer Res.* **49**, 1687-1692.

Jantzen, H-M., Strähle, U., Gloss, B., Stewart, F., Schmid, W., Boshart, M., Miksicek, R. and Schütz, G. (1987). Cooperativity of glucocorticoid response elements located far upstream of the tyrosine aminotransferase gene. *Cell* **49**, 29-38.

Jonat, C., Rahmsdorf, H. J., Park, K-K., Cato, A. C. B., Gebel, S., Ponta, H. and Herrlich, P. (1990). Antitumor promotion and antiinflammation: down-modulation of AP-1 (Fos/Jun) activity by glucocorticoid hormone. *Cell* **62**, 1189-1204.

Klebig, M. L., Russell, L. B. and Rinchik, E. M. (1992). Murine fumarylacetoacetate hydrolase (*Fah*) gene is disrupted by a neonatally lethal albino deletion that defines the hepatocyte-specific developmental regulation 1 (*hsdr-1*) locus. *Proc. Nat. Acad. Sci. USA* **89**, 1363-1367.

Kvittingen, E. A. (1986). Hereditary tyrosinemia type I: an overview. *Scand. J. Clin. Lab. Invest.* **46**, 27-34.

Laberge, C., Lescault, A. and Tanguay, R. M. (1986). Hereditary tyrosinemias (Type I): a new vista on tyrosine toxicity and cancer. In *Essential Nutrients in Carcinogenesis*. (ed. L. A. Poirier, P. M. Newbarne and M. W. Pariza), pp. 209-221. New York: Plenum Press.

Loose, D. S., Shaw, P. A., Krauter, K. S., Robinson, C., Englard, S., Hanson, R. W. and Gluecksohn-Waelsch, S. (1986). *Trans*regulation of the phospho*enol*pyruvate carboxykinase (GTP) gene, identified by deletions in chromosome 7 of the mouse. *Proc. Nat. Acad. Sci. USA* **83**, 5184-5188.

Lucibello, F. C., Slater, E. P., Jooss, K. U., Beato, M. and Müller, R. (1990). Mutual transrepression of Fos and the glucocorticoid receptor: involvement of a functional domain in Fos which is absent in FosB. *EMBO J.* **9**, 2827-2834.

McKnight, S. L., Lane, M. D. and Gluecksohn-Waelsch, S. (1989). Is CCAAT/enhancer-binding protein a central regulator of energy metabolism? *Genes Dev.* **3**, 2021-2024.

Morris Jr., S. M., Moncman, C. L., Kepka, D. M., Nebes, V. L., Diven, W. F., Dizikes, G. J., Cederbaum, S. D. and DeFranco, D. (1988). Effects of deletions in mouse chromosome 7 on expression of genes encoding the urea-cycle enzymes and phosphoenolpyruvate carboxykinase (GTP) in liver, kidney, and intestine. *Biochem. Genet.* **26**, 769-781.

Nebert, D. W., Petersen, D. D. and Fornace Jr., A. J. (1990). Cellular responses to oxidative stress: the [*Ah*] gene battery as a paradigm. *Environ. Health Perspect.* **88**, 13-25.

Niswander, L., Kelsey, G., Schedl, A., Ruppert, S., Sharan, S. K., Holdener-Kenny, B., Rinchik, E. M. R., Edström, J-E. and Magnuson, T. (1991). Molecular mapping of albino deletions associated with early embryonic lethality in the mouse. *Genomics* **9**, 162-169.

Nitsch, D., Ruppert, S., Kelsey, G., Schedl, A., Weih, F., Stewart, A. F., Strähle, U., Schmid, W., De Vack, C., Reik, A., Boshart, M. and Schütz, G. (1991). Hormonal and liver-specific control of expression of the tyrosine aminotransferase gene. In *The Hormonal Control of Regulation of Gene Transcription*, (ed. P. Cohen and J. G. Foulkes), pp. 223-234. Amsterdam: Elsevier.

Nitsch, D., Stewart, A. F., Boshart, M., Mestril, R., Weih, F. and Schütz, G. (1990). Chromatin structures of the rat tyrosine aminotransferase gene relate to the function of its *cis*-acting elements. *Mol. Cell Biol.* **10**, 3334-3342.

Park, E. A., Roesler, W. J., Liu, J., Klemm, D. J., Gurney, A. L., Thatcher, J. D., Shuman, J., Friedman, A. and Hanson, R. W. (1990). The role of the CCAAT/Enhancer-binding protein in the transcriptional regulation of the gene for phosphoenolpyruvate carboxykinase (GTP). *Mol. Cell. Biol.* **10**, 6264-6272.

Petersen, D. D., Gonzalez, F. J., Rapic, V., Kozak, C. A., Lee, J-Y., Jones, J. E. and Nebert, D. W. (1989). Marked increase in hepatic NAD(P)H:oxidoreductase gene transcription and mRNA levels correlated with a mouse chromosome 7 deletion. *Proc. Nat. Acad. Sci. USA* **86**, 6699-6703.

Rinchik, E. M. and Russell, L. B. (1990). Germ-line deletion mutations in the mouse: tools for intensive functional and physical mapping of regions of the mammalian genome. In *Genetic and Physical Mapping. Genome Analysis* vol. I (ed. K. E. Davies and S. M. Tilghman), pp. 121-158. New York: Cold Spring Harbor Laboratory.

Ron, D. and Habener, J. F. (1992). CHOP, a novel developmentally regulated nuclear protein that dimerizes with transcription factors C/EBP and LAP and functions as a dominant negative inhibitor of gene transcription. *Genes Dev.* **6**, 439-453.

Ruppert, S., Boshart, M., Bosch, F. X., Schmid, W., Fournier, R. E. K. and Schütz, G. (1990). Two genetically defined trans-acting loci coordinately regulate overlapping sets of liver-specific genes. *Cell* **61**, 895-904.

Ruppert, S., Kelsey, G., Schedl, A., Schmid, E., Thies, E. and Schütz, G. (1992). Deficiency of an enzyme of tyrosine metabolism underlies altered gene expression in newborn liver of lethal albino mice. *Genes Dev.* **6**, 1430-1443.

Russell, L. B., Montgomery, C. S. and Raymer, G. D. (1982). Analysis of the albino-locus region of the mouse. IV. Characterization of 34 deficiencies. *Genetics* **100**, 427-453.

Schmid, W., Müller, G., Schütz, G. and Gluecksohn-Waelsch, S. (1985). Deletions near the albino locus on chromosome 7 of the mouse affect the level of tyrosine aminotransferase mRNA. *Proc. Nat. Acad. Sci. USA* **82**, 2866-2869.

Shemshedini, L., Knauthe, R., Sassone-Corsi, P., Potron, A. and Gronemeyer, H. (1991). Cell-specific inhibitory and stimulatory effects of Fos and Jun on transcription activation by nuclear receptors. *EMBO J.* **10**, 3839-3849.

Talalay, P., De Long, M. J. and Prochaska, H. J. (1988). Identification of a common chemical signal regulating the induction of enzymes that protect against chemical carcinogenesis. *Proc. Nat. Acad. Sci. USA* **85**, 8261-8265.

Thaler, M. M., Erickson, R. P. and Pelger, A. (1976). Genetically determined abnormalities of microsomal enzymes in liver of mutant newborn mice. *Biochem. Biophys. Res. Comm.* **72**, 1244-1250.

Thorndike, J., Trigg, M. J., Stockert, R., Gluecksohn-Waelsch, S. and Cori, C. F. (1973). Multiple biochemical effects of a series of X-ray induced mutations at the albino locus in the mouse. *Biochem. Genet.* **9**, 25-39.

Tönjes, R. R., Xanthopoulos, K. G., Darnell Jr., J. E. and Paul, D. (1992). Transcriptional control in hepatocytes of normal and *c14CoS* albino deletion mice. *EMBO J.* **11**, 127-133.

Trigg, M. J. and Gluecksohn-Waelsch, S. (1973). Ultrastructural basis of biochemical effects in a series of lethal alleles in the mouse. *J. Cell Biol.* **58**, 549-563.

Yang-Yen, H-F., Chambard, J-C., Sun, Y-L., Smeal, T., Schmidt, T. J., Drouin, J. and Karin, M. (1990). Transcriptional interference between c-Jun and the glucocorticoid receptor: mutual inhibition of DNA binding due to direct protein-protein interaction. *Cell* **62**, 1205-1215.

Journal of Cell Science, Supplement 16, 123-127 (1992)
Printed in Great Britain

The role of the transcriptional activator protein DBP in circadian liver gene expression

JÉRÔME WUARIN, EILEEN FALVEY, DAN LAVERY, DALE TALBOT, ED SCHMIDT, VINCENT OSSIPOW, PHILIPPE FONJALLAZ and UELI SCHIBLER*

Department of Molecular Biology, Sciences II, 30 Quai Ernest Ansermet, CH-1211 Geneva-4, Switzerland

*Corresponding author

Summary

DBP, a liver-enriched transcriptional activator protein of the leucine zipper protein family, accumulates according to a very strong circadian rhythm (amplitude approx. 1000-fold). In rat parenchymal hepatocytes, the protein is barely detectable during the morning hours. At about 2 p.m., DBP levels begin to rise, reach maximal levels at 8 p.m. and decline sharply during the night. This rhythm is free-running: it persists with regard to both its amplitude and phase in the absence of external time cues, such as daily dark/light switches. Also, fasting of rats for several days influences neither the amplitude nor the phase of circadian DBP expression. Since the levels of DBP mRNA and nascent transcripts also oscillate with a strong amplitude, circadian DBP expression is transcriptionally controlled. While DBP mRNA fluctuates with a similar phase and amplitude in most tissues examined, DBP protein accumulates to high concentrations only in liver nuclei. Hence, at least in nonhepatic tissues, cyclic DBP transcription is unlikely to be controlled by a positive and/or negative feedback mechanism involving DBP itself. More likely, the circadian DBP expression is governed by hormones whose peripheral concentrations also oscillate during the day. Several lines of evidence suggest a pivotal role of glucocorticoid hormones in establishing the DBP cycle.

Two genes whose mRNAs and protein products accumulate according to a strong circadian rhythm with a phase compatible with regulation by DBP encode enzymes with key functions in cholesterol metabolism: HMG-coA reductase is the rate-limiting enzyme in cholesterol synthesis; cholesterol 7-α hydroxylase performs the rate-limiting step in the conversion of cholesterol to bile acid. DBP may thus be involved in regulating cholesterol homeostasis.

Key words: liver transcription factor, DBP, circadian rhythms.

Introduction

Our laboratory has chosen the liver as a paradigm for studying cell-type-specific gene expression. By identifying *cis*-acting albumin promoter elements and the transcription factors interacting with them, we hoped to gain access to key regulators of hepatic gene expression. A detailed functional dissection of the albumin promoter revealed six cognate sites (A, B, C, D, E and F) for sequence-specific DNA-binding proteins (Lichtsteiner et al., 1987). Two of these, sites B and D, proved to be particularly important for efficient cell-type-specific transcription (Maire et al., 1989; Lichtsteiner and Schibler, 1989). The proteins binding to these elements have been identified, purified and their cDNA and genes isolated by molecular cloning techniques (for review, see Crabtree et al., 1992). In this paper we concentrate on our results with a protein, DBP, which binds to the albumin D promoter element (Mueller et al., 1990). Studies on DBP expression have led to the surprising discovery that in adult rats, DBP accumulates according to a very strong circadian rhythm (Wuarin and Schibler, 1990). In the next paragraphs we will focus on the regulation of circadian DBP expression and speculate on its purpose.

DBP, a liver-enriched transcriptional activator protein

DBP cDNAs have been obtained by screening a lambda gt 11 liver cDNA expression library with a radiolabeled DNA probe encompassing the albumin promoter element D. Sequencing of close to full-length cDNAs revealed an open reading frame corresponding to 325 amino acids (Mueller et al., 1990; see corrected sequence, *Cell* **65**, 1991, page 915). DBP dimerizes via a carboxy-terminal leucine zipper and binds DNA via an adjacent basic region (Vinson et al., 1989). The three-dimensional structure of at least one leucine zipper, that of the yeast transcription factor GCN4, has been resolved by X-ray crystallography (O'Shea et al., 1991). Moreover, the amino acid residues located at func-

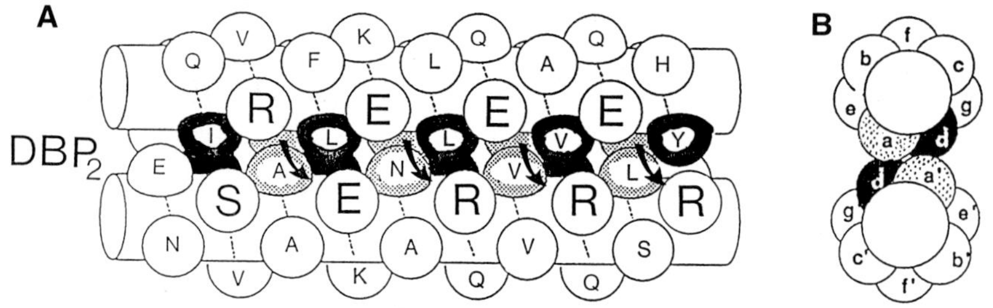

Fig. 1. Schematic representation (see O'Shea et al., 1992) of dimerized coiled coils (leucine zippers) within a DBP homodimer. The N- to C-polarity of the dimerization helices is from left to right. The right hand panel shows a vertical projection of the coiled coils. The D and A heptad positions containing the hydrophobic amino acids are shown in black and shaded, respectively. Hypothetical salt bridges between glutamic acids (E) and arginines (R) are indicated by arrows (four on each side of the paired amphipathic helices).

tionally strategic positions of leucine zippers have been determined by elegant genetic (Hu et al., 1990) and biochemical work (eg. Landschulz et al., 1989, and references therein; O'Shea et al., 1992). Leucine zipper regions form coiled-coil structures, consisting of parallel amphipathic α-helices that interact via hydrophobic side chains of amino acids located at positions A and D (Fig. 1B). While no particular D position must be occupied by a leucine, a total of at least two leucines appear to be required within this heptad repeat (Hu et al., 1990). However, not all leucine zippers can form stable interactions with each other. To a large extent, the specificity of dimerization is dictated by salt bonds between amino acids of opposite charges located in positions G and E (see O'Shea et al., 1992 and references therein). The leucine zipper of DBP (Fig. 1A) is somewhat unusual in two respects. Firstly, it contains only two leucines, the absolute minimum tolerated for dimerization (see above). Secondly, the G and E positions consist of four charged amino acids which, theoretically, could establish as many as eight salt bridges. However, the intrahelical repulsion created by the runs of three consecutive arginines and glutamic acids in the E and G heptad repeat positions, respectively, may also result in a destabilization of the dimerization helices. Indeed, our cross-linking experiments with purified recombinant proteins produced in *E. coli* suggest a three-fold higher dimerization dissociation constant for the DBP zipper as compared to the zipper of LAP, a C/EBP (Landschulz et al., 1989) related protein that also has affinity for the albumin D site (Descombes and Schibler, 1991). In keeping with this observation, replacement of the DBP zipper with the LAP zipper increases the affinity of the DBP chimeric protein to its DNA cognate site by about three-fold. Perhaps the dimerization properties of DBP have to be evaluated in the light of its role in diurnal transcription activation (see below). The circadian accumulation (see below) and action of this protein may call for a high off-rate of DNA binding to ensure a rapid equilibrium between specific and non-specific occupancy.

The N-terminal moiety of DBP contains the transcription activation domain(s). In cotransfection experiments with wild-type DBP and DBP target genes, over-expression of the N-terminal two thirds of DBP results in a strong attenuation of DBP-mediated transcription activation. This squelching is specific, since transcription activation through LAP is not impaired by an excess of N-terminal DBP sequences (E. F., unpublished results). Thus, the transcription activation domains of DBP and LAP appear to interact with different targets of the general transcription apparatus.

In liver, the majority of DBP is highly phosphorylated (V. O., F. Fleury and U. S., unpublished data). Dephosphorylation of DBP with acidic potato phosphatase does not appear to influence the equilibrium DNA-binding constant of DBP. We cannot exclude, however, that the dynamics of DNA binding (K_{on} and K_{off}) are modulated by phosphorylation, as has recently been suggested for the serum response factor SRF (Marais et al., 1992). Alternatively, phosphate groups may be involved in regulating the transcription activation potential of DBP.

The DBP gene

The rat gene specifying DBP has recently been isolated and characterized in our laboratory. It contains four exons and three introns and encompasses about 5.6 kb from cap– to polyadenylation sites (J. W. and E. F., unpublished data). The promoter is located within a CpG island and is devoid of a canonical TATA box. As has been observed for many promoters of this type, transcription initiation occurs at multiple start sites spread over about 50 nucleotides. The 5′ untranslated region of DBP mRNA is extraordinarily long (about 400 nucleotides) and contains three upstream AUGs. We remain uncertain whether these sequences are relevant for controlling translation initiation in the animal, but deletion of the sequences containing the three upstream AUGs increases the accumulation of DBP both in vivo (transient transfection) and in vitro (reticulocyte lysate) (S. Rufino and U. S., unpublished data).

Circadian and developmental DBP expression

DBP accumulates in parenchymal hepatocytes according to a very robust circadian rhythm. The DBP level measured at the time of maximal accumulation (8 p.m. in Lewis rats) exceeds the one determined at the time of minimal accumulation (8 a.m.) by about three orders of magnitude. This rhythm is transcriptionally controlled, free-running and independent of food or water uptake (Wuarin and Schibler, 1990). While DBP mRNA cycles with the same phase and a similarly strong amplitude in a variety of non-hepatic tissues (lung, spleen, kidney), the DBP protein accumulates to much lower levels in these cell types as compared to hepatocytes. As DBP mRNA appears to be associated with polysomes in all tissues (E. S., unpublished results), it is conceivable that protein stability rather than translation efficiency accounts for the differential DBP accumulation in various cell types.

In rodents, circadian rhythmicity commences shortly after weaning (for reviews on circadian rhythms, see Edmunds, 1988). As a consequence, DBP oscillation is only observed in rats older than three weeks (Mueller et al., 1990; Yano et al., 1992). In the liver of younger animals (birth to three weeks), DBP accumulates to very low levels and does not cycle during the day (D. L. and E. S., unpublished data).

How is the oscillation of DBP transcription regulated? As circadian mRNA expression is similar in tissues that do or do not accumulate the protein, DBP itself is unlikely to control its own transcription by positive and/or negative feedback loops. More likely, DBP transcription is governed by hormones whose secretion is also rhythmic. Several lines of evidence are compatible with glucocorticoid hormones participating in the "downphase" of circadian DBP expression. In rats, corticosterone shows highest peripheral levels at the dark/light switch (6 p.m. in our animal facility; for circadian glucocorticoid secretion, see Meier, 1975 and references therein), several hours after DBP transcription rates have climaxed (between 2 p.m. to 4 p.m.). Injection of dexamethasone around noon results in a strong attenuation of DBP mRNA accumulation during the afternoon. Moreover, in homozygous fa/fa Zucker rats, which exhibit higher corticosterone levels than their wild-type counterparts (Guillaume-Gentil et al., 1990), the DBP cycle is phase-shifted towards earlier hours and is reduced in amplitude. For the two following reasons, we consider it likely that DBP transcription is directly repressed by glucocorticoid hormones. (1) In hepatoma cells, dexamethasone addition leads to a rapid extinction of DBP mRNA accumulation (E. F., unpublished data). (2) The DBP promoter contains a GRE, as judged from DNase 1 footprinting studies with recombinant glucocorticoid receptor.

In an alternative attempt to localize *cis*-acting DBP promoter elements, DNase 1 hypersensitive sites in rat liver nuclei were mapped (D. T., unpublished data). Such sites have been observed exclusively within the 1 kb 5′ flanking region of the DBP gene. Some of these appear to be more efficiently cleaved in nuclei isolated during the afternoon, when the DBP gene is most active, than in nuclei isolated in the morning, when the DBP gene is not or very weakly transcribed. The importance of these elements for circadian DBP expression is currently being tested in transgenic mice.

Purpose of circadian DBP expression

In order to decipher the physiological significance of circadian DBP expression, one needs to identify bona fide DBP target genes. As mentioned earlier, DBP has been recognized as a transcription factor which binds to an albumin promoter element in vitro. Indeed, in adult rats, albumin transcription appears to fluctuate with a phase compatible with that of DBP accumulation (Wuarin and Schibler, 1990; J. W., unpublished results). However, this circadian transcription can hardly be physiologically relevant, since due to their long half-lives, neither albumin mRNA nor its protein product accumulate in a cyclic fashion. In order for a gene product to oscillate during the day, it must be unstable. The impact of the half-life on the accumulation of diurnally synthesized mRNAs is illustrated by the mathematical simulations shown in Figs 2 and 3. If one assumes that the transcription rates of DBP target genes parallel the circadian accumulation of DBP (shaded areas in Fig. 2) one can derive the accumulation profiles for the corresponding mRNAs of any given half-life (shown for $T_{1/2}$ between 1 hour and 20 hours). Even if the transcriptional amplitude of a gene were infinite, the level of mRNAs with half-lives of more than ten hours would fluctuate by two-fold or less (see Fig. 3, solid circles). However, the transcriptional fluctuation of most DBP target genes may only be five- to tenfold, since constitutively expressed activators in addition to DBP may be involved in controlling their expression. For mRNAs of such genes to cycle, their half-lives would have to be very short (see Fig. 3, open circles). Clearly then, DBP oscillation can be physiologically relevant only for those target genes whose mRNA and protein products are unstable.

Two particularly attractive candidates for DBP target genes are those encoding the liver-enriched enzymes HMG

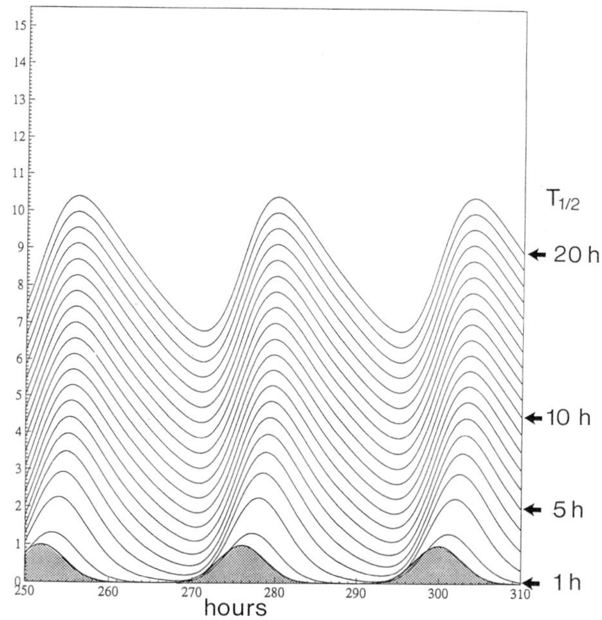

$$a_x = \int_0^x e^{\ln 2\,(t-x)/T_{1/2}} \cdot f(t) \cdot dt$$

Fig. 2. Circadian accumulation of mRNAs from putative DBP target genes as a function of their half-lives. As a first approximation, it is assumed that the transcription rates of DBP target genes (shaded areas) follow the accumulation profile of DBP. As demonstrated in Figs 2 and 4, the precise circadian transcription patterns of DBP target genes depends on the affinity of their DBP cognate sites and on the contribution of other, perhaps constitutively expressed transcription factors participating in the regulation of these genes. Accumulation profiles are shown for mRNA half-lives of one to twenty hours.

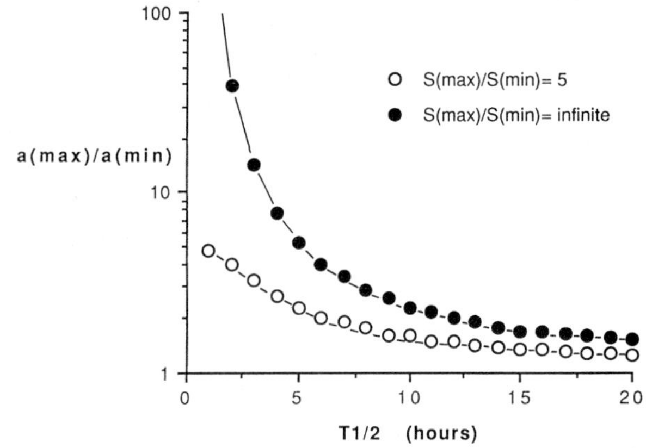

Fig. 3. Amplitudes of mRNA accumulation as a function of half-lives and amplitude of synthesis. Solid circles: Amplitudes of the accumulation (a) of mRNAs whose amplitude of synthesis (S) is infinite, such as shown in Fig. 2. Open circles: amplitudes of the accumulation (a) of mRNA whose amplitude of synthesis (S) is 5-fold. The amplitudes (maximal/minimal accumulation) were determined graphically from accumulation profiles such as shown in Fig. 2.

CoA reductase and cholesterol 7-α hydroxylase, which catalyze the rate-limiting steps in cholesterol synthesis and cholesterol conversion to bile acids, respectively. Enzyme activities as well as mRNAs from these genes accumulate according to a robust circadian rhythm with amplitudes of around 10-fold and a phase compatible with that of DBP accumulation (Clarke et al., 1984; Noshiro et al., 1990; D. L. and J. W., unpublished results). Furthermore, and in keeping with DBP accumulation, circadian expression of these enzymes is not detected until after weaning. Genomic clones encompassing the promoter regions of both of these genes have been obtained and are currently being examined for the presence of DBP-binding sites. Thus far, such elements have been discerned within the cholesterol 7-α hydroxylase promoter. Furthermore, in cotransfection experiments, a CAT reporter gene driven by this promoter is strongly activated by DBP. While none of the data thus far obtained prove that the circadian expression of cholesterol 7-α hydroxylase and HMG CoA reductase is governed by DBP, this transcription factor appears to be a likely player in the control of cholesterol homeostasis. In the model we are currently testing, DBP would simultaneously activate the diurnal production of bile acid and cholesterol. Such a coordination may be pivotal, since the massive cholesterol utilization during the activity period (evening and night hours for rats and mice) may call for a large, simultaneous increase in its synthesis.

Conclusions and perspectives

DBP, a liver-enriched transcriptional activator of the bZip protein family, accumulates according to an extraordinarily strong circadian rhythm. Preliminary results suggest that circadian DBP expression may be controlled, at least in part, by the cyclic secretion of glucocorticoid hormones. Like most (if not all) circadian activities in mammals, corticosterone synthesis and secretion is governed by the suprachiasmatic nucleus (SCN), a small brain structure located above the optical chiasma. Circadian outputs are entrained in the SCN by external time cues, such as light/dark phases. Once entrained, such outputs are free-running and persist for extended time periods, until the clock is reset by new time cues (for references on the circadian pacemaker SCN see Kornhauser et al., 1992; Meijer and Rietveld, 1989). In the case of adrenal hormone secretion, the SCN is believed to set the pace of CRF secretion in the hypothalamus. This in turn may dictate the rhythmic release of ACTH from the anterior pituitary gland, eventually resulting in a cyclic secretion of corticosterone in the adrenal gland (Meijer and Rietveld, 1989).

Deciphering of the physiological purpose of DBP oscillation will require the unequivocal identification of downstream genes. At present we are testing the hypothesis that one of the DBP functions is related to cholesterol homeostasis, because the rate-limiting enzymes in cholesterol synthesis (HMG coA reductase) and utilization (cholesterol 7-α hydroxylase) both fluctuate with a circadian phase compatible with that of DBP accumulation. Clearly, the possible role of DBP in cholesterol metabolism will remain hypothetical until scrutinized by a thorough genetic analysis. The tools required for this endeavor have recently become available. In the case of DBP, both gain-of-function (constitutive expression of DBP throughout the day) and loss-of-function (DBP gene knockout by homologous recombination) experiments should now be technically feasible.

We were surprised to observe that, while transcription of

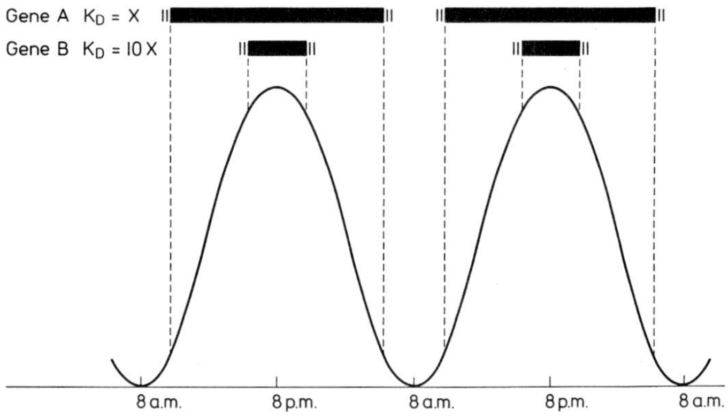

Fig. 4. Circadian transcription of two hypothetical DBP target genes containing DBP cognate sites with different affinities. The circadian DBP accumulation profile is schematically shown as a solid line. Target gene A is assumed to contain a 10-fold stronger DBP cognate site (equilibrium dissociation constant K_d=X) than gene B (equilibrium dissociation constant K_d=10X). As a consequence, half-saturation of the DBP site in gene B would require 10-fold higher DBP concentrations than that of its counterpart in gene A. Therefore, maximal transcriptional activity of gene A would be expected to persist for a longer circadian period than that of gene B. The periods of maximal transcription for genes A and B are depicted as black bars. The genes encoding cytochrome P450 CYP2C6 and cholesterol 7-α hydroxylase constitute putative candidates for the gene classes A and B, respectively.

certain genes, such as the one encoding serum albumin, cycles during the day, the accumulation of their products remains relatively constant. Apparently, diurnally expressed transcription factors like DBP may be used for genes whose products oscillate, such as HMG coA reductase and cholesterol 7-α hydroxylase, and genes whose products accumulate constitutively, depending on their stability.

The relative affinity of the DBP sites for their cognate factor may be another important parameter in determining the temporal pattern of putative DBP target gene expression. As illustrated by the cartoon shown in Fig. 4, DBP target genes with low affinity sites may be maximally expressed during a short circadian time window, since the critical DBP concentrations required for efficient occupancy may only be reached during a few hours during the day. In contrast, genes with high affinity sites are expected to be expressed during much longer time periods, as even low DBP levels would suffice to fill the respective recognition sequences. For example, the cytochrome P450 gene CYP2C6 contains a DBP site within its promoter that binds *E. coli*-derived recombinant DBP with a 17-fold lower K_d than the albumin D promoter element (Yano et al., 1992). Run-on transcription experiments with nuclei isolated at four hour intervals around the clock yielded more or less constant transcription rates for the CYP2C6 gene (D. L., unpublished data). While further experiments are required to identify this gene as a bona fide DBP target gene, these considerations exemplify the versatile potentials of oscillating transcription factors: circadian DBP accumulation may result in strong, weak or no oscillation of mRNAs and proteins issued from downstream genes, depending on the relative affinities of their DBP-binding sites and the stabilities of their products.

We would like to thank Prof. G. Wanner, Department of Applied Mathematics, University of Geneva, for preparing the computer program used in Fig. 2. This research was supported by the Swiss National Science Foundation and the State of Geneva.

References

Clarke, C. F., Fogelman, A. M. and Edwards, P. A. (1984). Diurnal rhythm of rat liver mRNAs encoding 3-hydroxy-3-methylglutaryl coenzyme A reductase. *J. Biol. Chem.* **259**, 10439-10447.

Crabtree, G., R., Schibler, U. and Scott, M. P. (1992). Transcriptional regulatory mechanisms in liver and midgut morphogenesis of vertebrates and invertebrates. In *Transcriptional Regulation* (ed. K. R. Yamamoto and S. L. McKnight). The Cold Spring Harbor Monograph (in press)

Descombes, P. and Schibler, U. (1991). A liver-enriched transcriptional activator, LAP, and a transcriptional inhibitory protein, LIP, are translated from the same mRNA. *Cell* **67**, 569-580

Edmunds, L. N. (1988). *Cellular and Molecular Bases of Biological Clocks*. 497pp. New York: Springer Verlag.

Guillaume-Gentil, C., Rohner-Jeanrenaud, F., Bestetti, G. E., Abramo, F., Rossi, G. L. and Jeanrenaud, B. (1990). Abnormal regulation of the hypothalamo-pituitary-adrenal axis in the genetically obese fa/fa rat. *Endocrinology* **126**, 1873-1879.

Hu, J. C., O'Shea, E. K., Kim, P. S. and Sauer, R. T. (1990). Sequence requirements for coiled coils: Analysis with lambda repressor-GCN4 leucine zipper fusions. *Science* **250**, 1400-1403.

Kornhauser, J. M., Nelson, D. E., Mayo, K. E. and Takahashi, J. S. (1992). Regulation of jun-B mRNA and AP-1 activity by light and a circadian clock. *Science* **255**, 1581-1584.

Landschulz, W. H., Johnson, P. F. and McKnight, S. L. (1989). The DNA-binding domain of the rat liver nuclear protein C/EBP is bipartite. *Science* **243**, 1681-1688.

Lichtsteiner, S. and Schibler, U. (1989). A glycosylated liver-specific transcription factor stimulates transcription of the albumin gene. *Cell* **57**, 1179-1187.

Lichtsteiner, S., Wuarin, J. and Schibler, U. (1987). The interplay of DNA-binding proteins on the promoter of the mouse albumin gene. *Cell* **51**, 963-973.

Maire, P., Wuarin, J. and Schibler, U. (1989). The role of *cis*-acting elements in tissue-specific albumin gene transcription. *Science* **244**, 343-346.

Marais, R. M., Hsuan, J. J., McGuigan, C., Wynne, J. and Treisman, R. (1992). Casein kinase II phosphorylation increases the rate of serum response factor-binding site exchange. *EMBO J.* **11**, 97-105.

Meier, A. H. (1975). Daily variations of concentration of plasma corticosteroid in hypophysectomized rats. *Endocrinology* **98**, 1475-1479.

Meijer, J. H. and Rietveld, W. J. (1989). Neurophysiology of the suprachiasmatic circadian pacemaker in rodents. *Physiol. Rev.* **69**, 671-707.

Mueller, C. R., Maire, P. and Schibler, U. (1990). DBP, a liver-enriched transcriptional activator, is expressed late in ontogeny and its tissue-specificity is determined post-transcriptionally. *Cell* **61**, 279-291.

Noshiro, M., Nishimoto, M. and Okuda, K. (1990). Rat liver cholesterol 7-alpha hydroxylase: Pretranslational regulation for circadian rhythm. *J. Biol. Chem.* **265**, 10036-10041.

O'Shea, E. K., Klemm, J. D., Kim, P. S. and Alber, T (1991). X-ray structure of the GCN4 leucine zipper, a two-stranded, parallel, coiled coil. *Science* **254**, 539-544.

O'Shea, E. K., Rutkowski, R. and Kim, P. S. (1992). Mechanism of specificity in the Fos-Jun oncoprotein heterodimer. *Cell* **68**, 699-708.

Vinson, C. R., Sigler, P. B. and McKnight, S. L. (1989). Scissor-grip model for DNA recognition by a family of leuzine zipper proteins. *Science* **246**, 911-916.

Wuarin J. and Schibler, U. (1990). Expression of the liver-enriched activator protein DBP follows a stringent circadian rhythm. *Cell* **63**, 1257-1266

Yano, M., Falvey, E. and Gonzalez, F. J. (1992). Role of the liver-enriched transcription factor DBP in expression of the cytochrome P450 CYP2C6 gene. *Mol. Cell. Biol.* **12**, 2847-2854.

Index